Synthese Library

Studies in Epistemology, Logic, Methodology, and Philosophy of Science

Volume 472

The aim of *Synthese Library* is to provide a forum for the best current work in the methodology and philosophy of science and in epistemology, all broadly understood. A wide variety of different approaches have traditionally been represented in the Library, and every effort is made to maintain this variety, not for its own sake, but because we believe that there are many fruitful and illuminating approaches to the philosophy of science and related disciplines.

Special attention is paid to methodological studies which illustrate the interplay of empirical and philosophical viewpoints and to contributions to the formal (logical, set-theoretical, mathematical, information-theoretical, decision-theoretical, etc.) methodology of empirical sciences. Likewise, the applications of logical methods to epistemology as well as philosophically and methodologically relevant studies in logic are strongly encouraged. The emphasis on logic will be tempered by interest in the psychological, historical, and sociological aspects of science. In addition to monographs *Synthese Library* publishes thematically unified anthologies and edited volumes with a well-defined topical focus inside the aim and scope of the book series. The contributions in the volumes are expected to be focused and structurally organized in accordance with the central theme(s), and should be tied together by an extensive editorial introduction or set of introductions if the volume is divided into parts. An extensive bibliography and index are mandatory.

Nora Mills Boyd • Siska De Baerdemaeker •
Kevin Heng • Vera Matarese

Editors

Philosophy of Astrophysics

Stars, Simulations, and the Struggle to Determine What is Out There

 Springer

Editors
Nora Mills Boyd
Department of Philosophy
Siena College
Loudonville, NY, USA

Siska De Baerdemaeker
Department of Philosophy
Stockholm University
Stockholm, Sweden

Kevin Heng
University Observatory Munich,
Faculty of Physics
Ludwig Maximilian University
Munich, Germany

Vera Matarese
Department of Philosophy
University of Perugia
Perugia, Italy

Center for Space and Habitability
Univeristy of Bern
Bern, Switzerland

This work was supported by the Center for Space & Habitability (CSH) at the University of Bern, Switzerland, as part of its interdisciplinary initiative between astrophysics and philosophy of science. The research of the second Editor, Siska De Baerdemaeker, was funded by the Swedish Research Council, via Project number 1598801.

ISSN 0166-6991   ISSN 2542-8292   (electronic)
Synthese Library
ISBN 978-3-031-26620-1   ISBN 978-3-031-26618-8   (eBook)
https://doi.org/10.1007/978-3-031-26618-8

This Springer imprint is published by the registered company Springer Nature Switzerland AG
The registered company address is: Gewerbestrasse 11, 6330 Cham, Switzerland

We dedicate this volume to the memory of Ian Hacking (1936–2023), whose work inspired much of the scholarship in it.

Acknowledgements

We, the editors, are grateful to all of the authors contributing to this volume. Their work helps to establish the young field of philosophy of astrophysics. We also want to thank Otávio Bueno for his support throughout the process of putting this volume together. This project has grown out of a biweekly discussion group at the Center for Space Habitability at the University of Bern—we thank participants there for stimulating conversations. The open access publication of this work was made possible thanks to generous funding from the Center for Space & Habitability (CSH) at the University of Bern, Switzerland, based on a long-standing collaboration between the then CSH director (Kevin Heng) and the Institute of Philosophy (Claus Beisbart), which was led by a CSH Fellow in philosophy of science (Vera Matarese). Siska De Baerdemaeker was partially supported by a grant from the Swedish Research Council (project number 1598801).

Contents

About the Editors

Nora Mills Boyd is Assistant Professor of Philosophy at Siena College. Her research focuses on empiricism in philosophy of science, philosophy of experiment, and the philosophy of astrophysics and cosmology. After working as a Research Engineer at the University of Washington Center for Experimental Nuclear Physics and Astrophysics, she received her PhD in the History and Philosophy of Science from the University of Pittsburgh in 2018. She has published journal articles in *Philosophy of Science* and *Studies in History and Philosophy of Modern Physics.* Her Cambridge University Press Element *Epistemology of Experimental Physics* was published in the Philosophy of Physics series in 2021.

Siska De Baerdemaeker is Researcher at Stockholm University. Between 2023 and 2025, she will work on a Project Grant from *Riksbankens Jubileumsfond* on the epistemology of experiment in context of dark matter research. Before, she worked as a Postdoctoral Researcher with Richard Dawid on theory confirmation in fundamental physics. She also has research projects on the history of relativistic cosmology in the 1920s and 1930s. In April 2020, she received her PhD in History and Philosophy of Science from the University of Pittsburgh. Her work has been published in *Philosophy of Science, Studies in History and Philosophy of Modern Physics*, and *HOPOS*. Her Cambridge University Press Element *Philosophy of Cosmology and Astrophysics* is under contract in the Philosophy of Physics series.

Kevin Heng is Chair Professor (German: *Lehrstuhlinhaber*) of *Theoretical Astrophysics of Extrasolar Planets* at the Ludwig Maximilian University in Munich, Germany. He is Honorary Professor in the Department of Physics at the University of Warwick in the United Kingdom. Previously, he was the Director of the interdisciplinary Center for Space and Habitability (CSH) at the University of Bern, Switzerland. He is the recipient of a 2017 European Research Council (ERC) Consolidator Grant and the 2018 Chambliss Astronomical Writing Award of the American Astronomical Society (AAS). He is mainly interested in the theory, simulation, and phenomenology of the atmospheres of exoplanets, including atmospheric radiative transfer, chemistry, fluid dynamics, and Bayesian data inversion methods. Recently, he has turned his attention to the geosciences, especially geochemistry, as it is indispensable in the understanding of biosignatures and their false positives. He is the author of *Exoplanetary Atmospheres: Theoretical Concepts and Foundations* (2017, Princeton University Press). He has published over 170 peer-reviewed papers, including in the *Astrophysical Journal (ApJ), Astronomical Journal (AJ), Monthly Notices of the Royal Astronomical Society (MNRAS), Astronomy & Astrophysics (A&A), Nature, Nature Astronomy, Science,* etc.; he has also published two modeling papers with epidemiologists.

Vera Matarese holds a tenure track Assistant Professorship (RTDb) in Philosophy of Science at the University of Perugia (Italy). Before, she was a postdoctoral researcher in the Institute of Philosophy and Research Fellow in the Center for Space and Habitability at the University of Bern. She was also a member of NCCR PlanetS, and she served on the Center for Space and Habitability Management Committee. She was a visiting fellow at the Center for Philosophy of Science at the University of Pittsburgh in fall 2019, and she was the recipient of an EPSA visiting fellowship to join the FraMEPhys project at the University of Birmingham in fall 2022. Her research focuses on the metaphysics of science as well as on the methodology of science. She is the single author of different journal articles in *Synthese, Foundations of Physics, International Studies in the Philosophy of Science*, of a monograph with De Gruyter, and of book chapters in *Synthese Library* and *Routledge*.

Chapter 1
Introduction

Vera Matarese, Siska De Baerdemaeker, and Nora Mills Boyd

Abstract This volume is the first edited collection of philosophy of astrophysics. In this introductory chapter, we provide a brief history of the rise of philosophy of astrophysics as a distinct subdiscipline in philosophy of science, brief summaries of the chapters in the volume and their interrelated themes, and a few suggestions for further work.

The volume you have before you is the first edited collection specifically devoted to philosophy of astrophysics. Our primary aims in producing this volume have been to gather contemporary research in philosophy of astrophysics together in one place as both a reference resource for scholars already working in this subdiscipline and as an introduction to curious newcomers. Several contributions in this volume will also likely be of interest to philosophers working on topics such as idealization, validation, and analogy, which extend well beyond the specificity of philosophy of astrophysics. This introduction provides some background on the rise of philosophy of astrophysics as a distinct subject area, brief summaries of the contributions, and closes with a few suggestions for future work.

V. Matarese (✉)
Department of Philosophy, University of Perugia, Perugia, Italy

Center for Space and Habitability, University of Bern, Bern, Switzerland
e-mail: vera.matarese@unipg.it

S. De Baerdemaeker
Department of Philosophy, Stockholm University, Stockholm, Sweden
e-mail: siska.debaerdemaeker@philosophy.su.se

N. Mills Boyd
Department of Philosophy, Siena College, Loudonville, NY, USA
e-mail: nboyd@siena.edu

© The Author(s) 2023
N. Mills Boyd et al. (eds.), *Philosophy of Astrophysics*, Synthese Library 472,
https://doi.org/10.1007/978-3-031-26618-8_1

1.1   Philosophy of Astrophysics Until Today

Astronomy, the observational science of the positions, motions and properties of celestial bodies, has a long and storied history, with roots going back even to prehistoric times.[1] Astrophysics 'proper', the scientific discipline that applies the laws of physics and chemistry to provide a dynamical explanation of these astronomical observations, originated in the unification of celestial and terrestrial physics from the Scientific Revolution, but found its proper start with the development of spectroscopy in the nineteenth century. Already by the early twentieth century, a relatively detailed theory of stellar evolution and stellar structure had been developed. Today, astrophysics is still progressing with leaps and bounds. For instance, in the 30 years since the first exoplanet discovery in 1992, over 5000 have been identified. LIGO detected its first gravitational wave signal in 2015, but by 2020 the collaboration was reporting candidate events at a rate of more than one per week (Abbott et al. 2021). With the expected launch of more advanced experiments like LISA, Euclid and the recently launched James Webb Space Telescope, as well as the development of better simulations, it is only to be expected that this process will continue.

Analytic philosophers of science have started to show interest in astrophysics since the 1980s, but philosophy of astrophysics has only properly come into maturity in the last decade or so. Indeed, Cameron Yetman's complete overview of all papers and books (in English) in philosophy of astrophysics (included at the end of this volume) lists only 87 entries, three quarters of which were published since 2010. Although this introduction won't go through every single one of these entries, it is worthwhile to review some of the history of the field to further explain why now is an especially salient time for an edited volume in philosophy of astrophysics.

Arguably the first philosophical writing in analytic philosophy of astrophysics was a series of remarks by Ian Hacking (1982, 1983, 1989). His was a negative take on the discipline. Hacking first observed a lack of experiments in astrophysics,[2] which led him to a negative conclusion about entity realism about astronomical objects (or, at least, astronomical objects not observable with the naked eye, like black holes and gravitational lenses). But the 1989 paper went much further: it claimed that the methodology of astronomy and astrophysics is merely one of saving the phenomena, and that, because of this methodology, "astronomy is not a natural science at all" (1989, 577).

A significant part of philosophy of astrophysics has been influenced by or directly responds to Hacking's initial dismissal. The first, most direct, and broadest

---

[1] See e.g. (North 2008) for a detailed history. Chanda Prescod-Weinstein's reading list on decolonizing science (available here: https://medium.com/@chanda/decolonising-science-reading-list-339fb773d51f) is an excellent resource, providing important corrections to a common western-centric historical narrative.

[2] This is most clearly summarized by the famous line: "galactic experimentation is science fiction, while extra-galactic experimentation is a bad joke" (Hacking 1989, 559).

response came from Shapere (1993). Shapere both defended the scientific status of astronomy and astrophysics and argued against the coherence of Hacking's entity realism more generally. But parts of the debate also percolate through in discussions about astronomy as a historical science (Anderl 2016; Cleland 2002), realism and astrophysics (Leconte-Chevillard 2021; Martens 2022), and the nature of direct or indirect observations in astrophysics (Elder 2020; Sandell 2010).

Nonetheless, the initial controversy about Hacking's gambit did not spur the development of philosophy of astrophysics as its own sub-field in philosophy of science. During the 1990s and 2000s, any philosophy of science engaging with astrophysics tended to remain limited to a few individual researchers using case-studies from astrophysics to engage with ongoing debates in philosophy of science. While such engagement is important—one goal of the current volume is to show how astrophysics presents a unique perspective on many debates in philosophy of science—these papers rarely explicated the unique philosophical opportunities posed by astrophysics.

Even if those papers do not make the unique philosophical interest of astro-physics explicit, they do so implicitly. For instance, Bailer-Jones (2000) uses the case of extragalactic radio sources to illustrate challenges arising in modeling novel phenomena, something especially prevalent in astrophysics, where novel physics (from types of supernovae to dark matter) is lurking around every corner. Cleland (2002) includes astrophysics as one example of a historical science in her discussion of the methodological differences between historical and experimental sciences. Insofar as astrophysics is reconstructing the past, there is an important contrast with areas like paleontology or evolutionary biology: there are strict constraints from established theories of physics. Ruphy (2010) shows how stellar kinds bear onto the natural kinds debate. And, finally, Salmon (1998) highlights the challenge of distinguishing between pseudo-processes and causal processes in astrophysics.[3]

Thus, a coherent body of work focusing specifically on philosophical questions arising in astrophysics remained wanting throughout the 1990s and 2000s. But the first seeds were already there. Aside from the aforementioned papers, it is also worth mentioning Bill Vanderburgh (2003, 2005) laying the groundwork for philosophy of dark matter. And in science and technology studies, the 2000s see an ongoing discussion about, e.g. the categorization of moons and planets (Messeri 2009; Metzger et al. 2022), and later on about the creation of telescope images (English 2017; Greenberg 2016) and of simulations (Sundberg 2010, 2012).

---

[3] The chapter is a hidden gem towards the end of Salmon's 1998 book on causation. It includes a full record of Salmon's correspondence with astrophysicists about a controversy about size measurements, as well as great personal anecdotes like the following: "In order to display [the shape of a spiral galaxy], I looked through our home collection of old LP records and serendipitously came upon "Cosmo's Factory" by the Creedence Clearwater Revival, a happy discovery given that we are interested in various types of engines in the cosmos, a clearing of the waters muddied by invalid arguments, and in a revival of credence in theories or models of such engines" (377–378).

From 2010 onwards, the philosophical literature shows a significant shift. Certain central themes in philosophy of astrophysics 'proper' started to crystallize out, this time in tandem with, but no longer solely in service of ongoing debates in philosophy of science. The annotated bibliography at the end of this volume gives a comprehensive overview divided into seven categories. Here, we close the historical overview by highlighting three themes that have garnered most attention since 2010—three themes that are also reflected by the contributions in this volume.

First, there is the question of how astrophysicists come to gather empirical evidence. No current philosopher of astrophysics would want to lapse into Hacking-style skepticism about the scientific status of astrophysics. But the question still stands of how astrophysical models are constrained by what Quine simply referred to as the 'tribunal of experience'. This becomes especially pressing when scientists are aiming to detect signals of novel physics that are buried in noise, like in the case of gravitational wave astronomy. A second theme is the epistemology of computer simulations. Astrophysicists use computer simulations to draw out empirical consequences of theoretical models, or to extend the epistemic reach of observations. But how is the reliability of simulations themselves established? Third, there is now quite an extensive literature on philosophy of black hole astrophysics. The aforementioned epistemology of gravitational wave astrophysics fits here, but also the recent debate about analogue gravity experiments.

From this brief historical overview, it is clear that philosophy of astrophysics has finally come to fruition. The contributions in this volume, which we summarize in the next section, represent how broad the discussion has become.

1.2   Philosophy of Astrophysics in This Volume

The contributions to this volume expand upon predominant themes of the extant philosophy of astrophysics literature. The book opens with a contribution by **Boyd**, which addresses the provocative challenge posed by Hacking mentioned above. While Hacking denies the empirical status of astrophysics due to the lack of experiments, Boyd attentively considers the field of laboratory astrophysics experimentation, which is often carried out by appealing to similarity arguments. In particular, she illustrates the case of laboratory supernova research carried out at the National Ignition Facility, which includes experiments studying the Rayleigh-Taylor hydrodynamic instability and based on the hydrodynamic similarity between terrestrial and celestial physics. While her conclusion cautiously warns against the purported epistemic significance of the particular experiment that is the subject of her case study, she also suggests that the division between experimental and non-experimental sciences is of little significance when evaluating the empirical status of astrophysics and, in general, any scientific discipline. Rather, according to her, attending to the empirical data and focusing on their causal chain can better illuminate the external validity of astrophysics research.

While Boyd convincingly underplays the significance of Hacking's challenge, she demonstrates a way of assessing the epistemic authority of astrophysics that nests complex epistemic challenges, which are beautifully addressed by the other contributions of this volume. In particular, the attention to how data is collected and to the enriched empirical evidence that they provide spurs questions on the reliability, validity and objective value of data, and on the connection between the results of astrophysical observations and theory.

Elder's and Patton's contributions richly exemplify the problem of data theory-ladenness and the hybridization of theoretical and empirical reasoning. **Elder**'s paper discusses both the vices and virtues of interdependence in theory testing by illustrating the case of the LIGO-Virgo experiments methods. Thanks to their very first direct detection of gravitational waves, the LIGO-Virgo experiments have opened the path not only for a successful observation of the universe, but also for a rigorous test of General Relativity (GR). The concern, however, is that the theory-ladenness of the LIGO-Virgo methods leads to a potentially vicious epistemic circularity, where GR assumptions and models may serve to interpret data that are actually inconsistent with GR as consistent. While the author clearly articulates the complex layers of theory-ladenness involved in the LIGO-Virgo experiments, and the threats of the vicious circularity involved, her conclusion is optimistic. Elder shows how the problem of circularity can be satisfactorily mitigated by leveraging improvements in modeling, simulation, or observation in one domain to place constraints in another. **Patton**'s contribution pursues a similar line of thought. She argues against the direct empiricist perspective according to which data are treated as windows on the world and as reflections of reality, by illustrating the case of population synthesis methods, which employ theories and models in analyzing data and in simulations. Stellar evolution theory is the foundation of population synthesis, and models recruit theoretical and empirical stellar libraries to generate simulations. The progressive and stunning observational development and the obtainment of higher-resolution and more precise empirical data are accompanied with a more numerous and more sophisticated theoretical and modeling resources. In particular, Patton shows that the stellar population synthesis methods not only use theories and models to interpret and analyze the data, but also necessarily need them to measure physical parameters: the physical variables that are the target of population synthesis cannot be even measured without employing significant theoretical resources.

The contribution by **Martens and King** examines another important problem concerning data, which is a case of underdetermination of data by two theories that are not perfectly empirically equivalent and are even not perfectly empirically adequate. The case presented is on dark matter and modified gravity. While the ΛCDM-model is still affected by small scale problems, modified gravity is unable to provide an accurate description of galaxy clusters and cosmological observables. Thus, they argue that in this case, the presumption of solving the underdetermination by an attentive examination of the empirical data is bound to fail. Martens and King provide a thorough theoretical discussion of both theories with regard to two theoretical virtues, which are unificatory power and simplicity.

Gueguen's contribution also deals with the problem of discord between different programs, in this case, of the Hubble constant controversy, which has been sometimes labeled as a crisis. Her approach is not theoretical but follows Boyd's suggestion of attending to the causal chain of empirical data and how they are collected: it provides an attentive analysis of how astrophysicists check for the errors affecting their measurements. Gueguen's contribution on the Hubble constant controversy showcases the intricate process of cross-checking different results in order to detect unknown systematic errors by the use of systematic replications and robustness analysis. While one well-trodden path would be to use robustness arguments to take the discrepancy of results of the Hubble constant value recently obtained as a clear sign of a crisis, after an attentive analysis of how the measurements are carried out, Gueguen warns against a precipitous evaluation of this case as a 'crisis' and endorses a more cautious approach that highlights the need for a better assessment of the presence of systematic errors. In this sense, her conclusion challenges those who have claimed there is a crisis in astrophysics: while the failure of systematic replications offered by the Tip of the Red Giant Branch and the Cepheids' teams and the consequent lack of robustness of results inform us of how their measurements can be further improved, it would be epistemically unjustified to use it to support a crisis of astrophysics.

While epistemic challenges are ubiquitous in astrophysics research, the extensive use of simulations is often regarded as a source of concerns that is more worrying than others. Indeed, simulations are often regarded as unsatisfactory to act as epistemic authorities in an empirical field, as they lack one of the most important components of experimental research, which is manipulation of the target system. In her contribution, **Abelson** challenges the common lore that simulations can play the role of experiments. Her thesis, however, is cautious. While she remains reluctant to regard simulations as experiments, she shows how a certain kind of astrophysical simulation can be regarded as conceptual experiments. These are dynamical simulations of temporal systems, which instantiate a significant amount of empirical temporal data and achieve a higher level of representational adequacy.

The contribution by **Kadowaki** concerns the epistemic justification of simulation as well. While a common practice to check for the reliability and trustworthiness of simulation would require the separation of the numerical/computational aspects of simulation from the relation of the simulation to its real-world target system, and separate the process of Verification from the process of Validation, the author argues that this is not epistemically advisable. Kadowaki supports his claim with a survey of the verification tests used in selected magnetohydrodynamics simulations. This case study shows that verification tests are not mere tests of numerical fidelity, as they also involve an exploration of the domain of possible real-world systems and of the space of simulation code types.

Another contribution that deals with the problem of the epistemic standing of simulations is the chapter by **Meskhidze**. She discusses code comparison, which is a method to check for the reliability and trustworthiness of computer simulations, and which has been criticized for relying on shaky grounds, as it is arguably not possible to achieve a good balance between difference and similarity to allow for a fair and informative comparison. Meskhidze presents a project she joined, which investigated two different implementations of self-interactions amongst dark matter particles in two computer simulation codes. In this case, the code comparison was epistemically informative, as the simulation outputs were diverse enough for an informative comparison and yet still comparable. Her conclusion is both optimistic and cautious: it shows that code comparisons, in cases where it is conducted as a part of eliminative reasoning, can be used to increase our confidence in computer simulations.

Along the same lines, **Gallagher and Smeenk** evaluate the reliability of simulation in spite of the challenge of 'uncomputed' alternatives, by examining the case of quasar formation. The problem of uncomputed alternatives is a type of selection effect that results from neglecting certain physically plausible scenarios because they are computationally intractable. In the case of quasar formation, some plausible explanations for the triggering of quasar activity have not been explored using simulations, and therefore have not been subjected to detailed observational evaluation, because of their computational intractability.

Another reason why the epistemic authority of simulations has been regarded with suspicion is its extensive use of idealization. **Jacquart and Arcadia**'s contribution deals with the problem of idealization. Their case study of Collisional Ring Galaxies simulations provides a perfect platform to analyze the nature of different kinds of idealization, their epistemic roles, and to discuss the delicate and sophisticated process of de-idealization involved in simulations. As their contribution shows, this process may involve different strategies, ranging from 're-composing' by adding back in features into a model that were at one point idealized, to 'reformulating', 'concretizing', and 'situating'. The authors highlight that these de-idealizations cannot be done as a simple reversal, and that they are processed according to the various aims and goals of the astrophysicist team.

The problem of idealization is very much connected to the problem of the extensive use of 'fictions' in models. **Suárez'** contribution summarizes the recent development of the field of asteroseismology and discusses its use of fictional posits, which are employed as effective means in allowing modellers to generate expedient predictions for observable quantities. While fictional assumptions have no further cognitive value beyond the convenience of their expediency, some of them have turned out to be just false idealizations. New asteroseismical methods, indeed, have produced knowledge regarding the energy transfer mechanisms inside a multitude of stars of different types, and this has shown that the equilibrium, spherical symmetry, and uniform composition assumptions do not operate as fictions, but are rather better understood as false idealizations.

The challenges that astrophysics has to face not only arise from the scientific tools used in their method, but also because of the nature of the objects of its investigation. This volume has dedicated one whole section to black holes, as the difficulties that hinder empirical access to black holes have encouraged the development of new epistemic techniques. These techniques range from indirect observation by the observation of their interaction with ordinary matter to analogical reasoning. **Mathie**'s chapter explores two uses of analogical reasoning regarding black holes, and the connections between them. On one hand, black hole thermodynamics relies on an analogical relationship between radiation from astrophysical black holes and radiation from ordinary thermodynamic systems. On the other hand, analog gravity experiments rely on arguments connecting analog systems displaying analog event horizons (in water, for example) to astrophysical black holes. Mathie argues that while physicists have generally been far more comfortable accepting the validity of black hole thermodynamics than analog gravity experiments, the analogical argument underpinning the former relies on input from the latter. In particular, black hole thermodynamics relies on the existence of astrophysical Hawking radiation, the evidence for which is only indirectly provided by the (to some, dubious) analog gravity experiments. Mathie considers, and ultimately rejects several strategies for avoiding this dependence.

Our final chapters by **Doboszewski and Lehmkuhl,** on the one hand, and by **Allzén** on the other hand, offer a complete overview of the epistemic challenges due to the nature of black holes and discuss several arguments for why our epistemic position with regards to black holes is problematic. Among other problems, they discuss that our epistemic access to black holes is not direct but indirect, and that black holes fail to be experimentally manipulable in a way that makes them deserving of a realist attitude, following not only Hacking's entity realism, but, as the paper by Allzén points out, also Cartwright's and Chakravartty's realist views. While Doboszewski and Lehmkuhl argue that all arguments supporting a failure of scientific realism are not convincing, Allzén's paper accepts that entity realism, as it stands, is not compatible with a realist attitude towards black holes. However, instead of supporting an anti-realist conclusion, the author seems to encourage a radical revisitation of our traditional realist criteria, according to the contemporary epistemic practices of astrophysics.

Following the contributed chapters, you will find a short essay titled "Reflections by a Theoretical Astrophysicist", written by our co-editor Kevin **Heng** in response to the contributions included in this volume. Heng's essay provides valuable insights for philosophers of science working in philosophy of astrophysics and more general topics such as modeling and simulation. He contrasts the practices and heuristics of working scientists, which are rarely explicitly mentioned in science publications, with the seemingly high standards of philosophers. Heng also notes several issues to which philosophers may wish to pay more attention, such as the inescapable influence of discretization in computer simulations and the unsolved problem of turbulence.

Taken together, we hope that the elements of this volume spark the further acceleration of valuable work in philosophy of astrophysics. From our vantage point, there are many fascinating avenues for future work. The ongoing engagement of philosophers with the Event Horizon Telescope is sure to produce additional illuminating scholarship on the relationships between theories, models, and empirical data, as well as the nature of astrophysical black holes. Exciting forays into philosophy of astrochemistry are currently underway. The domain of exoplanet research and the connections between planetary astrophysics, atmospheric, and climate science remain largely to be explored. Further case studies on the methodology and epistemology of laboratory astrophysics research, such as the formation of protoplanetary disks from low pressure dust, would undoubtedly enrich our understanding of the epistemology of experiment. While philosophers of astrophysics have investigated simulations of galaxies and galaxy clusters, and models of stars, the advanced stellar structure simulations have yet to receive due attention. Nurturing interdisciplinary collaborations between astrophysicists and philosophers will surely surface even further unforeseen questions and research topics and help strike the appropriate balance between fidelity to scientific practice and philosophical interest. Whatever directions the field ultimately takes, stars, simulations, and the struggle to determine what is out there will undoubtedly continue to inspire philosophical scholarship for many years to come.

References

Abbott, R., et al. 2021. GWTC-2: Compact Binary Coalescences Observed by LIGO and Virgo During the First Half of the Third Observing Run. *Physical Review X* 11 (2): 21053. https://doi.org/10.1103/PhysRevX.11.021053.

Anderl, Sibylle. 2016. Astronomy and Astrophysics in the Philosophy of Science. In *The Oxford Handbook of Philosophy of Science*. Oxford: Oxford University Press.

Bailer-Jones, Daniela M. 2000. Modelling Extended Extragalactic Radio Sources. *Studies in History and Philosophy of Science Part B – Studies in History and Philosophy of Modern Physics* 31 (1): 49–74.

Cleland, Carol E. 2002. Methodological and Epistemic Differences Between Historical Science and Experimental Science*. *Philosophy of Science* 69 (3): 447–451.

Elder, Jamee. 2020. *The Epistemology of Gravitational-Wave Astrophysics*. University of Notre Dame. https://curate.nd.edu/show/3f462517k8t.

English, Jayanne. 2017. Canvas and Cosmos: Visual Art Techniques Applied to Astronomy Data. *International Journal of Modern Physics D* 26 (4): 1730010.

Greenberg, Joshua M. 2016. Creating the "Pillars": Multiple Meanings of a Hubble Image. *Public Understanding of Science* 13 (1): 83–95.

Hacking, Ian. 1982. Experimentation and Scientific Realism. *Philosophical Topics* 13 (1): 71–87.

———. 1983. *Representing and Intervening*. Cambridge: Cambridge University Press.

———. 1989. Extragalactic Reality: The Case of Gravitational Lensing. *Philosophy of Science* 56 (4): 555–581.

Leconte-Chevillard, Gauvain. 2021. Experimentation in the Cosmic Laboratory. *Studies in History and Philosophy of Science* 90 (September): 265–274. https://doi.org/10.1016/j.shpsa.2021.10.005.

Martens, Niels C.M. 2022. Dark Matter Realism. *Foundations of Physics* 52 (1): 1–19. https://doi.org/10.1007/s10701-021-00524-y.

Messeri, Lisa R. 2009. The Problem with Pluto: Conflicting Cosmologies and the Classification of Planets. *Social Studies of Science* 40 (2): 187–214.

Metzger, Philip T., et al. 2022. Moons Are Planets: Scientific Usefulness versus Cultural Teleology in the Taxonomy of Planetary Science. *Icarus* 374: 114768. https://doi.org/10.1016/j.icarus.2021.114768.

North, John. 2008. *Cosmos: An Illustrated History of Astronomy and Cosmology*. Chicago: University of Chicago Press.

Ruphy, Stéphanie. 2010. Are Stellar Kinds Natural Kinds? A Challenging Newcomer in the Monism/Pluralism and Realism/Antirealism Debates. *Philosophy of Science* 77 (5): 1109–1120. https://doi.org/10.1086/656544.

Salmon, Wesley C. 1998. Quasars, Causality and Geometry: A Scientific Controversy That Did
    Not Occur. In *Causality and Explanation*, 369–384. Oxford: Oxford University Press.
Sandell, Michelle. 2010. Astronomy and Experimentation. *Techne* 14 (3): 252–269.
Shapere, Dudley. 1993. Astronomy and Antirealism. *Philosophy of Science* 60 (1): 134–150.
Sundberg, Mikaela. 2010. Cultures of Simulations vs. Cultures of Calculations? The Development
    of Simulation Practices in Meteorology and Astrophysics. *Studies in History and Philosophy of
    Science Part B – Studies in History and Philosophy of Modern Physics* 41 (3): 273–281. https:/
    /doi.org/10.1016/j.shpsb.2010.07.004.
———. 2012. Creating Convincing Simulations in Astrophysics. *Science, Technology & Human
    Values* 37 (1): 64–87.
Vanderburgh, William L. 2003. The Dark Matter Double Bind: Astrophysical Aspects of the
    Evidential Warrant for General Relativity. *Philosophy of Science* 70 (4): 812–832.
———. 2005. The Methodological Value of Coincidences: Further Remarks on Dark Matter and
    the Astrophysical Warrant for General Relativity. *Philosophy of Science* 72 (5): 1324–1335.

Part I
Theory, Observation, and the Relation Between Them

Chapter 2
Laboratory Astrophysics: Lessons for Epistemology of Astrophysics

Nora Mills Boyd

Abstract Astrophysics is often cast as an observational science, devoid of traditional experiments, along with astronomy and cosmology. Yet, a thriving field of experimental research exists called laboratory astrophysics. How should we make sense of this apparent tension? I argue that approaching the epistemology of astrophysics by attending to the production of empirical data and the aims of the research better illuminates both the successes and challenges of empirical research in astrophysics than evaluating the epistemology of astrophysics according to the presence or absence of experiments.

Keywords Experiment · Observation · Astrophysics · Dimensional analysis · External validity · Hydrodynamics

2.1 Introduction

If they mention astrophysics at all, philosophers of science often claim that experiments are impossible in astrophysics. The purported lack of experiments in astrophysics is usually taken to be a shortcoming of the field, an epistemic handicap. Indeed, the lack of experiments is painted as one of the most distinctive features of the epistemology of astrophysics in contrast to the so-called experimental sciences, thereby motivating special attention by philosophers of science. For example, Morrison (2015) and Jacquart (2020) have argued that, while lacking traditional experiments is a prima facie problem for astrophysics, astrophysicists successfully supplement their methodological toolbox by using computer simulations instead of experiments. Thus, the purported lack of experiment in astrophysics serves as a premise for arguments that simulation is an apt replacement for empirical research in astrophysics: "In the astrophysics case we may want to say that simulation is

N. M. Boyd (✉)
Department of Philosophy, Siena College, Loudonville, NY, USA
e-mail: nboyd@siena.edu

© The Author(s) 2023
N. Mills Boyd et al. (eds.), *Philosophy of Astrophysics*, Synthese Library 472,
https://doi.org/10.1007/978-3-031-26618-8_2

an acceptable source of experimental knowledge simply because we are unable to conduct materially based experiments in the way we can with other types of systems" (Morrison 2015, 214).[1]

Rather than take up the question of whether simulations can really serve as an apt replacement for empirical research here (for the record: I doubt they can), I want to focus on the prior issue already assumed in arguments such as those of Morrison and Jacquart, regarding the role of experiments in the epistemology of astrophysics. Is it really the case that there are no experiments in astrophysics?

However we ultimately want to answer that question, we must admit that it is certainly the case that there are many experimental physics laboratories that identify themselves as dedicated to astrophysical research. The University of Washington's Center for Experimental Nuclear Physics and Astrophysics (CENPA), the Compact Accelerator System for Performing Astrophysical Research (CASPAR) at the former Homestake Gold Mine, the Laboratory for Underground Nuclear Astrophysics (LUNA) at Gran Sasso, and the Laboratory Astrophysics branch of Harvard's Center for Astrophysics are just a few examples. This prevalence of 'laboratory astrophysics' in contrast to the philosophers' denial of experiments in astrophysics raises a bit of a puzzle. Do researchers at these laboratories conduct astrophysics experiments after all? And how does the answer to that question reflect back on the epistemology of astrophysics—on what we can hope to learn through empirical research in astrophysics?

This chapter will argue that powerful similarity arguments available in physics can sometimes span terrestrial laboratory experiments and celestial systems. In other words, there are indeed experiments in astrophysics. But, like all external validity arguments, these powerful similarity arguments have limitations and can break down. Care must therefore be taken to ensure the conditions that support the desired argument obtain in the intended domains. In Sect. 2.2, I briefly discuss some relatively straightforward examples of laboratory astrophysics that illustrate both its long pedigree and how manipulating material in a terrestrial laboratory can count as astrophysical research. These examples show that astrophysics is not a purely observational science. In Sect. 2.3, I present a more detailed case study of a laboratory research that I will eventually argue (in Sect. 2.4) does not quite succeed in attaining its astrophysical aspirations. The reasons for this particular shortfall are instructive—they demonstrate the crucial importance of establishing that the appropriate conditions obtain to support the intended similarity argument. The final section highlights the main methodological lesson for philosophers of science interested in understanding the epistemology of astrophysics in practice.

---

[1] Jacquart (2020) argues that simulations can be used for hypothesis testing in astrophysics: "because of the methodological challenges in astrophysics, comparison with observational data is extremely limited and in some cases impossible because there are no observations [. . .] I think it is clear that the simulations are not just testing a model but are playing the role of hypothesis testing in astrophysics [. . .] While a direct experiment would be helpful, as discussed above, for these kinds of systems in astrophysics this is the only means by which hypotheses can be tested" (1215).

The distinctions that we use to structure our inquiry can be fruitful for understanding science in practice or they can lead us astray. The distinction between observation and experiment has not served us well in appreciating the moves and arguments germane to empirical astrophysics. Instead, it is more fruitful to structure our inquiry by attending to what researchers in astrophysics are trying to study and to what in fact they have empirical access. In short: it's not whether it's an experiment that matters, it's how you use it.

2.2   Astrophysics as So-Called Observational Science

Astrophysics is often lumped under the description 'observational science' with fields like astronomy and cosmology. In the same breath, the lack of traditional experiments in astrophysics is taken to be an epistemic problem for astrophysics. The most extreme denigrator of astrophysics is undoubtedly Ian Hacking. In "Extragalactic Reality: The Case of Gravitational Lensing" Hacking quipped: "Galactic experimentation is science fiction, while extragalactic experimentation is a bad joke" (1989, 559). He explained, "the method of [astrophysics] is the same as that of astronomy in hellenistic times. Model, observe, and remodel in such a way as to save the phenomena" and in contrast, "[n]atural (experimental) science is a matter not of saving phenomena but of creating phenomena [...] But in astrophysics we cannot create phenomena, we can only save them" (577–578). Indeed he went so far as to say that "*astronomy* is not a natural science at all" and thus by implication, because it shares the same method, neither is astrophysics (577). This view of natural science is clearly too restrictive. Experiment, interference, and creation are not *necessary* for properly scientific research—surely at least *some* research in astronomy and astrophysics counts as bona fide natural science. However, Hacking is not alone in expressing the view that there's something wrong with astrophysics on account of the lack of experimentation in that field. We see this view reflected in the more recent work of some philosophers of astrophysics, as when Sibylle Anderl writes: "Astrophysics and cosmology share a common *problem* in that they both need to acquire knowledge of their objects of research without directly interacting, manipulating or constraining them" (2016, 653, my emphasis) and when Melissa Jacquart writes "Astrophysics faces methodological *challenges* as a result of being a predominantly observation-based science without access to traditional experiments" (2020, 1209, my emphasis). The common thought seems to be: experiments are impossible in astrophysics and astrophysics is epistemically poorer than it otherwise would be on that basis.

However, as I have already mentioned, it is not clear that astrophysics actually lacks experiments. In fact, astrophysics was born in the laboratory. The birth of astrophysics came with the application of physics to astronomy, in particular with the application of spectroscopy to light from the sun, and then to stars and nebulae (Becker 2011; Hearnshaw 2014). By comparing spectra thrown from elements committed to flame, arc, and spark in the laboratory to spectra from celestial sources, spectroscopists were able to match terrestrial sources with celestial ones. With this came the revolutionary possibility of determining the presence of particular elements in astronomical bodies (their chemical composition), and of determining the relative line-of-sight motion of such bodies via the determination of astronomical redshifts, thereby allowing for the addition of depth to our maps of cosmic structure.

Since those early days of astrophysics, the field has gained tremendous scope and embraced new aims and projects. Astrophysicists still use the chemical composition and redshift of celestial sources in their research, but also seek to understand the dynamical evolution of astronomical objects, processes, and systems, and the physical mechanisms in play. They investigate the causes and evolution of supernovae and their remnants, the formation of stars, planets, and galaxies, the flow of energy and material, the interactions of plasma, gravity, magnetic fields, and so on.

Still, in some ways just like in the early days of astronomical spectroscopy, astrophysics is about understanding the application of physics to astronomical targets and that application is often carried out in physics laboratories. There is a venerable branch of experimental physics devoted to accelerator-based nuclear astrophysics. With terrestrial accelerators, nuclear physicists can, and have, studied nuclear decay chains of astrophysical interest. Consider, for example, research on the second-forbidden beta decay of Boron-8. Solar neutrinos are produced by a combination of different nuclear reactions in the sun, and each of these needs to be carefully characterized in order to compare predictions to data from solar neutrino detectors. Although they are quite rare, some of the highest energy solar neutrinos originate from the second-forbidden beta decay of Boron-8 into the ground state of Berylium-8.[2] Nuclear physicists have studied the Boron-8 decay spectrum using terrestrial accelerators such as the University of Washington's Tandem Van de Graaff accelerator (Bacrania et al. 2007). In such nuclear physics experiments, researchers create conditions in the laboratory using ion sources, accelerators, and detectors to study the same kind of physical processes occurring elsewhere in the universe. Insofar as Boron and beta decays on Earth are of a kind with Boron and beta decays off-Earth, terrestrial accelerator experiments can study the very same kind of physical processes in the laboratory that are of astrophysical interest (see also Evans and Thébault 2020, Section 3). This is indeed 'experimental nuclear physics and astrophysics.'

There are also efforts to detect dark matter from our galactic halo in laboratory settings—that is, *not* waiting for celestial messengers to travel very long distances from their native environments to interact with detectors waiting to receive them on Earth, but rather capitalizing on the fact that our planet is swimming in a cosmic sea. For instance, some of these laboratory dark matter searches are being

---

[2] Characterizing these branching ratios is important because if their decays are numerous enough, they could serve as a significant background for solar neutrino research, which at present is one of the main empirical access points to physics beyond the Standard Model of particle physics.

conducted using instruments that have been called 'haloscopes' because they aim to detect dark matter from our Milky Way's own galactic halo right here on Earth. The Axion Dark Matter eXperiment (ADMX) is one example. ADMX aims to detect the signal of dark matter axions in a microwave cavity inside a powerful superconducting magnet housed in the basement of the Center for Experimental Nuclear Physics and Astrophysics at the University of Washington. The thought is that if galactic dark matter is composed of axions (undoubtedly a big "if"), then the magnetic field of the ADMX instrument will sometimes interact with these halo axions and produce a detectable signal. The dark matter axions would be expected in the laboratory microwave cavity, because we, the laboratory, and the cavity, are all riding along inside the Milky Way's dark matter halo—we're swimming in the stuff. In this research, the axions (if they exist) are not traveling from afar to be received by passive detectors. Rather, the experimental apparatus is intervening on the halo axions present in the laboratory via the strong magnetic field in the cavity. ADMX is just one example of laboratory research on an astrophysical target, from the relative comfort of our own planet. Empirical astrophysical research has also involved attempts at producing dark matter candidates using terrestrial particle accelerators (see e.g. Giagu 2019 and references therein).[3]

In short, astrophysics investigates the nature of celestial objects and processes using a suite of resources from physics, and some of that research—laboratory astrophysics—involves research in terrestrial laboratories. Laboratory astrophysics, even from its origins with astronomical spectroscopy, has involved studying conditions relevant to physics in space in laboratory settings, for instance by empirically investigating the spectra associated with different chemical elements and the spectra of decaying nuclei that occur throughout the universe. Thus, laboratory astrophysics includes investigation of physical phenomena that occur in both on-world and off-world settings. What makes astrophysics *astrophysics* is that it investigates the nature of celestial objects and processes using a suite of resources from physics. And what makes laboratory astrophysics *laboratory* astrophysics, is that it carries out such investigations using terrestrial experiments.

The very existence of laboratory astrophysics seems to undermine the 'no experiments in astrophysics' maxim we often see in philosophy of astrophysics. Moreover, the existence of laboratory astrophysics experiments might be surprising to those who conceive of astrophysics as a characteristically observational science (together with astronomy and cosmology). This surprise could lead us to expand our conception of astrophysics and to see the field as involving both observational and experimental research. Of course, someone like Hacking could still respond that the experiments employed in astrophysical research do not involve experimenting upon genuinely astrophysical targets—such as stars, black holes, supernovae and galaxies—and that it is this latter type of experimentation that would be relevant for

---

[3] The story regarding analog black hole experiments is related, but more complicated. See Unruh 1981, Dardashti et al. 2017, 2019, Evans and Thébault 2020, Crowther et al. 2021, Field 2021, and Field manuscript. See also the contribution from Alex Mathie, Chap. 14 in this volume.

being promoted to the status of 'experimental science' and thus for the epistemic status of astrophysical knowledge.[4] But I think that this response misses what is so fascinating about the examples of laboratory astrophysics I have highlighted. Accelerator-based nuclear astrophysics, haloscope experiments, and dark matter production experiments all experiment upon targets that are instances of physical types that occur both on Earth and in space. As I will discuss further below, if one is unwilling to countenance these experiments as astrophysical experiments, then one should also be unwilling to countenance most laboratory experiments as intervening on their targets in the relevant sense since in virtue of being conducted in the laboratory, laboratory experiments do not intervene on instances of their targets in the wild, but rather on instances of the relevant type located in the laboratory. This would be counterproductive to the project of someone like Hacking, who certainly would not want to undermine the epistemic usefulness of all laboratory experiments. Of course, arguments do need to be furnished to support the crucial claim that the instances in the laboratory belong to the relevant type, and these arguments are not always successful (as indeed my primary case study below will illustrate). This is a general challenge for scientific research however, not a specific handicap of astrophysics.

For my own part, I think that noting the fact that there are experiments in astrophysics and that thus astrophysics is not a purely observational science is not, in itself, particularly interesting. This is because I think that a field can be empirical without performing experiments.[5] Indeed, I claim that the existence of laboratory astrophysics betrays the unhelpfulness of the distinction between observation and experiment for philosophy of astrophysics.[6] Ignoring that distinction, and replacing it with another framework allows us to better notice and theorize the epistemically significant aspects of laboratory astrophysics. This alternative framework helps us to see where the 'epistemic action' really is, in a way that is obscured when we approach this field of research with questions about observations versus experiments. What is philosophically interesting about laboratory astrophysics is not the existence of astrophysics experiments simpliciter, but rather the method-ological and epistemological strategies that researchers use to study astrophysics in laboratory settings. Instead of attending to the distinction between 'observation' and 'experiment' (or 'observational science' and 'experimental science') in our

---

[4] An anonymous reviewer helpfully suggested this possible response on behalf of my interlocutors.

[5] Thank you to an anonymous reviewer for pressing me to clarify this point.

[6] In fact, I think the distinction between observation and experiment is largely unhelpful for the epistemology of science more broadly, not just in the context of astrophysics. In a separate manuscript, coauthor Dana Matthiessen and I argue against the usefulness of this distinction in general (manuscript). We argue that philosophy of science ought to shift its focus to other features of empirical research methods that better track the epistemic benefits of methods that researchers choose between in practice. Here, I want to come at these issues from a different angle: the framework premised on there being an important epistemic difference between observational and experimental sciences is unilluminating for the epistemology of much significant empirical research in astrophysics.

investigation of the epistemology of astrophysics, we should attend to the production of empirical data. When we pay attention to the production of empirical data against the backdrop of the aims of the research, we can better resolve the challenges and opportunities of the field, and we can better appreciate the continuity of astrophysics with other fields of empirical research while also remaining sensitive to any distinctive philosophically interesting features it may have. In the following section, I present a case study that clearly shows the advantages of attending to the production of empirical data rather than the presence or absence of experiment for understanding the epistemology of the research.

2.3 Laboratory Supernova Research and Physical Similarity Arguments

Some laboratory astrophysics research purports to investigate instances of physical phenomena that occur in both on-world and off-world settings via laboratory experiments. How is the epistemology of this research supposed to work exactly? To get some purchase on this question, I am going to consider a particular laboratory astrophysics experiment in detail, so that we can investigate what is involved in practice, and how it is all supposed to hang together.

The particular case I am about to describe is philosophically valuable because the premise of the experiment—that we can study supernovae by shining lasers on plastic and foam in an Earthly laboratory—is, on its face, peculiar enough to teach us something interesting about what doing empirical astrophysical research is like in practice. This is what drew me to the case in the first place. But as I worked deeper into the details, I was surprised to find that the interpretation of the results of this particular research that the scientists offer does not quite go through. So ultimately, I will also argue that this case sheds light on epistemic challenges in laboratory astrophysics.

Research at the National Ignition Facility (NIF) at Lawrence Livermore National Laboratory predominantly focuses on laser confinement for fusion. At peak power, NIF focuses 192 laser beams on a small volume of material ("about the size of a pencil eraser"), delivering more than 2 million joules.[7] Studying matter in such high-energy-density states also has applications beyond the energy sector. When they are not trying to advance fusion technology, NIF researchers use this laser facility to study nucleosynthesis in stars and supernovae, instabilities in supernovae, opacity of stars, black hole accretion, nuclear reactions in stars, and planetary interiors—in short: astrophysics.[8]

For present purposes, I want to focus on a particular paper, Kuranz et al. (2018a), which published some of the NIF laboratory astrophysics results. In this paper, the

---

[7] https://lasers.llnl.gov/about/what-is-nif

[8] https://lasers.llnl.gov/science/discovery-science

authors report results from a series of NIF experiments first designed in 2009 that aimed to study the Rayleigh-Taylor (RT) hydrodynamic instability, which occurs at the boundary between fluids of different densities, where the lower density fluid is somehow being pushed into the higher density one. At the interface between the fluids, characteristic finger-like shapes develop and then evolve mushroom-cap like tips that coil and expand. You may have seen something similar while pouring cream into coffee (if you had a clear cup). This instability is thought to occur in supernovae at the interface between the forward shock moving outwards into the relatively low density circumstellar medium around the exploding star, and the induced reverse shock in the relatively high density expanding stellar ejecta. The NIF researchers wanted to investigate the possible effects of high-energy fluxes on the structure of RT instabilities in supernovae. In particular, they were interested in whether or not material would be removed from the interface between the two shocks in cases where the instability is evolving under high-energy-flux conditions (Kuranz et al. 2018a, 2–3). Understanding the evolution of the remnants has implications for studying the timing of supernovae and, relatedly, supernova progenitors and the physical mechanisms that drive the explosions.

To study this phenomena in the laboratory, the NIF researchers use their powerful laser system to create such a shock in a test target: a little plug of plastic and foam. To do this, the NIF laser system is focused on a small holhraum (a cavity), which produces x-rays as it is energized by the lasers. These x-rays are then absorbed by the test target, producing a blast wave though relatively high density plastic into lower density foam. The experimenters report on two different conditions: a high flux case and a low flux case. By recording radiograph images of the test target material as it undergoes the blast wave, researchers can compare the structure of the instability as it evolves under the two conditions.

What they found was that in the high flux case, there were no mushroom caps on the characteristic finger-like shapes, and that the height of the region of mixed density was smaller than in the low flux case—in other words, the high flux conditions did seem to alter the shape of the instability. The researchers wanted to link this laboratory-generated data from x-ray blasted plastic and foam to astrophysical objects and processes and to draw conclusions about the evolution of the RT instability in high flux astrophysical conditions. In service of this aim, they consider a particular supernova (SN 1993J), where they suspect that the RT instability would have been subjected to high fluxes based on previous observations and modeling of that supernova (see Suzuki and Nomoto 1995; Fransson et al. 1996).[9] Various model parameters fitted to empirical data from this supernova suggest that in a dimensionless sense, i.e. comparing the relevant dimensionless numbers (more on this below), the energy fluxes present in the supernova would

---

[9] Note that the initial motivations for focusing on SN 1993J had to do with the fact that the researchers expected the RT instability to be present since, based on previous observations and modeling, they expected the interface of the two shocks. That alone is not enough to establish the similarity arguments ultimately necessary to support their conclusion due to the open question about how high fluxes would affect the dynamics of matter at this interface.

have been larger than those in the laboratory experiment. That is, the energy fluxes due to heat conducted from the shocked circumstellar medium back into the shocked stellar ejecta are evidently larger with respect to the astrophysical system than the fluxes present at NIF with respect to the experimental system (see Kuranz et al. 2018a, Table 1).

Insofar as the structure of the RT instability was affected in the experimental setup, the experimenters reason that the structure in the case of the supernova should have been affected too. Indeed, they suggest that insofar as the "energy fluxes are larger, in a dimensionless sense, in the emergent [supernova remnant] than they are in the lab experiment" and the fluxes "have a noticeable effect in the lab experiment", then the astrophysical fluxes "seem likely to have a larger effect in the [supernova remnant]" (Kuranz et al. 2018b, 9). They conclude: "realistic models of [supernova remnants] must account for the effects of thermal conduction to accurately predict their evolution at epochs immediately following the shock breakout" (Kuranz et al. 2018a, 5).

How is it, exactly, that conclusions about supernova remnants are supposed to have been drawn from terrestrial experiments? The short, but as we will see, not quite satisfactory answer is that one can argue that the physics, the RT instabilities under the influence of high energy fluxes, is the very same in both cases, such that in experimenting on x-ray-blasted plastic and foam, researchers are probing the very same kind of physics playing out in far distant supernovae.

The sort of reasoning exhibited here is not uncommon, especially in hydrody-namics. It is a powerful and widespread practice in physics and engineering to draw inferences about the behavior of physically similar systems by establishing that certain similarity criteria are met in the systems of interest (Sterrett 2009). Even without knowing which particular form the physical equations characterizing some system should take, if one knows which physical quantities a phenomenon or behavior of interest depend upon, then via application of the principle of dimensional homogeneity, it can be possible to determine a set of dimensionless ratios that pick out a class of systems that will be physically similar with respect to that phenomenon or behavior (ibid., 816–817).[10] The Reynolds number is perhaps a familiar example of an informative dimensionless number. The Reynolds number expresses the ratio of inertial to viscous forces in fluid flow and can be expressed as the local flow speed multiplied by the characteristic linear dimension of the system of interest, divided by the kinematic viscosity of the fluid. Certain values of the Reynolds number correspond physically to the transition between laminar and turbulent flow in a system (considered 2300 for a circular pipe, for example). Thus, by calculating Reynolds numbers for appropriate systems, one can predict

---

[10] The principle of dimensional homogeneity applies to dimensional equations. A dimensional equation is constructed by taking an equation relating physical quantities and replacing the symbol for the quantity with the associated dimension. When the dimensional equation is expressed using the dimensions of the basic quantities of a coherent system, dimensional homogeneity is achieved when the exponents of the dimensions of the basic quantities are the same on both sides of the equation (Sterrett 2009, 815–816).

if/when/where to expect turbulence, that is, a behavior of interest. This sort of reasoning to physical similarity based on dimensionless parameters is incredibly powerful where it can be achieved. The inferential payoffs do not come out of thin air of course, they are hard won via empirical knowledge, choices made in setting up the formalism, and finesse in characterizing the systems and phenomena of interest (ibid., 816). Moreover, in practice, researchers deploying similarity arguments via dimensional analysis rarely manage to (or aim to) capture physical similarity between systems of interest in *all* respects. The physical similarity established is circumscribed and often approximate (ibid., Section 6). Nevertheless, the fact that physical systems afford such similarity arguments at all may well constitute one of the most extraordinary epistemic resources of the physical sciences in comparison to the life and social sciences.[11]

In the National Ignition Facility experiment, the researchers attempt to deploy just such an argument from physical similarity via dimensionless parameters. Kuranz et al. (2018a) make use of a dimensionless parameter which they call the "Ryutov number" in their similarity argument, which can be interpreted as the ratio between pressure forces and inertial forces associated with a hydrodynamic system. We can trace the NIF researchers' use of this particular dimensionless number to their reference of a paper by Ryutov et al. (1999), titled "Similarity Criteria for the Laboratory Simulation of Supernova Hydrodynamics". Hydrodynamic systems well-described by the Euler equations can exist at vastly different scales. Ryutov et al. argue that as long as certain conditions they specify are met in the systems of interest—that is, viscosity and thermal conductivity are negligible, the energy density per unit volume of the fluid is proportional to pressure, dynamic influence of magnetic fields is absent, and the initial conditions are geometrically similar—the hydrodynamic behavior of the systems will be the same. Indeed, Ryutov et al. go so far as to state that if those similarity conditions are satisfied in an experimental and a natural system (i.e. in a laboratory and an astrophysical system), the two systems are *identical* with respect to their hydrodynamic physics (1999, 823). In particular, a laboratory system meeting these conditions should exhibit identical hydrodynamic behavior to an astrophysical system that also meets these conditions and has the same value of the dimensionless parameter mentioned above, which Kuranz et al. call the "Ryutov number" and which Ryutov et al. (and others) call the "Euler number." In attempt to avoid confusion, from now on, I will refer to this dimensionless number that is so crucial to the epistemology of the NIF experiments as the "Euler/Ryutov number".

Ryutov et al. caution that the similarity of the hydrodynamic behavior of systems can break down, however, when energy flow by particle heat conduction and/or energy flow by radiation flux are non-negligible: "The limit of applicability of this

---

[11] On the history of the concept of physically similar systems, see Sterrett (2017b). On universality arguments, see the work of Robert Batterman, e.g. Batterman (2002). For an analysis of the relationship between universality arguments à la Batterman and analog black hole arguments see Field (2021). Sterrett (2017a) contains a discussion of analog black hole experiments in relation to similarity arguments.

similarity is set by the validity of Euler's equations as an adequate description of the hydrodynamics" (1999, 826).

With this introduction to the experiment, let us attempt to unpack its epistemology. I take it that the NIF research I have just described may be readily identified as "experimental" without generating too much controversy. While I do not know of any characterization of what it means to be an "experiment" or "experimental" with any specificity that has been widely adopted in philosophy of science, experiments are often associated with a cluster of features that obtain in the NIF research. It is, after all, research in which the scientists prepare special conditions in their laboratory apparatus to test the outcome of varied conditions, which they manipulate themselves (in this case, by generating high and low flux conditions using the NIF laser system). However, noting this experimental character of the research does rather little to illuminate its epistemology. Is this laboratory research on plastic and foam (experiment or not) informative with respect to astrophysics? If so, how is that supposed to work? In the following section, I will argue that we make more headway in explicating and evaluating the epistemology of this empirical research if we attend instead to how the empirical data are produced and what the research target is supposed to be. To do this, it will be helpful to first have a view of what makes data empirical in general.

2.4   Attend to "Empirical" Not "Experimental"

Elsewhere I have argued that data, including astrophysical data, are *empirical* with respect to some target when there is an interpretation of the provenance of those data using the resources of an epistemic context, such that the data are products of causal interaction with that target (Boyd 2018). By 'epistemic context', I mean the collection of conceptual, theoretical and representational resources from the perspective of which the data is to be interpreted. It is important to note that data are empirical *relative to a target*. Without specifying a target it is impossible to say whether some particular dataset is empirical or not. Data are also empirical *relative to an epistemic context* and the epistemic context supplies the resources with which the data are interpreted. Data never speak for themselves, but rather always require interpretive resources. In particular, data require background theory to furnish a causal story connecting the worldly target of interest to the data collection and recording process.

An important feature of the view of what makes data empirical that I am defending is the causal production of data. As I said, to be properly empirical, data should have been produced by causal processes that connect the worldly target of research to the process of data collection and recording from the perspective of the epistemic context in which the data are to be interpreted. There is no perspective outside of an epistemic context from which the causal processes can be identified and traced. Indeed, there is no perspective outside of an epistemic context from which a worldly target can be identified in the first place. Yet, using the resources of an epistemic context, it can be possible to answer the question: were these data produced by causal interaction with the target?

Taking this view of what makes data empirical onboard, let us return to the National Ignition Facility Rayleigh-Taylor hydrodynamic instability experiment. How should we construe the data generated by the NIF experiment—are they empirical *and* astrophysical data? Following my view we should ask, first: what is the worldly target of the National Ignition Facility research and second: by causal interaction with what has the data been produced from the perspective of the relevant epistemic context?

Here are a few possible answers. First, we might say that the worldly target is SN 1993J, the particular supernova that NIF researchers highlighted as possibly displaying the RT instability under high flux conditions, yet the causal interaction producing the experimental data is with NIF plastic and foam targets, thereby ruining the *empirical* nature of the data with respect to the astrophysical target. The laboratory data was not produced by causal interaction with SN 1993J. Or we could say that NIF plastic and foam targets are the worldly targets of research too, but that would seem to ruin the *astrophysical* nature of the data.

To recover a sense in which this experiment produces empirical astrophysical data, we could construe the worldly target as the general class of RT hydrodynamic instabilities in high-energy-density states of matter. Then *insofar as laboratory systems and far removed astrophysical ones instantiate this very same physics*, investigating the effect of high energy fluxes on hydrodynamic instabilities in laboratory plastic and foam is just to investigate the very physics playing out in astrophysical contexts. NIF data are empirical with respect to high-energy-density states of matter and their behavior since there is an interpretation of the provenance of those data such that they are the products of the causal interaction of the matter energized and confined by the NIF lasers with the laboratory detectors systems. *Insofar* as such high-energy-density states are instantiated in faraway astrophysical systems also, the data gathered in NIF experiments can be used to constrain astrophysical theorizing. So, we should like to know, what is the justification for thinking that the same physics in instantiated in both contexts? As I have already alluded, the NIF researchers appeal to the hydrodynamic similarity of the two contexts.

At first glance, the move suggested in the previous paragraph might seem like slight-of-hand by mere redescription. Can it really make a difference to the epistemology of the research whether we think of the target as a plastic and foam target or as an instance of a class of physical systems? The answer is 'yes', but of course it is not the mere redescription that is doing all the heavy lifting, but rather the arguments and evidence in the background that justify treating the systems of interest as belonging to the relevant class.[12] In this case, the heavy lifting is

---

[12] I use the words "class", "kind" and "type" interchangeably, hoping any more metaphysically inclined readers will forgive me.

done by the justification of the relevant similarity criteria and the evidence for their applicability to the systems of interest.

The general epistemology of science issue here is that of external validity, which is both ubiquitous and absolutely crucial for the epistemology of empirical science (see e.g. Morgan 2003, Currie and Levy 2019, Leonelli and Tempini 2020, and Evans and Thébault 2020). 'Externally valid' experimental results are those that are valid outside of the local laboratory conditions (see e.g. Guala 2003, 1198). In much empirical research, arguing for external validity involves addressing features or conditions of the proximal and ultimate research target that could plausibly make a relevant difference. No two targets or experiments are identical in all respects. In practice, scientists must concern themselves with discerning (to the best of their abilities) features and conditions that might make a *relevant* difference, and then either providing arguments that the differences may be ignored for the limited purposes at hand, concocting circumstances so that the differences become negligible, or else modifying the way that they conceive of the scope of their research so as to responsibly accommodate those differences.

Without characterizing research targets as belonging to a type, we would be locked into an insufferably parochial epistemology, or, to borrow a delightful phrase that Alison Wylie cites from Bruno Latour: we want to avoid "tragically local" data (Wylie 2020, 285). Avoiding tragically local data can involve strategic characterization of the research target, and good arguments (backed by good auxiliary evidence) to support that characterization. I have highlighted a certain kind of similarity argument that can be made for some hydrodynamic systems. There are, of course, other types of arguments that can be made in other contexts. External validity claims need to be justified, they may be challenged by those clever enough to come up with physically plausible difference-makers that have not yet been taken into account, or by surprising empirical results. In general, scientists do well to take opportunities to empirically check their external validity arguments, and indeed to rigorously seek out such opportunities. This is an important part of what is involved in arguing for the epistemic significance of a result and, in particular, of what is involved in eliminating candidate sources of error, confusion, and alternative explanations.

I like to think about the work that goes into making good external validity arguments in terms of what I have elsewhere called "enriched lines of evidence" (Boyd 2018). The idea is that the epistemic utility of an empirical result depends on the details of its provenance. To use an empirical result in an epistemically responsible way, one has to know quite a bit about what assumptions have been baked into it. Some of those assumptions will cause epistemic problems for certain applications, for instance, in constraining a particular hypothesis or in attempting to combine the result with others that were generated by incorporating different assumptions. On this view, supplying good arguments for external validity involves arguing that the assumptions baked into an empirical result will not cause epistemic problems for the intended application. As it turns out, just such a problem seems to manifest in the NIF experiments with which we have been concerned here. If the argument from hydrodynamic similarity that links the laboratory and astrophysical systems as belonging to the same type requires that the energy flux from heat conduction in the systems be negligible, but the experiment in question is designed precisely to investigate how high energy flux from heat conduction influences the structure and evolution of the hydrodynamic instability, does the very aim of the experiment undercut the specified connection to astrophysical targets like SN 1993J? To address this worry, we need to know a bit more specifically what the criteria regarding heat conduction and radiation flux that would need to be met are, and then to check whether in fact those criteria are fulfilled in this context.

Following Ryutov et al. (1999), the source of the similarity criteria that the NIF researchers invoke, the criterion regarding heat conduction is that convective transport needs to dominate conduction in the systems of interest (828). Regarding radiation flux, convective transport ought to dominate the radiation contribution to thermal diffusivity, or, they explain, in cases where it is inconvenient to determine this due to difficulties evaluating the mean free path of photons, it is sufficient to show that the lower limit of the radiation cooling time is much larger than the characteristic hydrodynamic time (825). So, armed with these details, we can ask: do these conditions indeed obtain in the systems of interest in the NIF experiment and its astrophysical counterpart?

It seems these conditions are in fact *not* met. As the NIF researchers explain, for the supernova:

> The interface between the shocked ejecta and the shocked [circumstellar medium] thus arises hydrodynamically, and the transition across it will initially occur in a few ion-ion mean-free-paths [...] Because pressure is continuous across such an interface, the temperature is much higher in the shocked, less dense CSM than in the denser ejecta. This leads to the possibility of radiative or conductive transport of energy into the denser ejecta, which in turn can affect the evolution of the Rayleigh-Taylor (RT) instability at the interface by ablating material from it. In addition, there is a phase when radiation from matter heated by the reverse shock also might affect the RT. (Kuranz et al. 2018b, 3–4)

In particular, they state that "pure hydrodynamics may well be insufficient to accurately predict the structure of the young [supernova remnant]" and that, regarding heat conduction: "As the shock structure at the interface between the [circumstellar medium] and the dense ejecta forms, heat flow is possible by radiation and by electron heat conduction [...] the radiative losses from the shocked layer produced by the reverse shock are found to be large enough to cool it significantly ... [the energy flux associated with the cooling is much greater than the mechanical-energy flux] for a period of time, showing that the hydrodynamic model is not sufficient to accurately describe the behavior" (ibid., 5–6). Indeed, the authors state that "the incoming energy flux by heat conduction is larger than the incoming mechanical energy flux, by a factor of 1,000 at all times" and reason that this flux by heat conduction "is large enough that that this energy flux may be a dominant effect in establishing the structure of the layer" (ibid., 6).

We can see the failure to meet the necessary similarity conditions fairly directly by attending to the dimensionless parameters characterizing the laboratory and astrophysical systems (Kuranz et al. 2018b, Supplementary Table 2). For instance, consider the energy flux ratio R, which has the radiative energy flux in the numerator and the mechanical-energy flux in the denominator. For the similarity condition to be met, the denominator would have to swamp the numerator, yet R for the supernova is 10^3. Note also that the Euler/Ryutov numbers for the two systems are not the same: for SN 1993J at 0.1 years the Euler/Ryutov number is 4 and for the NIF experiment it is 5. On this latter point, the researchers explain that while the relevant sense of physical similarity between systems does not require the precise identity of all dimensionless parameters, Ryutov's et al.'s argument does require that the Euler/Ryutov number be the same for the systems whose similarity is to be established. In this case, the authors note that the Euler/Ryutov numbers for these two systems differ according to best estimates, but that the assumptions upon which those estimates have been made "could vary by at least an order of magnitude" (Kuranz et al. 2018b, 9). In response to these circumstances, they clarify in a supplementary note to the primary publication that the main goal of the work is not to scale their laboratory results to a specific astrophysical object, such as SN 1993J, but "rather to show the importance of energy fluxes in the evolution of young supernova remnants" more generally (Kuranz et al. 2018b, 9).

Here is the argument that I think these authors would need to make regarding the astrophysical relevance of these NIF experiments. Drawing on Ryutov et al., they suppose that in general, hydrodynamic systems that meet Ryutov et al.'s similarity criteria and have the same Euler/Ryutov number will exhibit the same hydrodynamic phenomena. Therefore, in particular, if the laboratory system and astrophysical systems of interest meet Ryutov et al.'s similarity criteria and have the same Euler/Ryutov number, then they will exhibit the same hydrodynamic phenomena. It then needs to be established that the laboratory system and the astrophysical systems of interest in fact meet the criteria and have the same Euler/Ryutov number. Then, supposing the NIF experiments show that the laboratory system displays a particular hydrodynamic phenomenon, the astrophysical systems can be expected to display that same phenomenon too.

The problem is that the similarity argument evidently fails in this instance since it is not the case that the laboratory system and the astrophysical ones meet the criteria and share the same Euler/Ryutov number. Instead, neither the laboratory system nor the astrophysical systems of interest meet the criteria necessary for the similarity argument to go through (and, at least for the particular supernova remnant the researchers considered) these systems have different Euler/Ryutov numbers. Therefore, while the experiments demonstrate that high flux conditions do influence the structure of the RT instability in the laboratory conditions, the argument for astrophysical relevance seems incomplete.

What *can* be gleaned from this case, what did we learn from the results? It seems clear that modeling these systems as evolving hydrodynamically is not appropriate, and therefore we do not have good reasons to suspect that the usual-shaped mushroom-cap RT fingers will show up in supernova remnants with high-energy flux conditions. Expecting such structures in supernovae remnants was premised on hydrodynamic modeling, but we have seen that the high-flux conditions make such modeling inappropriate. But this much might have already been clear before the NIF experiments on plastic and foam were ever performed. That is, reason to doubt the applicability of the usual hydrodynamic evolution of the RT instability in supernova remnants where high-energy flux conditions are present, could have been gleaned already from the conditions and argument set out in Ryutov et al. (1999). Combining these arguments with empirical data from SN 1993J supporting the presence of high-energy fluxes in that system, would have already been enough to cast doubt on the evolution of the structure of the RT fingers in the resulting supernova remnant. If the final epistemic payoff of the NIF experiment is supposed to be the relatively modest point that caution is warranted regarding predictions and interpretations of RT-like structures in supernova remnants in the presence of high-energy fluxes, then it is not clear that the experiment on plastic and foam adds anything new. One might consider the epistemic payoff as rather the demonstration that ablation of material from the RT fingers occurs in the experimental conditions. That is all well and good, but again, that result in itself does not speak to the astrophysical systems. Nevertheless, in some passages the NIF researchers seem to advance an epistemic payoff that goes beyond the modest point. They seem to argue that since the conditions that are responsible for the demonstrated phenomenon of interest in the laboratory setting are, in a dimensionless sense, larger in the astrophysical system, that ought to give us reason to expect the phenomenon in the astrophysical system as well—and that the effect would be larger in the astrophysical system (Kuranz et al. 2018b, 9). Unfortunately, the success of that argument depends on the soundness of the similarity argument, which in this case has evidently failed.

Let us take stock. I have suggested that properly empirical data must derive from a causal chain that has one end anchored in the worldly target of interest. In astrophysics, that does not necessarily mean that the target has to be outside of the terrestrial laboratory. Powerful similarity arguments are available to cast some laboratory targets and far removed astrophysical ones as instantiating identical physics. Insofar as phenomena that are identical with respect to the relevant physics can occur in laboratory conditions, then physics that occurs in astrophysical systems as well as laboratory ones can be studied on Earth. So laboratory astrophysics teaches us about astrophysics, by teaching us about kinds of physical phenomena that occur in laboratories and in astrophysical systems. Laboratory astrophysics teaches us astrophysics, by teaching us physics. Noticing this illuminates the continuity between empirical research in astrophysics and empirical research in other branches of science. Scientists conducting empirical research often need good external validity arguments to avoid producing tragically local data. This is certainly true in laboratory astrophysics. However, the unsurprising need for such good external validity arguments does not entail that astrophysics lacks experiments, is characteristically observational, or does not (sometimes) occur in terrestrial laboratories.

The powerful physical similarity arguments leveraged in laboratory astrophysics research can break down when the necessary conditions do not obtain. If the relevant conditions in the laboratory and astrophysical targets are not the same, then the crucial epistemic link will be broken. In such cases, other arguments would have to be furnished in order to justify couching data derived from terrestrial targets as both empirical and bearing on astrophysics. If such arguments cannot be made, then the research may still be construed as yielding empirical data, just not as bearing on astrophysics. Of course, even then, that does not mean that astrophysicists could not learn about their astrophysical targets using other methods.

2.5 Lessons for Epistemology of Astrophysics

As philosophers, our approaches to the epistemology of science can be more or less fruitful. Philosophers of science working in a normative mode sometimes deploy what we take to be informative distinctions to guide our inquiry. The most obvious example is Popper's falsifiability criterion for demarcating science from pseudo-science (1959). Methodologically, the falsifiability criterion tells us what the salient feature of a case is going to be. Approaching a case with the falsifiability criterion in mind tells us to pay close attention to whether or not it is met in that case, and to draw the associated normative judgements about it. A distinction such as falsifiable/unfalsifiable can thus structure how we approach the work of normative epistemology of science in practice by guiding our attention to certain features as salient for the epistemology of science.

Whatever you think about the utility of the falsifiable/unfalsifiable distinction in particular, there are other distinctions that play this sort of role in guiding attention in epistemology of science. A historically influential one has been the distinction between theory and observation (Boyd and Bogen 2021). Thinking within a traditional empiricist framework, in which pure observation was theory-free and thus suitable for confirming or disconfirming predictions from theory, philosophers of science would approach cases with questions such as: "Is this observation theory-laden in a way that would prevent its effective use for theory testing?" The theory/observation distinction generates an investigative framework in which the question of theory-ladenness becomes especially salient in the investigation of science in practice. From the perspective of twenty-first century philosophy of science, this distinction looks like a red herring. Philosophers of science largely agree that there is no empirical data that is totally theory-free, and furthermore that theory-laden empirical results can be perfectly useful for constraining theorizing. Indeed, it is *in virtue of* being imbued with theory that results can do the work of constraining theorizing (Boyd 2018). Some ways that theory can be integrated into empirical results do cause epistemic problems, but the interesting question is not whether the results are laden at all.

Similarly, normative epistemology of science in practice that approaches cases with the observation/experiment distinction in mind thereby operates within a framework that emphasizes the presence or absence of physical manipulation of the research target as especially salient for the epistemology of science. Is this a fruitful approach? Suppose we had approached the National Ignition Facility laboratory astrophysics research with the question "Is it an experiment?" in mind. Such an approach may have been natural from the perspective of a framework that prioritizes experiments in the epistemology of science. If, as we saw Hacking put it, natural science *is* experimental science, then the primary thing (or at least a very dominant thing) we want to know in investigating the epistemology of some science in practice is whether or not it rises to the standard of experimental science. On this approach, when we learn that a research project is observational, we learn that it is not experimental, and thus that the research is in this aspect, epistemically impoverished.

However, we are now in a position to clearly see just how unilluminating it would have been to ask if the NIF laboratory astrophysics research is experimental or involved experiments. As it happens, it did. And perhaps for those who thought astrophysics was characteristically observational, this (or indeed the existences of laboratory astrophysics research at all) would have been a surprise. But noting that the NIF research makes use of experiments does not, by itself, imply anything epistemically interesting. This is because the observation/experiment distinction does not make an epistemic difference in general. Like the theory/observation distinction, the observation/experiment distinction is largely a red herring for the epistemology of science (Boyd and Matthiessen manuscript). It sets philosophers of science up to attend to certain features of their cases as salient, but those features distract from the locus of epistemic action.

Nevertheless, there is something interesting going on in the NIF case for epistemologists of science. When we shift to a framework that foregrounds empirical data, we approach cases with questions such as "What is the worldly target?" and "By causal interaction with what had the data been produced from the perspective of the relevant epistemic context?" We saw that laboratory targets can instantiate the same physics as astrophysical targets, under the right description and with the right arguments. In the NIF case in particular, we encountered an instance in which it may have first seemed that powerful similarity arguments could be made to justify understanding the Rayleigh-Taylor instability in laboratory plastic and foam and in distant supernovae as instantiating the same physics. However, it turned out in the particular experiment that was the subject of my case study, that the aims of the experiment (namely, to study the evolution of this instability under high energy flux conditions) undermined the needed similarity argument. Notice how our choices in framing which features of scientific research count as salient for the epistemology of science serve to obscure or illuminate the site of epistemic action (or problems in the research). Thinking that astrophysics is an observational science, we might have simply dismissed the NIF research as not astrophysics, and missed the role of the powerful similarity arguments that can sometimes link together certain physical systems under one description. But once we pay attention to what the research target is and how the data are produced, we see that, as it happens, the needed powerful similarity argument does not go through in this case. One could easily miss what has gone wrong in the epistemology of this research insofar as one focuses on it as an experiment, or as purportedly an experiment relevant to a physically distant system. Instead, by tracking the aims of the research and the production of empirical data in it, we were able to notice how the conditions of the experiment affected the application of a crucial similarity argument.

I suggest that these lessons motivate a methodological shift in the epistemology of science in practice, the need for which we can see with particular clarity in the epistemology of astrophysics. Rather than attending to the presence or absence of experiments to investigate the epistemology of science, we ought to instead attend to the target of the research and the processes that produce the empirical data. Whereas attending to the former unhelpfully obscures the epistemology of astrophysics, and the epistemology of laboratory astrophysics in particular, attending to the latter is more helpful for highlighting features salient to the epistemology of science, both within astrophysics and in empirical science more broadly.

References

Anderl, S. 2016. Astronomy and Astrophysics. In *The Oxford Handbook of Philosophy of Science*, ed. P. Humphreys, 652–670. New York: Oxford University Press.

Bacrania, M.K., et al. 2007. Search for the Second Forbidden β Decay of ^{8}B to the Ground State of ^{8}Be. *Physical Review C* 76: 055806.

Batterman, R. 2002. *The Devil in the Details: Asymptotic Reasoning in Explanation, Reduction, and Emergence*. Oxford: Oxford University Press.

Becker, B.J. 2011. *Unravelling Starlight: William and Margaret Huggins and the Rise of the New Astronomy*. Cambridge: Cambridge University Press.

Boyd, N.M. 2018. Evidence Enriched. *Philosophy of Science* 85 (3): 403–421.

Boyd, N.M., and J. Bogen Theory and Observation in Science. *The Stanford Encyclopedia* (Winter 2021 Edition) ed. by E. N. Zalta. https://plato.stanford.edu/archives/win2021/entries/science-theory-observation/

Boyd, N.M., and D. Matthiessen. Manuscript. Observations, Experiments, and Arguments for Epistemic Superiority in Scientific Methodology.

Crowther, K., N.S. Linnemann, and C. Wüthrich. 2021. What We Cannot Learn from Analogue Experiments. *Synthese* 198 (S16): 3701–3726.

Currie, A., and A. Levy. 2019. Why Experiments Matter. *Inquiry: An Interdisciplinary Journal of Philosophy* 62 (9–10): 1066–1090.

Dardashti, R., K.P.Y. Thébault, and E. Winsberg. 2017. Confirmation via Analogue Simulation: What Dumb Holes Could Tell Us About Gravity. *The British Journal for the Philosophy of Science* 68 (1): 55–89.

Dardashti, R., S. Hartmann, K.P.Y. Thébault, and E. Winsberg. 2019. Hawking Radiation and Analogue Experiments: A Bayesian Analysis. *Studies in History and Philosophy of Science Part B: Studies in History and Philosophy of Modern Physics* 67: 1–11.

Evans, P.W., and K.P.Y. Thébault. 2020. On the Limits of Experimental Knowledge. *Philosophical Transactions of the Royal Society A: Mathematical, Physical and Engineering Sciences* 378 (2177): 20190235.

Field, G.E. 2021. Putting Theory in Its Place: The Relationship Between Universality Arguments and Empirical Constraints. Forthcoming in *The British Journal for the Philosophy of Science*. https://doi.org/10.1086/718276.

Field, G.E. Manuscript. *The Latest Frontier in Analogue Gravity: New Roles for Analogue Experiments*. http://philsci-archive.pitt.edu/20365/

Fransson, C., P. Lundqvist, and R.A. Chevalier. 1996. Circumstellar Interaction in SN 1993J. *The Astrophysical Journal* 461: 993–1008.

Giagu, S. 2019. WIMP Dark Matter Searches with the ATLAS Detector at the LHC. *Frontiers in Physics* 7: 75.

Guala, F. 2003. Experimental Localism and External Validity. *Philosophy of Science* 70 (5): 1195–1205.

Hacking, I. 1989. Extragalactic Reality: The Case of Gravitational Lensing. *Philosophy of Science* 56 (4): 555–581.

Hearnshaw, J.B. 2014. *The Analysis of Starlight: Two Centuries of Astronomical Spectroscopy.* New York: Cambridge University Press.

Jacquart, M. 2020. Observations, Simulations, and Reasoning in Astrophysics. *Philosophy of Science* 87 (5): 1209–1220.

Kuranz, C.C., et al. 2018a. How High Energy Fluxes May Affect Rayleigh-Taylor Instability Growth in Young Supernova Remnants. *Nature Communications* 9: 1564.

———. 2018b. How High Energy Fluxes May Affect Rayleigh-Taylor Instability Growth in Young Supernova Remnants: Supplementary Information. *Nature Communications* 9: 1564.

Leonelli, S., and N. Tempini, eds. 2020. *Data Journeys in the Sciences.* Cham: Springer Open.

Morgan, M.S. 2003. Experiments Without Material Intervention: Model Experiments, Virtual Experiments and Virtually Experiments. In *The Philosophy of Scientific Experimentation*, ed. H. Radder, 216–235. Pittsburgh: University of Pittsburgh Press.

Morrison, M. 2015. *Reconstructing Reality: Models, Mathematics, and Simulations.* Oxford: Oxford University Press.

Popper, K. 1959. *The Logic of Discovery.* London: Hutchinson & Co.

Ryutov, D.D., et al. 1999. Similarity Criteria for the Laboratory Simulation of Supernova Hydrodynamics. *The Astrophysical Journal* 518 (2): 821–832.

Sterrett, S.G. 2009. Similarity and Dimensional Analysis. In *Handbook of the Philosophy of Science. Volume 9: Philosophy of Technology and Engineering Sciences*, ed. A. Meigers, 799–823. Amsterdam: Elsevier.

———. 2017a. Experimentation on Analogue Models. In *Springer Handbook of Model-Based Science*, ed. L. Magnani and T. Bertolotti, 857–878. Cham: Springer.

———. 2017b. Physically Similar Systems – A History of the Concept. In *Springer Handbook of Model-Based Science*, ed. L. Magnani and T. Bertolotti, 377–411. Cham: Springer.

Suzuki, T., and K. Nomoto. 1995. X-Rays from SN 1993J and Structures of Ejecta and Circumstellar Medium. *The Astrophysical Journal* 455: 658–669.

Unruh, W.G. 1981. Experimental Black-Hole Evaporation? *Physical Review Letters* 46 (21): 1351–1353.

Wylie, A. 2020. Radiocarbon Dating in Archaeology: Triangulation and Traceability. In *Data Journeys in the Sciences*, ed. S. Leonelli and N. Tempini, 285–301. Cham: Springer Open.

Chapter 3
A Crack in the Track of the Hubble Constant

Marie Gueguen

Abstract Measuring the rate at which the universe expands at a given time–the 'Hubble constant'– has been a topic of controversy since the first measure of its expansion by Edwin Hubble in the 1920s. As early as the 1970s, Sandage and de Vaucouleurs have been arguing about the adequate methodology for such a measurement. Should astronomers focus only on their best indicators, e.g., the Cepheids, and improve the precision of this measurement based on a unique object to the best possible? Or should they "spread the risks", i.e., multiply the indicators and methodologies before averaging over their results? Is a robust agreement across several uncertain measures, as is currently argued to defend the existence of a 'Hubble crisis' more telling than a single 1% precision measurement? This controversy, I argue, stems from a misconception of what managing the uncertainties associated with such experimental measurements require. Astrophysical measurements, such as the measure of the Hubble constant, require a methodology that permits both to reduce the known uncertainties and to track the unknown unknowns. Based on the lessons drawn from the so-called Hubble crisis, I sketch a methodological guide for identifying, quantifying and reducing uncertainties in astrophysical measurements, hoping that such a guide can not only help to re-frame the current Hubble tension, but serve as a starting point for future fruitful discussions between astrophysicists, astronomers and philosophers.

M. Gueguen (✉)
Institut de Physique de Rennes 1, Université de Rennes 1, Rennes, France
e-mail: mgueguen@uwo.ca

© The Author(s) 2023
N. Mills Boyd et al. (eds.), *Philosophy of Astrophysics*, Synthese Library 472,
https://doi.org/10.1007/978-3-031-26618-8_3

3.1 Introduction

From the realization in the end of the 1920s by Edwin Hubble that a relation of proportionality exists between the recessional velocities of galaxies and their distance; to the crisis around the Hubble constant that currently undermines the standard model of cosmology, the history of this constant has been that of the chase of a fleeing number that kept escaping the scientists' net. Among the most remarkable episodes of this track: the *Hubble war* in the 1970s, opposing Sandage and de Vaucouleurs, arguing both about the correct methodology to adopt for measuring the Hubble constant and about its actual value;[1] the dispute between Sandage and his colleague Wendy Freedman at the Carnegie Observatories of Pasadena in the 1980s, the latter defending a much higher value than the former, probably lying in the middle of the range spanned by Sandage and de Vaucouleurs; the disagreement since 2014 between two opponents that nobody had seen coming: the distant and the local universe, the former with a Hubble value of $67.4 \, \text{km s}^{-1} \, \text{Mpc}^{-1}$, the latter one approaching 75, and finally the so-called and on-going Hubble 'crisis' that this persisting disagreement and its apparent confirmation by the publication of new local measures in 2019 have seeded.

The agitated history of the Hubble constant mirrors how fundamental this parameter has been for the development of our modern, precision cosmology. As one may remember from the famous words of A. Sandage, modern cosmology can be considered as the "search of two numbers": the values of the Hubble constant and of the q_0 parameter, which characterizes the deceleration in the expansion (Sandage 1970). But this troubled history certainly also reflects how tedious and delicate the task of measuring the Hubble constant is, and, as a result, how difficult it has been to assess the accuracy of its past measures. The determination of the Hubble constant requires one to find stellar objects (a) whose luminosity is known on theoretical grounds, (b) sufficiently far away from us to be freely moving (i.e., located in the so-called 'Hubble flow'), and (c) bright enough to be detected even that far away. But no single technique allows for the measurement of the distance satisfying all these properties. Hence, doing so necessitates deploying a 'cosmic ladder' from the nearby universe to the Hubble flow, where each of the rungs leads to a proliferation of systematic errors that must be constantly tracked and eliminated.

Yet, recent developments in astrophysics have led many scientists to consider that we know enough to consider the Hubble tension as a Hubble crisis. Such takes are grounded in the idea that (1) measurements of the Hubble constant have reached a sufficient precision for the discrepancy between early and late universe measures to become meaningful, and that (2) the robustness of the high value inferred from independent late universe techniques guarantees that no systematic errors will explain away this discrepancy. Here, I argue that in the context of highly uncertain measurements, methodologies that favor tracking the unknown unknowns

---

[1] See Guralp (2020).

always have epistemic priority over robustness arguments in the sense that they constrain the appropriate domain and timing for applying such arguments. On this basis, I contend that the current Hubble constant crisis is yet another avatar of a methodological confusion between the possible roles that different kinds of replication can play. The form of replication that robustness constitutes cannot be considered as evidence of a crisis when the necessary condition of systematic replication, which promotes tracking down unknown unknowns, is not successful.

Section 3.2 introduces the reader to the different ways of measuring the Hubble constant. In Sect. 3.3, I reconstruct the reasons that have been provided to justify the idea of a Hubble crisis and how they relate to the notion of robustness. Finally, Sect. 3.4 clarifies the roles that robustness and replication can play and contends that the use of robustness to establish a conclusion as dramatic as a Hubble crisis at this stage of the investigation is misguided: a day may come when a Hubble constant crisis arises and our cosmological model crashes down, but it is not this day.

3.2   How to Track the Hubble Constant

Different methods have been developed since Edwin Hubble's first attempt at measuring the Hubble constant *via* the cosmic ladder. One consists in inferring the Hubble constant from the early universe, for instance from the cosmic microwave background. The Santa Barbara conference of 2019 has seen new techniques based on the local universe blossom and reach a precision that makes their comparison with the most mature techniques genuinely informative. In this section, I briefly introduce some background for each of these techniques, such as to facilitate the philosophical interpretation of their concordance and of its signification for a cosmological crisis.

3.2.1   Jack and the Magic Bean: Building a Cosmic Distance Ladder in the Local Universe

At first sight, measuring the Hubble constant seems quite straightforward. The expansion law that must be solved in order to measure its value takes the following form:

$$c\frac{\delta\lambda}{\lambda_0} = H_0 D_0 \tag{3.1}$$

where c is the speed of light, $\frac{\delta\lambda}{\lambda_0}$ the redshift of the observed spectral lines of the galaxies compared to what would be expected only taking into account their distance, and D_0 their present distance. In other words, H_0 is known when the distance of an object and its redshift are known. Determining the redshift of a given stellar object is done by comparing the observed spectral lines of galaxies to the 'laboratory' ones. The distance, on the other hand, is determined on the basis of two pillars: the choice of a standard candle on one hand, its apparent magnitude and its absolute magnitude on the other hand. A standard candle is an object that has a known intrinsic luminosity, referred to as its 'absolute magnitude' M-the brightness we would measure if we were standing 10 parsec away from it.[2] The apparent magnitude m of an object corresponds to its brightness as it appears to us, taking into account its distance and the effects that interstellar dust or bright stars nearby could have on it. The distance modulus μ is equal to $m - M$ and is related to the distance d in parsecs as follows:

$$\mu = 5log_{10}d - 5 \qquad\qquad (3.2)$$

The problem, as we mentioned above, is that the relevant objects for measuring the Hubble constant must be located in the 'Hubble flow', and that determining the apparent magnitude of an object so far away requires a bundle of different methods that each comes with its own difficulties. There are, indeed, many phenomena that can alter the apparent magnitude of an object along our line of sight beyond its distance. It may, for instance, appear much fainter than it should, due to the absorption of part of its spectrum by the dust surrounding it –a phenomenon referred to as 'extinction'; or brighter than predicted, due to crowding effect by nearby stars. Such phenomena, among many others, must be accounted for and the apparent magnitude calibrated on this basis. Thus, objects in the Hubble flow cannot be *directly* probed. They require developing a ladder that will allow for the calibration of distances and magnitudes one rung at a time. Each rung is built on a different object, on a different technique, and on the information provided by the former rungs. Needless to say that in such a case, any error done in the first steps of the process has important repercussions on the final value found for the Hubble constant.

Cosmic distance ladders built to measure the Hubble constant are usually based on three rungs. The first rung 'anchors' the ladder in the sense that it serves as a zero-point calibration for extinction and crowding effects. Anchors must be sufficiently close to measure their distances with geometric methods, either through trigonometric parallaxes or Detached Binary Eclipses (DEBs). Anchors usually include the Large Magellanic Cloud, the Milky Way and NGC 4258. The second rung consists of determining the distances of galaxies known as 'calibrators', hosting both the selected standard candle and Type I Supernovae. Based on the calibration done in the first rung, the difference between apparent and absolute magnitudes allows for the determination of the distance of galaxies hosting both our

---

[2] Astronomers use the notion of "magnitude" to measure the brightness of an object. Magnitude is defined on a logarithmic scale, and the brighter an object is, the lower its magnitude. For instance, the absolute magnitude of the Sun is 4.8, but the faintest objects visible by the Hubble telescope have an apparent magnitude of 30.

standard candle and Type I Supernovae. Ideal standard candles consist of objects whose luminosity does not depend on their mass or composition. They usually fall into one of these two categories: either stars whose luminosity varies according to a known period-luminosity law, or extremely luminous objects whose brightness is due to a well-known and well-described phenomenon. They typically include, among many other examples, Cepheid stars, whose average intrinsic luminosity varies depending on the period at which they pulse, a relationship well-documented and empirically verified by Henrietta Levitt.[3] Another favored one are Type Ia Supernovae, which correspond to a rare but extremely bright explosion –around five million times the brightness of the Sun! –, that of a dying white dwarf star exceeding its critical mass. Their brightness is perfectly suited for exploring the Hubble flow. The combination of the two provides both the anchors needed for building the first rungs of the cosmic ladder, and an access to regions where galaxies are freely moving. One starts by calibrating distances to nearby Cepheids in the LMC, Milky Way or NGC 4258, before gauging distances to much farther away Cepheids. Finally, the distance of Type I Supernovae in the Hubble flow is determined, on the basis of the second rung calibration.

This picture of a "Jack and the Magic Bean" astronomer climbing the cosmic ladder to catch supernovae, as beautiful as it is, is however anything but simple. As we mentioned above, each rung comes with many traps and errors propagating from one rung to the others. Maybe surprisingly, the zero-point calibration of the ladder is one of the trickiest part of the process and the largest source of systematic uncertainties. Uncertainties associated with the anchors carry over as systematic errors and have a huge possible impact on the determination of H_0. Yet, these sources of uncertainties are not only important, but impossible to reduce otherwise than by improving the accuracy of observational tools. Among them, one can include the distance to these anchors, extinction, but also difficulties related to converting the I-band photometric system of space telescopes to ground-based telescopes,[4] both needed. The second major source of uncertainties comes from the fact that it is usually difficult to find a statistically significant sample of galaxies that host both the relevant standard candles and Type I Supernovae. In the case of the Cepheids, four decades of research have allowed astronomers to build a sample of only 19 host galaxies[5]. But more generally speaking, standard candles are rarely really 'standard', as their luminosity may actually depend on their age or metallicity.[6] Random velocities of specific galaxies can be perturbed by local gravitational perturbations, thereby complicating the task of determining their redshift if the statistical sample of standard candles is not big enough to average

---

[3] Leavitt (1908).

[4] See Freedman et al. (2019, p. 11).

[5] Riess et al. (2022) have succeeded in more than doubling this sample in 2022, with a Hubble value now at 72.53 ± 0.99 km s^{-1} Mpc^{-1}.

[6] In astronomy, the metallicity of a star corresponds to the heavy elements it contains, a 'heavy' element being any element other than hydrogen or helium.

away these perturbations. In sum, each rung comes with its ensemble of systematic and statistical proliferating errors, each of which could significantly distort the final value inferred for H_0. Significant progress has allowed for the improvement of the measurement of the Hubble constant based on Cepheids up to 1%,[7] after many years of rigorous investigation to explore and reduce the systematics associated with this technique.

3.2.2 Hubble Constant in the Early Universe

What characterizes measures of the Hubble constant based on the primordial universe is their model-dependence. One cannot infer the value of the Hubble constant from the early universe without already assuming a cosmological model. This feature holds both for measures inferred from the cosmic microwave background (henceforth CMB) or from the Baryon Acoustic Oscillations (BAO). For space reasons, I chose to limit the introduction to the early universe measures of H_0 to the CMB measure, but a detailed and accessible introduction to the BAO determination of H_0 can be found for instance in Fong (2011).

A couple words on the CMB first. When the primordial universe got cold enough for the first neutral hydrogen atoms to form –the epoch of 'recombination', photons decoupled from matter and started to free-stream across the universe. These photons have been propagating ever since, and the relic of this radiation is what is referred to as the 'cosmic microwave background' (henceforth CMB). This fossil electromagnetic radiation offers an extraordinary window into the early universe, as it provides a map of how matter was distributed across the universe at the time of decoupling and thus informs us about fundamental parameters of the ΛCDM model, included its matter density Ω_m. Assuming the standard cosmological model, and the dark energy density Ω_λ and spatial curvature k that characterise this model, secondary parameters such as the Hubble constant can be derived through the following equation:

$$H_z = H_0 \sqrt{\Omega_m (1 + z)^3 + \Omega_\Lambda + \Omega_k (1 + z)^2} \qquad (3.3)$$

A nearly exact geometrical degeneracy exists however, both for Ω_Λ and k, that make different cosmological models based on different Ω_Λ and spatial curvature k compatible with the anisotropies mapped by the CMB.[8] Hence the need and importance of a model-independent measure of H_0 that could waive this degeneracy, such as measures in the local universe.

---

[7] Riess et al. (2019).

[8] See e.g. Fong (2011) or Efstathiou (2020). The Planck 2018 results released in Collaboration (2020) seem to possibly break this degeneracy however.

3.3 A Tale of Two Values: The Hubble Crisis

Now, here lies the problem: the results delivered by the two techniques, the one based on the early universe and the Cepheids-based one, do not agree. In other words, the expansion of the universe as measured from our local universe is much faster than that predicted on the basis of the CMB, by almost 8%. The difference between the two is significant, close to 5σ: the value announced by the SH0ES team in 2019, led by A. Reiss and working with Cepheids, was $74.03 \pm 1.42\,\mathrm{km\,s^{-1}\,Mpc^{-1}}$,[9] when the value obtained from the CMB after the last release of Planck results[10] is $67.4 \pm 0.5\mathrm{km\,s^{-1}\,Mpc^{-1}}$. Until now, this difference was not considered as too alarming. As we saw, the measure of the Hubble constant using the cosmic ladder with Cepheids as standard candles is a delicate task. Cepheids are young stars, thereby living in the dusty and crowded center of galaxies–an environment that maximizes extinction and crowding effects, whose period-luminosity depends on their age and metallicity, and whose nature itself is a problem, inasmuch as variable stars necessitate many exposures during several observational campaigns, adding new sources of systematic errors to account for. No wonder then that the first measure of H_0, based on Cepheids, was off by an order of magnitude: Edwin Hubble estimated the constant value around $500\,\mathrm{km\,s^{-1}\,Mpc^{-1}}$. The complexity of the first technique, the multiple systematic and statistical errors that could affect each rung of the ladder, and the degeneracy associated with H_0 in the early universe context could legitimately lead us to think that, as the accuracy of the measures improve in the future, the results would have converged, especially as these values did progress a lot over the last two decades, and globally toward a possible convergence.[11]

The conference that was held in Santa Barbara in 2019 has however shaken this confidence and consolidated the gap that one was hoping to bridge. Over the last decade, many new techniques have indeed been developed to measure the Hubble constant independently of the Cepheids and the CMB, in order to break the tension between the two–especially as it becomes less and less clear how the precision of these measurements could further be improved. Four new measures were released during the conference that did not resolve the tension, but on the contrary corroborated the high value found by the SH0ES team, turning an apparent disagreement into a genuine problem, and generating a strong feeling of crisis among cosmologists.[12] Part of the goal of this paper is to elucidate the reasons that underlie such a feeling and whether these reasons are justified.

---

[9] See Riess et al. (2019).

[10] See Collaboration (2020).

[11] See Di Valentino et al. (2021) for an exhaustive plot of the current Hubble constant situation, and figure 17 of Freedman et al. (2019) to see the evolution of these values over the last two decades.

[12] A nice overview of the different reactions heard during this conference can be found in the paper "Cosmologists debate how fast the universe is expanding"(https://www.quantamagazine.org/print), written for *Quanta Magazine* by N. Wolchover.

3.3.1 The Blossoming of New Measurement Techniques

The main reason driving this sense of crisis is the fact that these new techniques are considered as independent measurements, that is, measurements that differ enough from the Cepheids measure to exclude possible common sources of systematic errors. Let us briefly review three of these techniques[13] to assess the extent to which this claim is justified.

- H0LiCOW: $H_0 = 73.3 \pm 1.7 \pm 1.8 \mathrm{km\,s^{-1}\,Mpc^{-1}}$
  The H0LiCOW[14] project uses strong gravitational lensing, i.e., the distortion of spacetime produced by supermassive objects, to measure the Hubble constant. This lensing phenomena permits the researchers to obtain multiple images of a same object, based on the different paths that the electromagnetic radiation follows given the curvature of spacetime. The idea of H0LiCOW is to study the light emitted by 5 quasars, i.e, extremely luminous active galactic nuclei whose magnitude varies. Massive galaxies between us and these objects act as magnifying and distorting lenses that multiply the images of the lensed target. Since light takes a different path for each of the images, the oscillation of the luminosity is delayed for each of these pictures. Thus, given that the distance travelled by light depends on the expansion of the universe, the time delay between each image allows to calculate the Hubble constant.
- MIRA variables:[15] $H_0 = 73.3 \pm 3.9 \mathrm{km\,s^{-1}\,Mpc^{-1}}$
  This method is a variant of the cosmic distance ladder, but using red pulsating stars called MIRAs as standard candles. The main sequence of the life of a star consists of converting the hydrogen contained in its heart into helium through nuclear fusion. At some point of the life of the star, after hydrogen and helium in the core are fully exhausted, the fusion starts in the outer shell which will expand, sometimes up to 1 AU; then cool down and shrink, before expanding again. This cycle is what makes MIRA-type stars good standard candles, because this period-luminosity relationship is fully determined by the mass and radius of the star. Note that this cycle, that lasts at minimum 100 days, is even longer than the Cepheids', whose fluctuations usually range from 1 to around 50 days.
- The Megamasers Cosmology Project:[16] $H_0 = 73.9 \pm 3.0 \mathrm{km\,s^{-1}\,Mpc^{-1}}$
  Megamasers and gravitational lensing are especially interesting techniques, as both offer the possibility of a direct measure of the Hubble constant, skipping the rungs of the ladder altogether. The characteristic of interest of masers is the equivalent of the laser effect in the microwave domain. The rough idea is

---

[13] I leave aside the result obtained by the Surface Brightness Fluctuations method, whose error bar is so large that its agreement/disagreement with other results is not meaningful. See more on this in Potter et al. (2018).

[14] See Wong et al. (2020).

[15] See Huang et al. (2020).

[16] See Pesce et al. (2020).

the following: the spontaneous emission of a photon generated by an atom's transition to its fundamental state triggers a cascade of similar emissions. The incident photon provokes the deexcitation of another atom; and thus the emission of another photon. Hence, the initial photon is so to speak photocopied up to a very powerful electromagnetic beam. Megamasers are typically located around 1pc of the center of a galaxy, close to the active galactic nuclei that can stimulate the surrounding gas clumps or water clumps (in the cases of water masers). Hydrogen and oxygen atoms composing the water maser absorb the galactic nuclei energy and radiate it in the form of a microwave 22 Hz beam that can be detected by Very Long Baseline Interferometry (VLBI). From this stable radiation can be inferred the velocities of gas clumps and water clouds orbiting the nuclei, their radius from the nuclei and distance to the galaxy, and their host galaxy's redshift.

These new techniques are all very promising, but also very recent. Beyond the many known sources of errors and unknown unknowns to uncover associated with them, their youth comes with a high price, that of relying on limited (and so possibly biased) statistical sample. Take for instance the H0LiCOW or the Megamaser Cosmology Project results: the former was based on only 7 lensed quasars in their 2019 paper, and the latter had only 4 megamasers– including NGC4258 which is used as an anchor for cosmic ladder techniques. One has to grant however that they are based on totally different physical assumptions, a fact that renders the possibility of unveiling common plausible sources of errors explaining their convergence toward a high value very unlikely.

This makes the fourth result published during the Santa Barbara conference even more disturbing. Indeed, the Carnegie-Chicago-Hubble program, led by W. Freedman, presented their Hubble value during this exact same conference, based on a new version of the cosmic ladder using the Tip of the Red Giant Branch stars (TRGBs) as a standard candle. Their announcement, far from solving the issue, created a new puzzle, since their value lies right in the middle between the low value based on the CMB and the high value of the local universe methods, at $H_0 = 69.8 \pm 0.8 \, \text{km} \, \text{s}^1 \, \text{Mpc}^{-1}$.

3.3.2 Houston, We have a Rogue Measure

TRGBs offer an excellent standard candle to calculate the distance of supernovae. This phase of the evolution of red giants corresponds to the moment when a star of around 1 to $2M_\odot$ has exhausted the hydrogen of its core and started the fusion in the outer shell. Unlike what happens for stars of higher mass though, the core does not contract but is entirely sustained by the electron degeneracy pressure. As a result, the temperature increases in the core as the helium piles up there, without any corresponding dilatation of the star, until the temperature reaches 100,000 K. At this point a triple α reaction is triggered. Under the combined effects of the temperature and the pressure exerted in the core, the fusion of helium into beryllium and into carbon turns into a runaway reaction and creates an extremely violent flash of helium, with a release of energy superior to the entire output of a whole galaxy. The infrared luminosity of stars going through an helium flash is independent from their mass and composition, which makes them eligible for the title of standard candles.

Probably one of the most interesting features of this technique is that it constitutes the perfect counterpart to the Cepheids technique. This means that, even though we currently have no way to decide whether one of these results, that obtained based on the Cepheids or that based on TRGBs is correct, a comparison between them is both very telling and very informative, as each one fills the lacuna of the other. Whereas Cepheids are short lived-stars that live in dusty environments, TRGBs are old stars living in isolation, in the outskirts of galaxies. As such, they are not as exposed to extinction and crowding effects as Cepheids are. Likewise, while the period-luminosity relationship of the latter depends on their metallicity, that of TRGBs can be accurately accounted for in two different ways: first, the infrared I-passband is almost not affected by their metallicity. Second, this metallicity manifests itself in the color of the star, by a widening of the RGB color that has been well-studied and calibrated empirically. Finally, TRGBs are not variable stars and do not require multiple exposures. As we will see in the next section, these complementary features constitute an ideal investigation path and offer a fecund scenario not only to test the robustness of the Hubble value, but also to discover new sources of uncertainties not necessarily accounted for in the report of the accuracy of these measurements. The question is thus the following: how can we explain the fact that, TRGBs excluded, the different methods based on the local universe agree on a high value and the methods based on the early universe on a low value, whereas at the same time the two methods that are the closest to each other, the more complementary and the more likely to agree fail to do so? How can we account for this success on one hand and this failure on the other?

3.4   Should We Call it a Crisis?

How should we thus react to these diverging measures? For many cosmologists, as reflected in the number of papers attempting to resolve the Hubble problem published since 2019, the convergence of the SH0ES, the H0LiCOW, the MIRAs and the Megamaser techniques towards a high value of H_0 is taken to indicate that the standard model of cosmology is undergoing a crisis. Although it is difficult to reconstruct the exact argument supporting a crisis, the papers that endorse the idea of a 'Hubble tension' or a 'Hubble crisis' tend to agree on the following statements: first, the discrepancy is much higher than the error estimates associated with each of these results. Put differently, the error bar is sufficiently small for a meaningful assessment of the discrepancy. Second, the independence of the techniques converging toward a high value for H_0 excludes that one single, shared, source of systematic errors could waive the tension. Hence, the latter will likely resist an improvement of the accuracy of these techniques and a subsequent decrease in the size of their error bar.[17] The quote below summarizes this view:

> Given the size of the discrepancy and the independence of routes seeing it, a single systematic error cannot be the explanation. [...] Moreover, a suite of low redshift, different, truly independent measurements, affected by completely different possible systematics, agree with each other; it seems improbable that completely independent systematic errors affect all these measurements by shifting them all by about the same amount and in the same direction" (Verde et al. 2019, 7).[18]

3.4.1   From Robustness to Reliability

Although the word is never explicitly stated in the discussion, this line of argument captures what constitutes the core of robustness analysis as theorized within the tradition starting with Levins (1966), Levins (1993) and Wimsatt (2012).[19] Robustness analysis has been famously suggested by Levins as a way to assess the trustworthiness of models in the absence of a background theory providing analytically solvable equations. Since models have to be simplified to get predictions suseptible to being measured against nature, a method must be developed in order to evaluate the impact of these simplifications on the predictions of the model and to determine "whether a result depends on the essentials of the model or on the details of the simplifying assumptions" (Levins 1966, 423). One way to do so, to Levins' eyes, is to compare different models M_1, M_2, ..., M_N of the same target system, where each model is conceived of as the intersection of a common, plausible core C and of an unshared, variable part V_1, V_2, ..., V_N, and to look for a connection between C and a predicted property R (Levins 1993). The plausible core includes the biological or physical assumptions that are undergoing the test, while the unshared part corresponds to different idealizations or simplifications used to make the problem tractable. If one can show that the intersection of C with the union of the V_i implies R, then one can establish under certain conditions[20] that C alone implies R-put differently, that R does not depend on the V_i, but on the common core of all models whose adequacy is under test:

---

[17] See notably Verde et al. (2019), Efstathiou (2020), and Riess (2020).

[18] See also (Riess 2020, p. 2).

[19] See also Weisberg (2006) and Weisberg (2012) for a more recent take on this version of robustness analysis. Of course, many schools of thoughts have arisen since Levins' and Wimsatt's accounts of robustness analysis. The goal here, however, is not to address whether robustness is a sound tool, but whether the kind of robustness allegedly displayed in the Hubble context supports the existence of a crisis. Therefore, the paper focuses on their version of robustness, which seems to capture the line of argument defended by cosmologists. For a more up-to-date account of how robustness is used in the actual practice of scientists, see among many others (Soler et al. 2012).

[20] For instance, the condition that the set of V_i exhausts the space of admissible possibilities.

(...) Thus the search for robustness as understood here is a valid strategy for separating conclusions that depend on the common [...] core of a model from the simplifications, distortions and omissions introduced to facilitate the analysis, and for arriving at the implications of partial truths. (Levins 1993, 554).

Levins, however, remained rather careful about what can be learned through robustness; at least in his 1993 piece:

(...) the more inclusive the set of V_i's, the more we can have confidence that C implies R. If we feel that the set of V_i's spans a wide enough range of possibilities, then we may generalize to claim that C usually implies R, a result that is not very exciting as a mathematical theorem but may be good biology. (Levins 1993, 554)

It was Wimsatt in 1981 who tightened the bond between robustness and reliability. According to him, robustness analysis is defined through the following three principles: first, it is a procedure aiming at distinguishing the "reliable from the unreliable"; second, it requires one to show the invariance of that which reliability is scrutinized over independent[21] processes or models, in order to build confidence in their independence from these; and finally it requires determining the scope of this invariance. Hence, within this framework, establishing reliability is no longer a mere possible goal of robustness, but one of its core and definitional tenets–robustness and reliability go hand in hand, and where one is to be found, the other is expected:

[A]ll the variants and uses of robustness have a common theme in the distinguishing of the real from the illusory; the reliable from the unreliable; the objective from the subjective; the object of focus from artifacts of perspective; and, in general, that which is regarded as ontologically and epistemologically trustworthy and valuable from that which is unreliable, ungeneralizable, worthless, and fleeting. (46)

Since Wimsatt, robustness has been generally accepted as an indicator of reliability, that is, as evidence that a prediction is not an artifact of specific modelling assumptions.[22] Orzack and Sober have however convincingly undermined this claim, in a beautiful paper that seeded many of the questions about robustness that have been debated ever since. As they see it, there are three possible scenarios resulting from a Levins-like robustness reasoning:

- Scenario 1: We already know that one of the M's among M_1, M_2, ..., M_N is true.[23] In that case, if for all i, M_i implies R, then R must be true. As emphasized in Justus (2012, 797), this inference is unproblematic but also relatively uninteresting, given that robustness is precisely needed in those cases

---

[21] For space reasons, I will not dwell on how to characterize what 'independence' means here, given that models have to be of the same target and thus presumably share some core assumptions to be even comparable. This term is present in both Wimsatt's and Levins work, but never fully elucidated and is subject to controversy. See for instance Schupbach (2015).

[22] See for instance Weisberg (2006), Weisberg and Reisman (2008) or Soler et al. (2012), but also Parker (2011) for a criticism.

[23] Although we do not endorse the terminology in terms of 'truth' that Orzack and Sober use, we will keep it here in order to remain faithful to the authors.

where no observations or no analytic solutions are available that could establish the truth of one of the M_i.

- Scenario 2: We know that all the models are false. In this case, we have no reasons to believe that the fact that each M_i implies R is evidence that R is true. Their simple example illustrates this point beautifully: if all models we compare in population biology admit natural selection as the only force acting on the size of the population, then all models will predict populations with infinite size. That they agree does not say anything about the truth of the prediction, but only about the convenience of the assumption (Orzack and Sober 1993, 538).

- Scenario 3: We do not know whether one of the models is true. This is the most common situation in astrophysics and cosmology, and robustness is precisely used in these contexts to help to establish the reliability of predictions converging across different models. According to Orzack and Sober, we have no more reasons yet in this situation to infer that R is true than we have in the second scenario: "if we do not know whether one of the models is true, then it is again unclear why a joint prediction should be regarded as true (ibid., 538–539).

From a purely inferential point of view, I think that Orzack and Sober make a valid logical point here in emphasizing that we have no reason to consider R as true in any of the last two scenarios. Nonetheless, I am still willing to grant that the last scenario can correspond to very different epistemic situations, and that for each of them the degree of confidence possibly supported by the robustness of R or its value as a heuristic guide could vary a lot. A comparison where recently developed techniques, with many shared assumptions, and from which little is understood, are compared, does not support a high degree of confidence in R. But it seems reasonable to say that a much higher degree of confidence in R would be justified–in the words of Levins, would be "good astrophysics"– if the comparison is performed across mature and independent models, that have been rigorously examined and studied such that systematics and statistical errors have been identified and reduced to the best of our knowledge. The question is thus the following: which one of these situations corresponds to the Hubble constant crisis's scenario? Are we in the epistemic position to apply robustness and infer conclusions with a high degree of confidence on this basis? Or are we putting the cart before the horse?

3.4.2   Temporary Discrepancy vs. Residual Discrepancy

When are we justified in thinking that a discrepancy is symptomatic of a crisis? My –presumably uncontroversial– answer to this question would be: when the discrepancy is not a temporary one, but a *residual* discrepancy. That is, when we find ourselves in the case where known sources of systematic and statistical errors have been quantified and sufficiently reduced for a comparison between the two values thus obtained and their error bars to be significant, and when enough efforts have been done to chase unknown unknowns, i.e., new sources of yet undiscovered errors. A *residual* discrepancy, in other words, is the discrepancy that remains after adequate efforts have been deployed to identify, quantify and reduce all possible sources of uncertainties that could explain away the disagreement.

Of course, this does not mean that we need to be absolutely certain that all sources of errors have been excluded. It does mean, however, that the significance of the tension is directly related to the assumption that all sources of uncertainty have been identified and accounted for. If we have strong evidence that there are still errors, which are not accounted for but which could resolve the tension, the robustness of the tension does not have sufficient epistemic strength to justify a call for a crisis.

Up to the Santa Barbara conference, one can see that the discrepancy between the early and the late universe measures was interpreted as a temporary one, thus merely seen as a tension rather than as a crisis. What changed after the Santa Barbara conference is that the robustness of the values resulting from the local universe measurements[24] was interpreted as an indication that the discrepancy had gone from temporary to residual. Now, do we have good reasons on the basis of the new results to think that the epistemic situation switched from temporary to residual? And is robustness the appropriate tool to decide whether this is the case, that is, if systematic and statistical errors have been sufficiently purged from our measurements?

In the remainder of the paper, I content that we have clear evidence that not only the discrepancy is not residual, but that robustness has so far been unsuccessful in detecting unknown systematic errors in our case study, notably because of the emphasis on comparing measurements *as independent as possible*. I illustrate the latter point by showing that the robustness of the high late-universe value is blind to the systematics since acknowledged (notably) in time-delay cosmography. Next, I show that the tool that could diagnose these errors is actually in competition with robustness and thus often neglected despite its epistemic priority.

3.4.2.1 The Example of Time-Delay Cosmography

Announced in 2019, the H0LiCOW result was considered as one of the most important evidence of a Hubble crisis. As mentioned above, a measure based on gravitational lensing is a *direct* measure of the Hubble constant, in that it does not require appealing to the ladder technique. Furthermore, the H0LiCOW and the Cepheids' measurements are as independent as two measures of the Hubble constant can be. They rely on completely different objects and different physics, whereas the Mira project is a version of the cosmic ladder which might suffer from the same

---

[24] This robustness holds only at the price of excluding the TRGBs' result of course. The SH0ES team (Yuan et al. 2019) has justified such an exclusion on the basis of, as they argued, a calibration error on the TRGBs side. This claim has now been debunked several times (Freedman et al. 2020; Freedman 2021; Mortsell et al. 2021).

issues as the Cepheids or the TRGB technique (due to common anchors for instance) and the Megamaser Project involves the maser located in NGC4258, which is a common zero-point for the Cepheids and the TRGB measures. Thus, an agreement between the two independent measures released by H0LiCOW and by SH0ES was considered as particularly exciting and telling, and a major reason to interpret the Hubble tension as a Hubble crisis.

But this technique is a really young one, and a lot of work still needs to be done to understand how the different assumptions that enter this measurement might distort the result. Note that this is not pure speculation about possible future developments for time-delay measurements of gravitational lensing. The effects of relaxing assumptions about the mass density profile of the deflector have already been carefully studied, with surprising results. Indeed, one of the most important sources of uncertainties in time-delay lensing is the mass profile of the deflector. If no assumption is made about it, the precision of the measurement, based on the 7 H0LiCOW lenses, drops from 2% precision to 8%. Such an error budget is far too important to resolve the Hubble tension. So where do the mass assumptions used in this context come from?

A lens model should ideally be able to reproduce the observables associated with the lens with as few unconstrained parameters as possible. With respect to quasar astronomy however, there are too few observational constraints to reach this standard, which means that different models can reproduce the same set of observations but give different $H_0\delta t$ product and thus different values for H_0. This degeneracy can be broken by relying on stellar kinematics–from which the above 8% uncertainties estimate is obtained– or, for an improved precision, by further constraining the mass distribution of the lensing system. Traditionally, the two main solutions adopted are that of a power-law, or of a constant mass-to-light ratio plus the so-called Navarro-Frenk-White dark matter halo density profile[25] inferred from simulations (Navarro et al. 1997). These mass assumptions are not however chosen on theoretical grounds or observational constraints. To be sure, surveys did show that the mean slope of the density profile of lenses is nearly isothermal. But this slope is an average, and thus need not be adequately described by a power law (Schneider and Sluse 2013). On the contrary, there are good reasons to think that the central regions of the lens would significantly depart from a perfect power law. Yet, the studies published by Birrer et al. (2020) and Birrer and Treu (2021) have shown that, with a sample of lenses increased from 7 to the 33 lenses from the SLAC-TDCOSMO collaboration, if the mass modelling assumptions are relaxed to be maximally degenerated, then the value obtained for the Hubble constant is no longer of $H_0 = 73.3\pm1.7\pm1.8\,\mathrm{km\,s^{-1}\,Mpc^{-1}}$, but of $H_0 = 67.4\pm4.1\pm3.2\,\mathrm{km\,s^{-1}\,Mpc^{-1}}$, no longer in significant tension with the CMB measure. Such a result demonstrates how much we need to further improve our understanding of the systematics for time-delay cosmography before we can claim a 2% precision measurement that does not

---

[25] Note that the NFW profile is challenged, as it fails to reproduce the observations especially for low-surface-brigtness galaxies. See for instance Bullock and Boylan-Kolchin (2017).

involve unjustified assumptions when it comes to gravitational lensing. Clearly, the high Hubble value with a small error bar highly depends on an assumption that has no strong physical justification. Acknowledging what we still do not know, while not taking any bets, leads to a low value with a much larger error bar. But more importantly, it shows that the robustness of the high Hubble value is no guarantee that this high value is not an artefact from systematic errors. Had the robustness argument for a crisis been taken at face value, the track for these unknown facts would have stopped and the importance of the mass distribution assumption not been properly understood.

3.4.2.2   Systematic Replication and Unknown Unknowns

Now, what about the Cepheids's claim of 1% precision measurement, grounded in more than four decades of refining the ladder technique? Can we legitimately believe that new systematics could remove the discrepancy at this stage of the scientific investigation?

Before addressing this question, a short detour through another toolbox, that of replication, is necessary. The Replication Crisis,[26] according to which many findings in social, behavioral and biomedical sciences have failed to replicate at alarming rates, has led to many interesting developments when it comes to understand what is a replication and what purpose it can serve. I will briefly present a typology of replication adapted from Schmidt and Oh (2016), Schmidt (2016), Zwaan et al. (2018), and Fletcher (2021) before going back to our current issue. The typology I suggest orders replication along four categories. It is important to note that these categories are better conceived of as covering a spectrum and revealing different aspects of replication than as clean-cut separations between different types of replication:[27]

- *Direct replication*: direct replication is an attempt to reproduce exactly the original study, on a different statistical set. It is especially useful to exclude errors related to the statistical sample or to contextual factors. In the case of the Cepheids' technique for instance, a successful direct replication would consist of obtaining the same Hubble constant value based on different Cepheids calibrators.
- *Methodological replication*: this kind of replication is a simple re-analysis of an experiment, ideally by another team. As Fletcher puts it, methodological replication "ensures that the results of a scientific study are not due to data-entry, programming or other suchlike technical errors", or in general to human errors.

---

[26] An introduction to the replication crisis and to its importance can be found in Romero (2019).

[27] One way to think about this spectrum is that the typology aims to capture situations going from: nothing is changed in the replication, only one thing is changed, to several if not all are changed. How you quantify the number of variables that vary also depends on how fine-grained your perspective is.

The re-analysis of the SH0ES data found in Javanmardi et al. (2021) is such an example, among many ones, of such methodological replication.

- *Systematic replication*: it consists in systematically varying one of the variables of the experiment or measure while maintaining the others fixed. The goal of systematic replication is to help to identify which variables causally contribute to the final outcome, but more importantly to better understand and circumscribe the causal contribution of a given variable to the result. Contrary to the other three, systematic replication is more about understanding a protocol, measurement, experiment or model than about assessing the reliability of its prediction.
- *Conceptual replication*: here, the goal is to measure the same phenomenon or test the same hypothesis as in the initial study, but by using different methods, techniques or models. Conceptual replication subsumes robustness analysis: an agreement between different models or different types of measurements amount to a successful replication. But given that the more independent the measurements under comparison are, the stronger the link between robustness and reliability is, robustness is on the far-end of the spectrum–it is a form of conceptual replication that insists on the fact that the original and the replicated study ideally have nothing in common but the targeted value. In other words, the most extreme form of conceptual replication is needed to warrant strong robustness-based conclusions. The alleged agreement between the four local measures detailed above is supposedly of such a nature: it is a successful conceptual replication, allegedly between fully independent measures, inasmuch as the TRGB result is excluded.

The comparison between the TRGB and the Cepheids results exemplifies how the replication spectrum goes from systematic replication to conceptual replication. As the reader may remember, TRGB and Cepheids have complementary weaknesses and strengths, as well as common and independent zero-point anchors: on the Cepheids side, the zero-point calibration is based on Milky Way parallax, on the LMC or on NGC4258, and the sample of Type I SNe used is the Supercal sample. On the TRGB side, the zero-point calibration has been done on the basis of the distance modulus to the LMC based on DEBs + Hubble Space Telescope parallax calibration in 2019, and on the basis of the LMC, NGC 4258 and the Milky Way globular clusters in 2020. The sample of supernovae can be that of the Carnegie-Chicago Program or the Supercal SNIa sample used by the SH0ES team. Galaxies where the calibrators (either TRBG or Cepheids) can be found as well as SNe include 18 host galaxies on the TRGB side, 19 on the Cepheids's side, 11 of which are common to the two groups. In other words, the comparison between the two results can be constructed such as to maximally overlap and leave only the choice of standard candle as the variable explored–which amounts to the perfect picture of a systematic replication, or to be fully independent, which would amount in the case of an agreement to a perfect conceptual replication, inasmuch as the standard candle used is no longer considered a mere variable but a method. As it turns out, neither the systematic replication nor the conceptual replication are successful replications in this context: recently updated TRGB and Cepheids measurements result in differing values of $H_0 = 69.6 \pm 1.9 \, \text{km s}^{-1} \, \text{Mpc}^{-1}$ (Freedman et al. 2019) for the TRGB and 73.04 ± 1.4 (Riess 2020) for the Cepheids. But the conceptual replication does not teach us anything about how to locate the problem, as differences in several variables do not allow us to pin down the most probable culprit. Differences in one variable only, as it the case with systematic replication, do not fall prey to this problem. If the two measurements only differ in the sample of supernovae chosen, then the calibration or a possible bias in the sample of supernovae is most likely responsible for the disagreement. It is only because the Carnegie Chicago Hubble Program led by W. Freedman proceeded to a detailed systematic comparison between TRGB stars and Cepheids that we now have a better idea about where to look for possible unidentified unknown unknowns. While the two methods show excellent agreement on the distance modulus to 28 galaxies for instance, the study shows that this agreement no longer holds when comparing the distance to the 10 SNIa host galaxies that the two have in common. Future observational campaigns with much higher resolution, notably thanks to the JWST telescope, might be in a position to elucidate this disagreement.

The failure of systematic replication not only indicates with no possible doubts that new systematics are yet to be discovered, but informs us about where to look for them: if one wants to test an hypothesis about a possible source of systematics (e.g., the distance modulus to the LMC), the overlap between these two techniques easily allows one to design a crucial test that permits one to verify such an hypothesis—for instance by comparing the TRGB result to the value inferred from Cepheids only on the basis of the Milky Way and of NGC 4258. Likewise, the fact that the metallicity can be constrained for TRGB stars allows one to decouple the problem of metallicity and of extinction, given that the TRGB I-band is not affected by metallicity effects. Metallicity and extinction can thus be individually solved, and the measure of the extinction obtained from TRGB stars can inform the calibration of Cepheids for common objects.[28] Hence the epistemic superiority of systematic replication in contexts of highly uncertain measurements, and the need to wait for successful systematic replication before applying robustness arguments:

- Given that the focus of robustness is on comparing models or measurements that are as independent as possible, arguments drawn on its basis remain mute and offer no explanations and no guide to locate the problem when the robustness analysis fails. A good example of this is a comparison between the success of the robustness strategy when excluding the TRGB result and its failure when including it. Once the claim of poor calibration on the TRGB side is excluded, how ought we to account for this failure? How can scientists decide where to start to explain it? Systematic replication, on the other hand, is maximally informative and is in a position to identify possible sources of failure. Conceptual replication does not have in principle to be mute about such possibles sources of errors, but

---

[28] For a detailed comparison between TRGB stars and Cepheids for the different replications performed, see section 4 of Freedman et al. (2019).

the part of the conceptual replication spectrum that corresponds to robustness does, inasmuch as it necessitates the maximization of the independence of different measurements.

- As we have seen above, a successful robustness analysis or conceptual replication does not logically establish the reliability of a given prediction. But the failure of systematic replication demonstrates unequivocally that new systematics have yet to be identified. Hence, if we grant the claim made in 3.4.1 that a high degree of confidence is better supported when the comparison holds between mature and independent models, free as much as possible of unknown systematic and statistical errors, then systematic replication does not only have the epistemic priority to assess a discrepancy, but also the chronological priority. Indeed, it is the role of systematic replication to diagnose whether such unknown unknowns have still to be identified. It is only when successful systematic replications are performed that one can be confident that the most important sources of systematic and statistical errors have been accounted for. In other words, if systematic replication cannot be performed successfully, it demonstrates that the conditions for applying conceptual replication understood as robustness are not yet met, at least in the conditions that would allow for a high degree of confidence in conclusions drawn from it.[29] In the case of the Hubble constant, the failure of the systematic replication performed on the Cepheids and TRGB results show that the precision of these measurements, though by far the most mature techniques for determining the value of the Hubble constant, has not reached a sufficient level for robustness arguments to be telling and/or trusted.

Two remarks are needed to qualify the claim made here. First, one does not have to reject robustness analysis altogether on these grounds. Robustness analysis and systematic replication are complementary tools and can work together very well to address different problems. They cannot however be deployed at the same stage of the inquiry, as the latter indicates when the conditions for justified inferences based on robustness are met. The appeal to robustness has no epistemic grounds if systematic replication is not successfully achieved. If systematic replication fails, no meaningful robustness-based conclusions about the reliability of the measure can be drawn. Second, when used too early, robustness analysis is actually an obstacle to its companion and leads to neglect of the chase for unknown unknowns. This happens because robustness, as an extreme form of conceptual replication, focuses on developing independent techniques that have, ideally, absolutely nothing to do with each other–e.g., time-delay cosmography and Cepheids-based distance ladder. On the other hand, systematic replication requires measurement techniques sufficiently close to each other to be mutually informative, as Cepheids and TRGB

---

[29] Although space reasons prevent me to expand on this point, it would be interesting to analyze the different scenarios that can arise: failure of CR vs success of SR, failure of both, and so on and so forth. Here we only address the claim that a specific kind of conceptual replication, that usually referred to as robustness analysis, is sufficient to warrant the reliability of the late-universe measurements and thus the problematic nature of the discrepancy.

can be. Robustness deployed too early leads to developing techniques that are too independent from each other to offer the grounds needed for systematic replication. Hence the need to understand their roles and places in the scientific investigation, so as to not let robustness becomes the crack in the track of the Hubble constant. The failure of systematic replication offered by the TRGB and the Cepheids' measurements tells us that the discrepancy between early universe and local universe measurements is not a residual discrepancy, contrary to what the defenders of a crisis would like to see on the basis of their robustness strategy. Moreover, it informs us about how measurements can be further improved and where to start doing so. Robustness cannot be used to justify such as crisis if it is not established that the track for unknown unknowns has gone far enough, and that we are indeed comparing mature and well-understood techniques. There might be a cosmological crisis to come, but such a crisis is not justified by the current epistemic situation, and certainly not on the basis of robustness arguments made at this stage of the scientific investigation.

3.5   Conclusion

Recent developments in astrophysics have seen the community working on the Hubble constant shaken by the robustness of the high value found by local universe measurements, and subsequently by the significant discrepancy between this value and the one obtained from early universe measurements. Some have gone as far as claiming that these developments prove that the standard model of cosmology is undergoing a crisis. The robustness of the high-values, they contend, shows that the discrepancy between the early and late universe will hold, and thus is the long-wished for evidence that new physics is needed to amend the standard model. I hope to have shown here that we do not have good reasons to interpret the discrepancy as residual on the basis of robustness arguments, and to think that we are currently facing a crisis. The Hubble debate does not offer the conditions that would warrant a strong degree of confidence in robustness-based inferences. The fact that we seem to have reached 1% precision measurements is not yet a sign that systematic errors have been almost eliminated, but that the unknown systematics are getting harder and harder to track. The blossoming of new techniques is an opportunity not to establish the robustness of the discrepancy, but to use their overlap to deploy systematic replication and refine our understanding of where the skeletons could still be hiding.

Acknowledgments The author is very grateful to Chris Smeenk, Barry Madore and Wendy Freedman for their helpful feedback. They would also like to thank the Bonn History and Philosophy of Physics Seminar, the APC colloquium (Paris 7), and the audience of the "What can Astrophysics teach us about Replicability?" workshop organized by the CSH center (Bern) for insightful discussions. Finally, the author would like to thank the two anonymous referees for their helpful feedback. This paper is based on work done while funded under the John Templeton Foundation grant: "New Directions in Philosophy of Cosmology" (grant number 61048).

References

Birrer, S., A. Shajib, A. Galan, M. Millon, T. Treu, A. Agnello, M. Auger, G.-F. Chen, L. Christensen, T. Collett, et al. 2020. TDCOSMO-IV. Hierarchical time-delay cosmography–joint inference of the Hubble constant and galaxy density profiles. *Astronomy & Astrophysics* 643: A165.

Birrer, S., Treu, T. 2021. TDCOSMO-V. Strategies for precise and accurate measurements of the Hubble constant with strong lensing. *Astronomy & Astrophysics*, 649: A61.

Bullock, J.S., M. Boylan-Kolchin. 2017. Small-scale challenges to the ΛCDM paradigm. arXiv:1707.04256.

Collaboration, P. 2020. Planck 2018 results. VI. Cosmological parameters. *Astronomy and Astrophysics* 641(A6): 1–56.

Di Valentino, E., O. Mena, S. Pan, L. Visinelli, W. Yang, A. Melchiorri, D.F. Mota, A.G. Riess, J. Silk. 2021. In the realm of the hubble tension—a review of solutions. *Classical and Quantum Gravity* 38(15): 153001.

Efstathiou, G. 2020. A Lockdown Perspective on the Hubble Tension (with comments from the SH0ES team). Manuscript. arXiv preprint arXiv:2007.10716. Accessed on 12 May 2023.

Fletcher, S.C. 2021. The role of replication in psychological science. *European Journal for Philosophy of Science* 11(1): 1–19.

Fong, Y.C. 2011. Measuring the hubble constant. https://www.imperial.ac.uk/media/imperial-college/research-centres-and-groups/theoretical-physics/msc/dissertations/2011/Yick-Chee-Fong-Dissertation.pdf

Freedman, W.L. 2021. Measurements of the hubble constant: Tensions in perspective. *The Astrophysical Journal* 919(1): 16.

Freedman, W.L., B.F. Madore, D. Hatt, T.J. Hoyt, I.S. Jang, R.L. Beaton, C.R. Burns, M.G. Lee, A.J. Monson, J.R. Neeley, et al. 2019. The Carnegie-Chicago hubble program. VIII. An independent determination of the hubble constant based on the tip of the red giant branch. *The Astrophysical Journal* 882(1): 34.

Freedman, W.L., B.F. Madore, T. Hoyt, I.S. Jang, R. Beaton, M.G. Lee, A. Monson, J. Neeley, J. Rich. 2020. Calibration of the tip of the red giant branch. *The Astrophysical Journal* 891(1): 57.

Guralp, G. 2020. Calibrating the universe: The beginning and end of the hubble wars. *Standardization in Measurement. Philosophical, Historical and Sociological Issues* 125–138.

Huang, C.D., A.G. Riess, W. Yuan, L.M. Macri, N.L. Zakamska, S. Casertano, P.A. Whitelock, S.L. Hoffmann, A.V. Filippenko, D. Scolnic. 2020. Hubble space telescope observations of Mira variables in the SNIa host NGC 1559: An alternative candle to measure the hubble constant. *The Astrophysical Journal* 889(1): 5.

Javanmardi, B., A. Mérand, P. Kervella, L. Breuval, A. Gallenne, N. Nardetto, W. Gieren, G. Pietrzyński, V. Hocdé, S. Borgniet. 2021. Inspecting the Cepheid distance ladder: The hubble space telescope distance to the SN Ia host galaxy NGC 5584. *The Astrophysical Journal* 911(1): 12.

Justus, J. 2012. The elusive basis of inferential robustness. *Philosophy of Science* 79(5): 795–807.

Leavitt, H.S. 1908. 1777 variables in the magellanic clouds. *Annals of Harvard College Observatory* 60: 87–108.

Levins, R. 1966. The strategy of model building in population biology. *American Scientist* 54(4): 421–431.

Levins, R. 1993. A response to Orzack and Sober: Formal analysis and the fluidity of science. *The Quarterly Review of Biology* 68(4): 547–555.

Mortsell, E., Goobar, A., Johansson, J., Dhawan, S. 2021. Sensitivity of the Hubble Constant Determination to Cepheid Calibration. Manuscript. arXiv preprint arXiv:2105.11461. Accessed on 12 May 2023.

Navarro, J.F., C.S. Frenk, S.D. White. 1997. A universal density profile from hierarchical clustering. *The Astrophysical Journal* 490(2): 493.

Orzack, S.H., E. Sober. 1993. A critical assessment of Levins's the strategy of model building in population biology (1966). *The Quarterly Review of Biology* 68(4): 533–546.

Parker, W.S. 2011. When climate models agree: The significance of robust model predictions. *Philosophy of Science* 78(4): 579–600.

Pesce, D.W., Braatz, J.A., Reid, M.J., Condon, J.J., Gao, F., Henkel, C., Kuo, C.Y., Lo, K.Y., Zhao, W. 2020. The Megamaser Cosmology Project. XI. A Geometric Distance to CGCG 074-064. *The Astrophysical Journal*, 890, pp. 340–398.

Potter, C., J.B. Jensen, J. Blakeslee, P. Milne, P.M. Garnavich, P. Brown. 2018. Calibrating the type Ia supernova distance scale using S brightness fluctuations. In *American astronomical society meeting abstracts# 232*, vol. 232, 319–02.

Riess, A.G. 2020. The expansion of the universe is faster than expected. *Nature Reviews Physics* 2(1): 10–12.

Riess, A.G., S. Casertano, W. Yuan, L.M. Macri, D. Scolnic. 2019. Large magellanic cloud cepheid standards provide a 1% foundation for the determination of the Hubble constant and stronger evidence for physics beyond λCDM. *The Astrophysical Journal* 876(1): 85.

Riess, A.G., W. Yuan, L.M. Macri, D. Scolnic, D. Brout, S. Casertano, D.O. Jones, Y. Murakami, G.S. Anand, L. Breuval, et al. 2022. A comprehensive measurement of the local value of the Hubble constant with 1 km s^{-1} mpc^{-1} uncertainty from the Hubble space telesope and the SH0ES team. *The Astrophysical Journal Letters* 934(1): L7.

Romero, F. 2019. Philosophy of science and the replicability crisis. *Philosophy Compass* 14(11): e12633.

Sandage, A.R. 1970. Cosmology: A search for two numbers. *Physics Today* 23(2): 34–41.

Schmidt, F.L., I.-S. Oh. 2016. The crisis of confidence in research findings in psychology: Is lack of replication the real problem? Or is it something else? *Archives of Scientific Psychology* 4(1): 32.

Schmidt, S. 2016. Shall we really do it again? The powerful concept of replication is neglected in the social sciences. In *Methodological issues and strategies in clinical research*, ed. A.E. Kazdin, 581–596. Washington: American Psychological Association.

Schneider, P., D. Sluse. 2013. Mass-sheet degeneracy, power-law models and external convergence: Impact on the determination of the Hubble constant from gravitational lensing. *Astronomy & Astrophysics* 559: A37.

Schupbach, J.N. 2015. Robustness, diversity of evidence, and probabilistic independence. In *Recent developments in the philosophy of science: EPSA13 Helsinki*, 305–316. Berlin: Springer.

Soler, L., E. Trizio, T. Nickles, W. Wimsatt. 2012. *Characterizing the robustness of science: After the practice turn in philosophy of science*, vol. 292. Berlin: Springer.

Verde, L., T. Treu, A.G. Riess. 2019. Tensions between the early and late Universe. *Nature Astronomy* 3: 891–895.

Weisberg, M. 2006. Robustness analysis. *Philosophy of Science* 73(5): 730–742.

Weisberg, M. 2012. *Simulation and similarity: Using models to understand the world*. Oxford: Oxford University Press.

Weisberg, M., K. Reisman. 2008. The robust Volterra principle. *Philosophy of Science* 75(1): 106–131.

Wimsatt, W.C. 2012. Robustness, reliability, and overdetermination (1981). In *Characterizing the robustness of science*, 61–87. Berlin: Springer.

Wong, K.C., S.H. Suyu, G.C. Chen, C.E. Rusu, M. Millon, D. Sluse, V. Bonvin, C.D. Fassnacht, S. Taubenberger, M.W. Auger, et al. 2020. *H0LiCOW*–XIII. A 2.4% measurement of H_0 from lensed quasars: 5.3 σ tension between early-and late-universe probes. *Monthly Notices of the Royal Astronomical Society* 498(1): 1420–1439.

Yuan, W., A.G. Riess, L.M. Macri, S. Casertano, D.M. Scolnic. 2019. Consistent calibration of the tip of the red giant branch in the large magellanic cloud on the hubble space telescope photometric system and a redetermination of the hubble constant. *The Astrophysical Journal* 886(1): 61.

Zwaan, R.A., A. Etz, R.E. Lucas, M.B. Donnellan. 2018. Making replication mainstream. *Behavioral and Brain Sciences* 41: e120.

Chapter 4
Theory Testing in Gravitational-Wave Astrophysics

Jamee Elder

Abstract The LIGO-Virgo Collaboration achieved the first 'direct detection' of gravitational waves in 2015, opening a new "window" for observing the universe. Since this first detection ('GW150914'), dozens of detections have followed, mostly produced by binary black hole mergers. However, the theory-ladenness of the LIGO-Virgo methods for observing these events leads to a potentially-vicious circularity, where general relativistic assumptions may serve to mask phenomena that are inconsistent with general relativity (GR). Under such circumstances, the fact that GR can 'save the phenomena' may be an artifact of theory-laden methodology. This paper examines several ways that the LIGO-Virgo observations are used in theory and hypothesis testing, despite this circularity problem. First, despite the threat of vicious circularity, these experiments succeed in testing GR. Indeed, early tests of GR using GW150914 are best understood as a response to the threat of theory-ladenness and circularity. Each test searches for evidence that LIGO-Virgo's theory-laden methods are biasing their overall conclusions. The failure to find evidence of this places constraints on deviations from the predictions of GR. Second, these observations provide a basis for studying astrophysical and cosmological processes, especially through analyses of populations of events. As gravitational-wave astrophysics transitions into mature science, constraints from early tests of GR provide a scaffolding for these population-based studies. I further characterize this transition in terms of its increasing connectedness to other parts of astrophysics and the prominence of reasoning about selection effects and other systematics in drawing inferences from observations.

Overall, this paper analyses the ways that theory and hypothesis testing operate in gravitational-wave astrophysics as it gains maturity. In particular, I show how these tests build on one another in order to mitigate a circularity problem at the heart of the observations.

J. Elder (✉)
Black Hole Initiative, Cambridge, MA, USA
e-mail: jelder@fas.harvard.edu

© The Author(s) 2023
N. Mills Boyd et al. (eds.), *Philosophy of Astrophysics*, Synthese Library 472,
https://doi.org/10.1007/978-3-031-26618-8_4

4.1 Introduction

The LIGO-Virgo Collaboration achieved the first 'direct detection' of gravitational waves in 2015, a discovery that marked a new epoch for gravitational-wave astrophysics—one in which gravitational waves provided a new "window" for observing the universe. Since this first detection ('GW150914'), dozens of detections have followed, most produced by binary black hole mergers.

Merging black holes offer us unique access into the 'dynamical strong field regime' of general relativity (GR), due to the high speeds and strong gravitational fields involved with these events. Such events are of interest for learning about strong-field gravity, as well as about black hole populations and formation channels.

However, the LIGO-Virgo methods for observing these events also pose some interesting epistemic problems. Parameter estimation and other inferences about the source system are highly theory- or model-laden, in that all such inferences rely on assumptions about how source parameters determine merger dynamics and gravitational wave emission. This leads to a potentially vicious circularity, where general relativistic assumptions may serve to mask phenomena that are inconsistent with GR. Under such circumstances, the fact that GR can 'save the phenomena' may be an artifact of theory-laden methodology.

In this paper I examine the ways that the LIGO-Virgo Collaboration engages in theory testing (and model and hypothesis testing) despite this circularity problem.

I begin, in Sect. 4.2, by rehearsing some of the key epistemic challenges of gravitational-wave astrophysics, with an emphasis on issues of theory- or model-ladenness and circularity as they arise in the context of theory testing.

In Sect. 4.3 I examine several ways that the LIGO-Virgo observations act as tests of GR. I argue that the tests of GR using individual events (such as GW150914) are best understood as a response to the circularity problem described above; they test whether the circularity in their methodology is problematic, through searching for evidence that these methods are biasing the overall conclusions. This allows the LIGO-Virgo Collaboration to place constraints on the bias introduced by the specific assumptions being tested. While this does much to mitigate the circularity problem, I also discuss how degeneracies of various kinds make it difficult to constrain all sources of bias introduced by the use of GR models.

In Sect. 4.4 I then describe the further tests that become available as gravitational-wave astrophysics transitions into mature science. This involves a shift in focus from individual events to populations. I show how the earlier, event-based tests of GR provide a foundation on which these population-level inferences may be built. Specific examples include (in Sect. 4.4.1) inferences about the astrophysical mechanisms that produce binary black hole mergers (e.g., van Son et al. 2022), and (in Sect. 4.4.2) inferences about cosmological expansion (e.g., Chen et al. 2018). The use of populations helps reduce some remaining sources of uncertainty (e.g., due to distance/inclination degeneracy) but not others (e.g., due to 'fundamental theoretical bias'). Either way, hypothesis testing using populations must continue to grapple with the issues raised in Sects. 4.2 and 4.3.

Finally, in Sect. 4.5, I discuss the themes that emerge from my examination of theory testing in earlier sections. First, I characterize a transition that is occurring in gravitational-wave astrophysics as it gains maturity. This involves a shift to populations; an increasing interdependence or connectedness with electromagnetic astrophysics; and a greater resemblance to other parts of astrophysics. This includes, for example, grappling with a 'snapshot' problem of drawing inferences about causal processes that occur over long timescales based on observations of a system at a single time. Second, I argue that theory-ladenness, circularities, and complex dependencies between hypotheses are important—but not insurmountable—challenges in gravitational-wave astrophysics. Furthermore I suggest the progress in this field will come where these vices can be made virtuous, by leveraging improvements in modeling, simulation, or observation in one domain to place constraints in another.

4.2 Epistemic Challenges for Theory Testing

The LIGO-Virgo Collaboration uses a network of detectors (specifically, gravitational-wave interferometers) to detect the faint gravitational wave signals produced by compact binary mergers, such as binary black hole mergers. The data produced by the interferometers are very noisy, so an important challenge of gravitational-wave astrophysics is that of separating the signals from the noise. This is most efficiently done through a modeled search, using a signal-processing technique called 'matched filtering'.[1] Having extracted a gravitational wave signal, features of the source are inferred using a Bayesian parameter estimation process. This produces posterior distributions for values such as the masses and spins of the component black holes, the distance to the binary system, etc.

In the 'discovery' paper announcing GW150914, the LIGO-Virgo Collaboration describe this event as the first 'direct detection' of gravitational waves and the first 'direct observation' of a binary black hole merger (Abbott et al. 2016b). Elder (In preparation) provides an analysis of these terms—with a focus on what is meant by 'direct' in these cases—drawing connections to recent work in the philosophy of measurement (e.g., Tal 2012, 2013; Parker 2017).[2] However, these descriptions

---

[1] For details about this technique, see e.g., Maggiore (2008).

[2] Roughly, the 'directness' has to do with the nature of the inferences needed to make the detection claim; for a direct detection, this is based on the model of the measuring process (i.e., the understanding of the measuring device and how it couples to its environment) while an indirect detection also relies on a model of a separate target system. Note also that 'detection' and 'observation' have a range of meanings across different scientific contexts but are often used interchangeably to describe gravitational-wave detection/observation (Elder In preparation). In this paper, I use the term 'observation' in a broad, permissive way, encompassing measurements with complex scientific instruments (e.g., gravitational wave interferometers). Something like Shapere (1982)'s account of observation in astronomy will do to capture what I have in mind. I use the term

have the potential to obscure the fact that these observations are also *indirect* in the sense that they are mediated by models, such as those in the 'EOBNR' and 'IMRPhenom' modeling families (Abbott et al. 2016d). These models take source parameters and map them to gravitational-wave signals, via a description of the dynamical behaviour of the binary system. The success of the LIGO-Virgo experiments depends on the availability of accurate models spanning the parameter space for systems that the LIGO and Virgo interferometers are sensitive to.

Theory-ladenness comes in many forms and degrees, and is not inherently problematic. However, the theory-ladenness of experimental observations may be problematic when it leads to a vicious circularity. Such a circularity arises when the theoretical assumptions made by the experimenters—either in the physical design of the experiment or in subsequent inferences from the empirical data—guarantee that the observation will confirm the theory being tested.[3] In the case of the LIGO-Virgo experiments, the concern is that the use of general relativistic models to interpret the data guarantees that the results will be consistent with general relativity.[4]

For the LIGO-Virgo experiments, there are two main layers of theory-ladenness to consider. These are due to the two main roles of the experiments: detecting gravitational waves (via 'search' pipelines) and observing compact binary mergers (via 'parameter estimation' pipelines).[5]

Detecting gravitational wave signals in the noisy LIGO-Virgo data is done using both modeled and unmodeled searches. The two modeled search pipelines, 'GstLAL' and 'PyCBC', are targeted searches for gravitational waves produced by compact binary coalescence (e.g., by the merger of two black holes). Both searches use 'matched filtering.' This involves correlating a known signal, or *template*, with an unknown signal, in order to detect the presence of the template

---

'detection' to include any empirical investigation (measurement, observation, etc.) that purports to establish the existence or presence of an entity or phenomenon within a target system. A successful detection meets some threshold of evidence (relative to the background knowledge and the standards of acceptance of the relevant scientific community) such that we accept the existence of that entity (though precisely what this acceptance constitutes will differ for realists and anti-realists).

[3] See Boyd and Bogen (2021, section 3) for an overview of theory-ladenness (and value-ladenness) in science.

[4] See Elder (Forthcoming) and Elder and Doboszewski (In preparation) for discussion of theory-ladenness and (vicious) circularities in the context of the LIGO-Virgo experiments. My brief exposition here draws on these more detailed discussions.

[5] Here 'pipeline' refers to the set of processes used to generate the particular product from the data. Thus 'search pipelines' are alternative pathways for processing data to detect gravitational wave signals and 'parameter estimation pipelines' are alternative pathways for estimating the values of source parameters (e.g., masses and spins of the binary components).

within the unknown signal.[6] The unmodeled (or 'minimally modeled'[7]) 'burst' search algorithms, 'cWB' and 'oLIB', look for transient gravitational-wave signals by identifying coincident excess power in the time-frequency representation of the strain data from at least two detectors.[8] The modeled search pipelines, using matched filtering, are heavily theory- or model- laden, while the unmodeled search is less so. However, the modeled searches are more efficient at extracting signals from compact binary mergers, regularly reporting detections at higher statistical significance than the unmodeled searches.

Once a detection has been confirmed through the search pipelines the next step is to draw inferences about the properties of the source system on the basis of the signal. This involves assigning values to parameters describing features of interest. This process, 'parameter estimation,' is performed within a Bayesian framework. The basic idea is to calculate posterior probability distributions for the parameters describing the source system, based on some assumed model M that maps parameters about the source system to gravitational-wave signals. The sources of the gravitational waves detected so far are compact binary mergers. Such events are characterized by a set of intrinsic parameters—including the masses and spins of the component objects—as well as extrinsic parameters characterizing the relationship between the detector and the source—e.g., luminosity distance, and orientation of the orbital plane. The observation of compact binary mergers relies on having an accurate model relating the source parameters to the measured gravitational waveform.[9]

These two ways that the LIGO-Virgo observations are theory-laden lead to some specific concerns about how the use of general relativistic models may be systematically biasing the LIGO-Virgo results. These issues have been discussed in detail elsewhere (see e.g., Elder Forthcoming, Sect. 5.3) so I limit my discussion to a brief summary.

In both cases, inaccuracies of the models may be due to either the failure of the models to accurately reflect the full general relativistic description of the situation, or the failure of GR to accurately describe the regimes being observed (or both). Following Yunes and Pretorius (2009), call these 'modeling bias' and 'fundamental theoretical bias' respectively.

In the first case—the observation of gravitational waves—the main concern is that any inaccuracy of the models might lead to a biased sampling of gravitational wave signals. Here, by 'inaccuracy of the models' I mean a lack of fit between the morphology of actual gravitational waves and template models. This may be due

---

[6] The details of the modeled searches and their results for GW150914 are reported in Abbott et al. (2016a).

[7] The unmodeled searches are often (and more accurately) called 'minimally modeled' searches. However, I will use the 'unmodeled' terminology to highlight the contrast between search pipelines, since this search doesn't use general relativistic modeling to detect gravitational waves.

[8] The details of this search and the results for GW150914 are reported in Abbott et al. (2016c).

[9] Examples include the previously mentioned EOBNR and IMRPhenom modeling families.

to either modeling bias or fundamental theoretical bias. There are several ways that the sampling may be biased. This includes both false positives (falsely identifying noise as a gravitational wave signal) and false negatives (failing to detect genuine signals). While these are both genuine problems to be overcome, a more insidious concern lies in between: imperfect signal extraction. If the models used for matched filtering are inaccurate, but still adequate to make a gravitational wave detection, the extracted signal may not accurately reflect the real gravitational waves. In this case, we can think of there being some residual signal left behind once the detected signal is removed. Thus the observation of gravitational waves may be systematically biased by any inaccuracies in the models used to observed them.

In the second case—observation of compact binary mergers—the main concern is that inaccurate models used in parameter estimation will systematically bias the posterior distributions for the parameters representing the properties of the source. Here, both modeling bias and theoretical bias are important sources of potential error, but the latter requires some explanation. The final stages of the binary black hole mergers observed by the LIGO-Virgo Collaboration occur in the dynamical strong field regime, because the two black holes are orbiting each other closely enough to be moving through strong gravitational fields with high velocities. At this stage, approximation schemes (such as post-Newtonian approximation) are no longer adequate for describing the dynamics of the binary and the full general relativistic description is needed. Such regimes have not been observed before; our empirical access to such regimes comes only from the LIGO-Virgo observations. The final stage of a binary black hole merger (including its gravitational wave emission) is thus (for all we know, at least prior to the LIGO-Virgo observations) a place where the description offered by GR may be inadequate. Thus the LIGO-Virgo Collaboration is performing theory-laden observations of a process where the theory itself is in question.

The two layers of theory-ladenness just described lead to circularity in the following sense: justifying confidence in the LIGO-Virgo observations relies on justification for the applicability of the theory in this context, while justification for the applicability of the theory in this context can only be based on the LIGO-Virgo observations.[10] Put another way, general relativistic models are used in the detection and interpretation of gravitational waves in such a way that detected signals will inevitably be well-described by general relativistic models, and interpreted as being produced by systems exhibiting general relativistic dynamics. In short, GR seems guaranteed to 'save the phenomena' of binary black hole mergers that are observed this way, since it is this theory that tells us what is being observed.

The theory-ladenness, and even the circularity I have described so far may seem unremarkable—indeed, some degree of theory-ladenness is a generic feature of scientific experiments. However, there are other features of the epistemic situation

---

[10] For the purposes of this paper, I neglect the additional complication of validating models and simulations with respect to theory, which is made challenging by the lack of exact solutions to the Einstein field equations for such systems (Elder Forthcoming).

that render this circularity problematic in that they make it particularly difficult to uncover and circumvent any bias introduced by the theory- and model-laden methodology.

First, the LIGO-Virgo observations probe new physical regimes—the 'dynamical strong field regime'—where it is possible that target systems will deviate from the predictions of GR in terms of their dynamical behaviour and gravitational-wave generation.

Second, the LIGO-Virgo observations are the only line of evidence into the source system. There are no earlier empirical constraints on the behaviour of such systems in the dynamical strong field regime, and no other independent access to these regimes is possible.[11] This limits the prospects for corroboration through coherence tests or consilience.

Third, no interventions on these systems are possible, given the astrophysical context. Controlled interventions to choose source parameters and then observe the resulting gravitational waves would allow the source properties and the resulting dynamics to be disentangled. Such a process would allow for a comparison between independent predictions and observations. In astrophysics, where such controlled interventions are generally impossible, it is common to use simulations as a proxy experiment in order to explore causal relationships and downstream observational signatures. However, in this case, it is the dynamical theory governing these systems that is in question. Increasing the precision of the simulation cannot overcome the problem of 'fundamental theoretical bias' introduced by the need to assume that GR is an adequate theory in these regimes.

These features of the epistemic situation mean that the theory-ladenness of the LIGO-Virgo methodology is a potential problem. If the observations are biased (e.g., distance estimates were consistently underestimated) this bias may go unchecked.

4.3 Testing General Relativity

Despite the epistemic challenges highlighted in Sect. 4.2, the LIGO-Virgo Collaboration does claim to use their observations to test general relativity. On the surface, the epistemic situation just described might seem to render this impossible. However, I will show that the theory testing done with individual events such as GW150914 goes beyond merely 'saving the phenomena'. Rather, the tests performed probe specific ways that the theory-laden methodology might be biasing results and masking discrepancies with general relativity, by testing specific hypotheses about the signal and noise. In doing so, they place constraints on the bias

---

[11] For GW170817, the first multi-messenger observation of a binary neutron star merger, there was independent access—but not of a kind that placed strong independent constraints on the dynamical strong field regime (Elder In preparation).

introduced by the model-dependent methods being employed.[12]

For a start, the direct detection of gravitational waves itself constitutes a kind of test of general relativity. After all, gravitational waves are an important prediction that distinguishes GR from its predecessor, Newtonian gravity. This is one test of GR that doesn't suffer from the kind of circularity problem I described above. This is largely because confidence in the detection of gravitational waves depends primarily on confidence in the detector and the modeling of the measuring process, rather than confidence in a general relativistic description of the source (Elder and Doboszewski In preparation; Elder In preparation). However, this test largely serves as corroboration of what was already known from observations of the Hulse-Taylor binaries (Hulse and Taylor 1975; Taylor and Weisberg 1982). In short, this test of GR avoids the circularity objection, but it also fails to provide new constraints on possible deviations from this theory.

Other tests of GR by LIGO-Virgo depend on the properties of the detected waves and what these properties tell us about the nature and dynamical behavior of compact binaries. Abbott et al. (2016e) discusses a number of such tests performed using the data for GW150914. In addition, Abbott et al. (2019c) and Abbott et al. (2021d) extend these tests to the full populations of events from O2 and O3. The tests considered include what I will call the 'residuals test,' the 'IMR consistency test,' the 'parameterized deviations test,' and the 'modified dispersion relation test.'[13] None of these tests have yet found evidence of deviations from general relativity.

First, the 'residuals test' tests the consistency of the residual data with noise (Abbott et al. 2016e). This involves subtracting the best-fit waveform from the GW150914 data and then comparing the residual with detector noise (for time periods where no gravitational waves have been detected). The idea here is to check whether the waveform has successfully removed the entire gravitational-wave signal from the data, or whether some of the signal remains. This process places constraints on the residual signal, and hence on deviations from the subtracted waveform that might still be present in the data. However, this doesn't constrain deviations from GR *simpliciter*, due to the possibility that the best-fit GR waveform is degenerate with non-GR waveforms for events characterized by different parameters (Abbott et al. 2016e). That is, the same waveform could be generated by a compact binary merger (described by parameters different from those that we think describe the GW150914 merger) with dynamics that deviate from general relativistic dynamics. In this case, we could be looking at different compact objects than we think we are, behaving differently than we think they are, but nonetheless producing very

---

[12] For the purposes of this paper, I use the term 'test' fairly loosely to encompass any empirical investigation where the outcome is taken to have a bearing on (or provide evidence relevant to) the acceptability of a theory or hypothesis. Thus I will consider something to be a test of GR if it makes use of empirical data and has an outcome that counts as evidence for or against the acceptance of GR.

[13] The first two of these, the residuals and IMR consistency tests, are also briefly discussed in Elder (Forthcoming).

similar gravitational wave signatures. Thus the residuals test could potentially show inconsistency with general relativity, but not all deviations from GR will be detectable in this way.

Second, the 'IMR consistency test' considers the consistency of the low-frequency part of the signal with the high-frequency part (Abbott et al. 2016e). This test proceeds as follows. First, the masses and spins of the two compact objects are estimated from the inspiral (low-frequency), using LALInference. This gives posterior distributions for component masses and spins. Then, using formulas derived from numerical relativity, posterior distributions for the remnant, post-merger object are computed. Finally, posterior distributions are also calculated directly from the measured post-inspiral (high-frequency) signal, and the two distributions are compared. These are also compared to the posterior distributions computed from the inspiral-merger-ringdown waveform as a whole. Every step of this test involves parameter estimation using general relativistic models. Thus general relativistic descriptions of the source dynamics are assumed throughout the test. Nonetheless, this test could potentially show inconsistency with general relativity. This would occur if parameter estimation for the low and high frequency parts of the signal did not cohere. The later part of the signal in particular might be expected to exhibit deviations from GR (if there are any). In contrast, previous empirical constraints give us reason to doubt that such deviations will be significant for the early inspiral. In the presence of high-frequency deviations, parameter estimation based on GR models will deviate from the values of a system that is well-described by general relativity. Hence (in such cases) we can expect the parameter values estimated from the low frequency part of the signal to show discrepancies with the parameter values estimated from the high frequency part of the signal.

Third, the 'parameterized deviations test' checks for 'phenomenological deviations' from the waveform model (Abbott et al. 2016e). The basic idea of this test is to consider a family of parameterized analytic inspiral-merger-ringdown waveforms and to treat the coefficients of these waveforms as free variables (Abbott et al. 2016e, 6). This means that a new family of waveforms gIMR is generated by taking the frequency domain IMRPhenom waveform models and introducing fractional deformations $\delta\hat{p}_i$ to the phase parameters p_i. To test the theory, the p_i are fixed at their GR values while one or more of the $\delta\hat{p}_i$ are allowed to vary. The physical parameters associated with mass, spin, etc. are also allowed to vary as usual. Within this new parameter space, GR is defined as the position where all of the testing parameters $\delta\hat{p}_i$ are zero. The values of all of the varying parameters are then estimated through a LALInference analysis and the resulting posterior distributions are compared to those generated using only standard GR waveform models. Although there are a range of alternatives to GR on the table, models of the gravitational waves we can expect to observe according to these theories are not available. Without detailed knowledge of the predictions of such theories (i.e., a library of models spanning the parameter space) it is difficult to say with confidence that any given gravitational wave observation favors GR over the alternatives. The parameterized deviations test is a way of overcoming this lack of alternative models. It does so by generating a more general set of models (gIMR) that do not presuppose general relativity. Evidence of deviations would thus not support any particular theory, but it would provide evidence against the hypothesis that GR is uniquely empirically adequate. Abbott et al. (2019c) extends this test to consideration of individual events from the LIGO-Virgo catalog GWTC-1 as well as ensembles of particularly strong events.[14]

Fourth, the 'modified dispersion relation' test specifically considers the possibility of a modified dispersion relation, including that due to a massive graviton. Since such a modification would alter the propagation of gravitational waves, this allows us to place constraints on the mass of the graviton by using a similar method to that of the third test: the post-Newtonian terms for both the EOBNR and IMRPhenom waveform models are altered (according to the modified dispersion relation) and the Compton wavelength λ_g is then treated as a variable. As with other tests, the posteriors generated in this analysis are consistent with general relativity—in this case, meaning consistency with a massless graviton.

Abbott et al. (2019c) classifies the first two of these tests as *consistency* tests, while the latter two are described as *parameterized* tests of gravitational wave generation and propagation respectively. This reflects the fact that the latter two use a modified set of models—models that have been altered to allow for the possibility of waveforms that deviate from the predictions of general relativity. Despite this distinction, all of these tests can be understood as consistency tests in the sense that they demonstrate that GR 'saves the phenomena' with respect to the dynamical behaviour of compact binary systems.

However, we have seen that the theory- or model-ladenness of the LIGO-Virgo observations has the potential to guarantee the consistency of their empirical results with the predictions of general relativity. The fact that general relativistic models 'save the phenomena' in this case may be an artifact of the role such models play in the observation process (e.g., parameter estimation).

With this problem of theory- or model-ladenness (and the related problem of circularity in validating both results and models) in mind, I offer an alternative interpretation of these 'tests of general relativity'. Rather than testing high-level theory itself, these four tests probe specific ways that the model-ladenness of the observations could be biasing results, and thus masking inconsistencies with general relativity. The residuals test looks for evidence of imperfect signal extraction; the IMR consistency test looks for evidence of non-GR behaviour in the final stages of the merger that is obscured by the consistency of the early signal with general relativity; and the two parameterised tests look for evidence that a non-GR model might also 'save the phenomena', perhaps better than the general relativistic model. In each case, the tests place constraints on ways that the model-laden methodology might obscure deviations from general relativity.

The IMR consistency test is a particularly interesting case, searching for evidence of deviations from GR *despite assuming GR at every individual step*. Here, the possibility of degeneracies between GR and non-GR signals (noted above as a

---

[14] See Patton (2020) for an insightful discussion concerning this test.

problem) is leveraged to search for deviations. Using this testing method, these deviations would manifest as discrepancies in parameter estimates. But such discrepancies could indicate that the overall gravitational wave signal we observe was inconsistent with general relativity, even though the two sections of the signal (taken separately) were consistent with general relativity.

Overall, these tests take a potentially vicious circularity and make a virtue of it; improving either models or the sensitivity of the detector will reap rewards in terms of both improved model validation and tighter empirical constraints on deviations from GR.

However, these tests do not place constraints on all possible deviations from general relativity. One important reason for this is that the possibility of degeneracies remains. The IMR consistency test and the two parameterised tests do place some constraints on the kind of deviations that might be present, but it remains an area of ongoing concern. This is especially true when we consider how any undiagnosed deviations might further bias the inferences made on the basis of biased parameter estimates. Yunes and Pretorius (2009) discuss this issue in their broader discussion of 'fundamental theoretical bias':

> For a second hypothetical example, consider an extreme mass ratio merger, where a small compact object spirals into a supermassive BH [black hole]. Suppose that a Chern-Simons (CS)-like correction is present, altering the near-horizon geometry of the BH [...] To leading order, the CS correction reduces the effective gravitomagnetic force exerted by the BH on the compact object; in other words, the GW emission would be similar to a compact object spiraling into a GR Kerr BH, but with smaller spin parameter a. Suppose further that near-extremal (a ≈ 1) BHs are common (how rapidly astrophysical BHs can spin is an interesting and open question). Observation of a population of CS-modified Kerr BHs using GR templates would systematically underestimate the BH spin, leading to the erroneous conclusion that near-extremal BHs are uncommon, which could further lead to incorrect inferences about astrophysical BH formation and growth mechanisms. (Yunes and Pretorius 2009, 3)

This provides an example of how certain deviations from general relativistic dynamics might not be detected as differences in gravitational waveforms, since these deviations are indistinguishable from a change in source parameters. Thus there is an underdetermination of the properties of the source by the observed gravitational wave signal. In this passage, Yunes and Pretorius also note that this 'fundamental theoretical bias' has consequences for further inferences based on the estimated properties of the binary. As I discuss in Sect. 4.4, managing biases in gravitational wave observations becomes a prominent feature of reasoning using populations of compact binary mergers.

Overall, the tests of GR performed by the LIGO-Virgo Collaboration provide a variety of constraints on ways that their model-dependent methodology might be biasing their results. Nonetheless, the theory- or model-ladenness remains an ongoing problem, especially when it comes to disentangling variation in source parameters from (possible) deviations from the predictions of general relativity.

4.4   Theory-Testing Beyond Individual Events

Since the announcement of the first gravitational wave detection (Abbott et al. 2016b) the LIGO-Virgo Collaboration has made nearly one hundred further detections.[15] These occurred over three observing runs (O1-O3), with upgrades to the interferometers being undertaken between runs to increase sensitivity. Details of these detections are available in gravitational wave transient catalogs GWTC-1, GWTC-2, and GWTC-3 (Abbott et al. 2019b, 2021b,c)

As gravitational wave events accumulate, gravitational-wave astrophysics is undergoing a transition in the scope of their targets of inquiry: from individual events to ensembles. With this comes the further possibility of probing astrophysical and cosmological processes, including the astrophysical mechanisms responsible for producing the population of compact binaries ('formation channels'), and cosmic expansion (as measured by the Hubble constant).

Parameter estimation for individual events provides the foundations on which further inferences can be built. What I mean by this is that inferences about populations of binary black hole mergers (for example) are based on prior inferences about the properties of individual events. If these parameter estimates are systematically biased then these biases will be passed on to further inferences about the population. For this reason, the tests of GR discussed in Sect. 4.3 are vital for controlling systematic error beyond the original context of those tests.

In what follows, I sketch two examples of research programs that can be built on the initial scaffolding afforded by the observation of compact binary mergers. These include: testing models of binary black hole formation channels (Sect. 4.4.1) and performing precise measurements of the Hubble constant that are independent of the 'cosmic distance ladder' (Sect. 4.4.2).[16]

These examples illustrate how the transition of gravitational wave astrophysics into mature science is characterized by a move towards learning about the causal processes responsible for shaping the observable universe, as well as building bridges with other areas of astrophysics. This building of connections with the rest of (electromagnetic-based) astrophysics means that gravitational wave astrophysics is becoming increasingly unified with astrophysics as a whole as it gains maturity. The picture that emerges is one of hypothesis-testing that leverages this unification;

---

[15] The collaboration has also expanded to become the LIGO-Virgo-KAGRA Collaboration, reflecting the expansion of the network to include KAGRA in Japan. However, KAGRA came online only shortly before the end of O3 and has not yet been involved in any detections.

[16] For the reader unfamiliar with these terms: 'formation channel' refers to the set of causal processes involved in the evolution of (in this case) a binary black hole system. Different formation channels are different pathways for producing the same kind of system; the 'Hubble constant', H_0, is a measure of the rate of expansion of the universe; and the 'cosmic distance ladder' refers to a set of methods for measuring distances to celestial objects. Since different methods have different domains of applicability, earlier 'rungs' on the ladder, measuring distances to closer objects, help form the basis for using methods corresponding to later 'rungs', measuring distances to more distant objects.

by placing an increasing number of constraints from multiple directions, an increasingly narrow region of the possibility space is isolated where the overall picture of astrophysical processes holds together.

4.4.1 Binary Black Hole Formation Channels

Once binary black hole mergers have been detected, one astrophysically important question is how such binaries formed in the first place. Some familiar methodological challenges of astrophysics make inferences about this evolution difficult. First, the evolution of a binary black hole system involves long timescale, meaning that we cannot watch such a process unfold. Instead, as is often the case in astrophysics, dynamical processes must be inferred from 'snapshots' (Jacquart 2020, 1210–1211, see also Anderl 2016). Second, there are no terrestrial experiments that we can perform that replicate the conditions relevant for the formation of these objects. Thus we are limited to what data are made available by nature. In the case of binary black holes, this is very sparse: while it is possible to make observations pertaining to the conditions of star formation at high redshifts, binary black hole systems are currently only observable in the very final stages of their inspiral and merger. As in the case considered by Jacquart (2020)—the formation of ring galaxies—computer simulations are vital for inferring long-timescale dynamics from snapshots. However, unlike Jacquart's example, the available data here do not include snapshots from different stages in the process. Instead, the only 'traces' of these processes are the observed properties of the binary black hole merger.

Despite these challenges, some early progress has been made in seeking evidence about binary black hole formation channels. This work depends on developing models of the processes by which binary black holes form and evolve and determining the signatures of these different processes in terms properties of the observed populations of binary black hole mergers. For example, Belczynski et al. (2020) reviews several of the dynamical processes proposed for explaining the formation of the binaries observed by LIGO-Virgo, including the near-zero effective spins observed among binaries from O1 and O2.[17]

Recent work by van Son et al. (2022) nicely illustrates the kind of reasoning involved in making inferences about formation channels for the binaries observed by LIGO-Virgo. This paper investigates the formation channels for binary black holes using simulations of different channels to determine signatures in (current and future) observations of binary black hole mergers. They find that the redshift evolution of the properties of binary black hole mergers encodes information about the origins of the binary components. However, decoding this information is complicated by what is called the delay time, t_{delay}: the time between the

---

[17] See also Vitale et al. (2017) and Farr et al. (2018) for similar discussions relating binary black hole formation channels and measured spins.

formation of the progenitor stars and the final binary black hole merger.[18] Factoring in differing delay times, binary black hole mergers at the same redshift do not necessarily have shared origins.

Through simulations, van Son et al. (2022) identify two formation channels that contribute to the overall population of binary black hole mergers. These are the common envelope (CE) and stable Roche-lobe overflow (RLOF) channels. These simulations incorporate modeling of astrophysical processes relating to stellar evolution and binary interactions, such as stellar wind mass loss, mass transfer between the binary components, the role of supernova kicks, etc.

They find that the CE channel preferentially produces low mass black holes ($<$ $30 M_\odot$) and short delay time ($< 1 Gy$), while the stable RLOF channel preferentially produces black holes with large masses ($> 30 M_\odot$) and long delay times ($> 1 Gy$). These differences mean that the channels exhibit different redshift evolution; the binary black hole merger rate $R_{BBH}(z)$ is expected to be dominated by the CE channel at high redshifts, while there is significant contribution ($\sim$40%) from the stable RLOF channel at low redshifts. Thus van Son et al. (2022) predict a distinct redshift evolution of $R_{BBH}(z)$ for low and high component masses.

Finding observational evidence of these signatures is challenging at present. As of O3, the gravitational-wave interferometer network was only observing out to redshifts of $z \sim 0.8$ for the highest mass primary black holes. For smaller primary black holes ($M_{BH,1} \sim 10 M_\odot$), this network is only probing out to redshifts of $z \sim 0.1 - 0.2$. However, this is expected to change in the coming decades, as upgrades to existing detectors (Advanced LIGO, Advanced Virgo, and KAGRA) and the addition of new detectors (e.g., the Einstein telescope and Cosmic Explorer) allow for the observation of binary black hole mergers out to higher redshifts (Abbott et al. 2017a; Maggiore et al. 2020).

With future observations, it may be possible to observe how a range of properties of the observed binaries change with redshift. Of course, doing so accurately requires having a good handle on the biases in the observations, from selection effects (e.g., the limited sensitivity of the detector means that lower-mass binaries are less likely to be observed at greater distances) to parameter degeneracies (e.g., distance/inclination of the binary), to the potential bias introduced by using general relativistic models. Nonetheless, the use of modeling to account for these biases in the data may result in data that are reliable enough to draw inferences about the evolution of the observed binary population.

Correcting data via models may naively seem problematic—after all theory- or model-ladenness is sometimes (as in this paper) a cause for concern. However, I am inclined to think that such treatment of data is both common and (usually) virtuous, resulting in 'model-filtered' data that is a better basis for inferences as a result of mitigating biases in the data. Here, I borrow the term 'model-filtered' from Alisa

---

[18] Note that the delay time is the sum of two timescales: first, the lifetime of the binary stars up until they both become compact objects; and second, the inspiral time of the black holes, up until the binary black hole merger (van Son et al. 2022, 2).

Bokulich, who defends a view of 'model-data symbiosis' and discusses a range of ways that data can be model-laden—in a way that is beneficial (Bokulich 2020, 2018).[19] Bokulich's examples include the correction of the fossil record to account for known biases in this record. The resulting model-corrected data forms a better basis for inferences about biodiversity in the deep past. I take this view to also be a good fit for astrophysics. The gravitational wave catalog resembles the fossil record; while there are important biases in the observed data, modeling can help correct for these in order to build a timeline of events and learn more about the processes driving change over long timescales. On the flip side, the reliability of such data depends on the reliability of the models used to correct it.

4.4.2   Measuring the Hubble Constant

As the population of detected gravitational wave transients grows, it also becomes possible to use gravitational waves to probe cosmology. In particular, compact binary mergers (especially binary neutron star mergers) can be used to measure the Hubble constant (Abbott et al. 2017b). The Hubble constant H_0 is a measure of the rate of expansion of the universe at the current epoch. For distances less than about $50 Mpc$, H_0 is well-approximated by the following expression:

$$v_H = H_0 d \qquad (4.1)$$

where v_H is the local 'Hubble flow' velocity of the source (the velocity due to cosmic expansion rather than the peculiar velocities between galaxies) and d is the (proper) distance to the source.

To measure the Hubble constant one needs measurements of both v_H and d. Gravitational wave signals encode information about the distance to their source, through the amplitude of the waves. Thus the parameter estimation process described above provides a measurement of d that is independent of the 'cosmic distance ladder' (and the electromagnetic observations used to calibrate this). This leads to compact binary mergers being called 'standard sirens' (analogous to the 'standard candles' provided by Type Ia supernovae). In contrast, measurements of v_H rely on electromagnetic radiation. In particular, v_H is inferred using the measured redshift of the host galaxy. Measurements of v_H from the redshift must account for the peculiar velocity of the host galaxy through analysis of the velocities of the surrounding galaxies.

Distance measurements are complicated by the fact that there is a degeneracy between the distance to, and inclination of the source in terms of the amplitude of the measured gravitational waves. Face on binaries are "louder", radiating gravitational

---

[19] See also Boyd (2018), Bokulich and Parker (2021), and Leonelli (2016) for related views of data.

waves at higher amplitude (hence SNR) than a binary that is viewed side-on. Altering the inclination of the binary mimics the effect of moving the source closer or farther away. Thus (as with standard candles) the measurement of distance provided by standard sirens is a little more complicated than it initially appears.

So far, one multi-messenger event has been reported by the LIGO-Virgo Collaboration.[20] For this event, an optical transient (AT 2017gfo) was found to coincide with the source of gravitational waves, leading to an identification of the source of both signals as a binary neutron star merger. Using gravitational waves to measure d and the electromagnetic counterpart to measure v_H, Abbott et al. (2017b) obtain a measurement for H_0: $70.0^{+12.0}_{-8.0}\,\mathrm{kms}^{-1}\,\mathrm{Mpc}^{-1}$.[21]

It is also possible to perform standard siren measurements with binary black hole mergers, fittingly called 'dark sirens', even in the absence of an electromagnetic counterpart. In such cases, the velocity v_H is inferred statistically from the redshifts of possible host galaxies. The first such measurement was performed using GW170814, giving an estimated value of $H_0 = 75^{+40}_{-32}\,\mathrm{kms}^{-1}\,\mathrm{Mpc}^{-1}$ (Soares-Santos et al. 2019).[22]

These measurements of the Hubble constant based on individual events both have large uncertainties, in part due to the distance-inclination degeneracy and (in the latter case) the lack of an electromagnetic counterpart.

However, as gravitational-wave astrophysics transitions into a mature field, with a substantial population of events, the precision of measurements will improve. This is in part due to the projected increase in information about the polarization of the gravitational waves with an expanded network of detectors. This should lead to better measurements of the source inclination, and therefore distance (i.e., it should help break the degeneracy between these).

Additionally, the precision of the measurement is expected to increase with a larger population of detections, as uncertainties from peculiar velocities and distance should decrease with the sample size. Revised estimates of H_0 have already been made using populations of events, following O2 (Abbott et al. 2019a), and O3 (Abbott et al. 2021a).

The precision of H_0 measurements does not increase uniformly with each new event. Events with an identifiable counterpart and host galaxy contribute the most (since these have the best estimated peculiar velocities and hence redshift). Strong contributions also come from compact binary mergers in the 'sweet spot' where uncertainties associated with the distance and the peculiar velocities are comparable

---

[20] 'Multi-messenger' means that the event was observed using at least two cosmic 'messengers' (electromagnetic radiation, neutrinos, gravitational waves, and cosmic rays). In this case, GW170817 was observed with gravitational waves as well as electromagnetic radiation across a broad range of frequencies.

[21] Note that all values quoted are based on Bayesian analysis and report the maximum posterior value with the minimal-width 68.3% credible interval.

[22] GW170814 was a strong signal that gave a well localized source region within a part of sky that is thoroughly covered by the Dark Energy Survey. This made it highly amenable to a statistical treatment of the redshifts of possible host galaxies.

(Chen et al. 2018).[23] Chen et al. (2018) provide calculations showing how standard siren measurements of H_0 should increase in precision according to the number and properties of compact binary mergers detected over time. For example, for a population of 50 binary neutron star mergers with associated electromagnetic counterparts, the fractional uncertainty in the H_0 measurement would reach 2%. With this precision, gravitational-wave-based measurements of H_0 would be precise enough to perform stringent coherence tests with existing measurements—a potentially important step toward resolving the 'Hubble tension'.[24] However, the prospects are less good for a population of binary black hole mergers.

In the absence of a large population of events with counterparts, much recent work has focused on maximizing the cosmological information that can be inferred from the increasingly large population of binary black hole mergers. Such populations can be effective probes of cosmic expansion when analyzed together with known astrophysical properties of the overall compact binary population. For example, Abbott et al. (2021a), provides improved estimates of the Hubble constant using 47 events from GWTC-3. They use two methods to do this. The first makes assumptions about the redshift evolution of the binary black hole population—in particular, that the mass scale *does not* vary with redshift—then fits the population for cosmological parameters (Abbott et al. 2021a, 11–12). In essence, this works by assuming that redshift evolution of the population properties is due to the cosmic expansion as opposed to intrinsic properties of the binary population. This approach is especially useful in the case that the population has sharp cut-off features (such as the 'mass gap' produced by pair-instability supernovae) (Ezquiaga and Holz 2022). The second method that they use involves associating gravitational wave sources with a probable host galaxies using existing galaxy surveys. This approach also makes assumptions about the binary black hole merger source population—including a fixed source mass distribution and fixed-rate evolution of the binaries—as well as assumptions about the selection effects that lead to incompleteness in the galaxy surveys (Abbott et al. 2021a, 13).

Overall, these methods yield estimates of $H_0 = 68^{+12}_{-8} \, \mathrm{kms}^{-1} \, \mathrm{Mpc}^{-1}$ (first method) and $H_0 = 68^{+8}_{-6} \, \mathrm{kms}^{-1} \, \mathrm{Mpc}^{-1}$ (second method). The latter is a significant improvement in precision over the measurement from O2, which was $H_0 = 69^{+16}_{-8} \, \mathrm{kms}^{-1} \, \mathrm{Mpc}^{-1}$ (Abbott et al. 2019a). However, Abbott et al. (2021a) note that their estimate strongly depends on assumptions about the binary black hole source mass distribution: 'if the source mass distribution is mismodeled, then the cosmological inference will be biased' (Abbott et al. 2021a, 27).

---

[23] This sweet spot changes with the detector sensitivity since the fractional distance uncertainty scales with SNR. In addition to having a counterpart, GW170817 fell close to the sweet spot for the detector at the time (Chen et al. 2018, 'Methods').

[24] The 'Hubble tension' refers to the disagreement between existing current-best estimates of H_0 based on local and high-redshift measurements. See Matarese and McCoy (In preparation) for a detailed philosophical discussion of the Hubble tension.

We can now see that the two different investigations discussed in Sects. 4.4.1 and 4.4.2 are intertwined. There is a kind of degeneracy between population properties and cosmology; if we know about the population properties, we can learn about cosmology, and if we know something about cosmology, we can learn about the properties of the binary black hole population. This has the potential to introduce a new circularity when it comes to justifying inferences about either. However, as with the case of testing general relativity (Sect. 4.3), the circularity can also be virtuous; increasing constraints on one helps to constrain the other.

4.5   Conclusion

Gravitational-wave astrophysics has begun to transition from new to mature science. This transition is partly characterized by a shift from individual events (such as GW150914 or GW170817) to populations of events as the target of investigation. In studying populations of events, gravitational waves may be used to to probe new phenomena, including astrophysical processes (e.g., binary black hole formation channels) and cosmology (e.g., Hubble expansion).

As for studies of individual events, studying the properties of populations is based on theory-laden methodology. Parameter estimation using general relativistic models forms the foundation on which further inferences about populations can be built. The reliability of these inferences thus depends on the success of the LIGO-Virgo Collaboration in constraining the bias introduced by their use of general relativistic models. The tests I described in Sect. 4.3 provide crucial scaffolding for proceeding with inferences like those described in Sect. 4.4.1.

However, working with populations can also help with some sources of error. For example, the distance/inclination degeneracy limits the precision of measurements of H_0 from a single event. But this uncertainty is largely washed out with a large sample size, since binaries are not expected to have any preferred orientation with respect to us (though this must be corrected for selection effects, since the 'louder' face-on orientations are more likely to be detected).

Another feature of gravitational wave astrophysics' transition to maturity is its increasing unification with other areas of astrophysics. This began with the multi-messenger observations of a binary neutron star merger, where electromagnetic observations combined with gravitational wave observations to give a more complete picture of the processes involved. However, just as importantly, bridges are built between gravitational wave and electromagnetic astrophysics when methods from these different fields are brought to bear on the same processes. This is true in the case of independent measurements of the Hubble constant, which offer the possibility of coherence tests between different measurements—and perhaps the eventual resolution of the Hubble tension. It is also true in the studies of binary black hole formation channels, where constraints from other areas of astrophysics (e.g., concerning star formation) can be brought to bear on plausible channels for the evolution of binary black holes. Thus connections are being forged via observational targets, coherence tests across independent measurements, and by importing constraints from one domain to inform investigations in another.

Alongside this increasing connectedness to other areas of astrophysics, gravitational-wave astrophysics is also increasingly resembling electromagnetic astrophysics in a few ways.

First, it is now facing the 'snapshot' problem of trying to infer causal processes that occur over long timescales based on temporal slices. This is a characteristic challenge not only for astrophysics at large, but also for historical science in general.

Second, the dependency relations between the target phenomenon and observational traces are increasingly complex, due to parameter degeneracies, degeneracies between source properties and cosmological effects, and other uncertainties concerning the astrophysical processes by which binary stars are thought to form and collapse, accrete matter and dissipate angular momentum, etc. There is thus a kind of holism that impedes hypothesis testing due to the difficulty in isolating a single hypothesis to test independently.

These complex relationships are gradually disentangled with the help of simulations. This often proceeds as a search for clear signatures to act as 'smoking guns'. However, as with the case of standard sirens, progress in studying relationships between observations and target phenomena seems to be characterized more by increasing appreciation of the complexities than by finding true unambiguous smoking guns. Given all of this, controlling for selection effects and other systematics becomes a prominent component of reasoning about the populations being studied using gravitational waves.

Overall, testing hypotheses about astrophysical and cosmological processes with gravitational waves proceeds iteratively through improving the precision of measurements and further exploring dependency relationships, especially through simulations. This has some resemblance to what (Chang 2004, 45) calls 'epistemic iteration', a process where 'successive stages of knowledge, each building on the preceding one, are created in order to enhance the achievement of certain epistemic goals'. This process, which Chang says could also be called a kind of 'bootstrapping', is one of progress through self-improvement, in the absence of secure epistemic foundations. This seems like a fruitful picture for making sense of progress in gravitational-wave astrophysics, where we have seen that the foundational theory (GR) is in question at the same time as the observations made utilizing that theory.[25] Furthermore, intertwined lines of inquiry about cosmology

---

[25] However, the case of gravitational-wave astrophysics is rather different from the examples that Chang considers. First, the theoretical and experimental developments in this field proceeded in parallel for decades; both strands of development had to reach an advanced stage before such measurement was possible. Thus the main focus of my analysis has not been on iterative improvement in the measurement of gravitational waves by the LIGO and Virgo interferometers, but rather on downstream inferences on the basis of these. Second, the phenomena of temperature Chang describes were accessible by a variety of means, including sensory experience and a range of measuring instruments. In contrast, gravitational waves are much harder to measure (to put it mildly) and accessible only by a small number of similar detectors. This provides a very

and black hole populations and formation channels proceed simultaneously, feeding back into one another without any of them providing an independently secure foundation for the others.

This process of iterative improvement gradually places tighter constraints on viable models and parameter values, while teasing apart complex dependencies between populations of binary systems and observations of their final moments. Even when these constraints are conditional, depending on various other assumptions in order to place a constraint on the phenomenon of interest, the accumulation of constraints across the entire web of hypotheses can gradually reduce the space of viable possibilities. Here again there are resonances with Chang (2004)'s coherentist epistemology, in which epistemic iteration allows for progress by building on previous knowledge despite its lack of secure foundations. Instead, previous knowledge enjoys a kind of tentative acceptance so that it can be built upon in the pursuit of epistemic values.

Overall, this paper has exhibited both the vices and virtues of interdependence in theory testing. We have seen how theory-laden methodology can lead to a vicious circularity, but also how this need not be an insurmountable hurdle for theory testing. Indeed, Sect. 4.3 showed how the LIGO-Virgo Collaboration perform tests of GR by placing constraints on different ways that their methods could be masking deviations from GR. The confidence in both theory and observations based on these tests provides the confidence needed to base further inferences on the LIGO-Virgo observations. The examples of Sects. 4.4.1 and 4.4.2 show how interdependence can lead to a kind of holist underdetermination in the sense that changes in cosmology are degenerate with changes in the properties of black hole binary populations. However, this also means that improved constraints in one domain can do double work, simultaneously introducing further constraints to related domains.

With the increasing interconnectedness between electromagnetic and gravitational wave astrophysics come more constraints on how it all must fit together. Between the increasing precision of gravitational wave detection, the improved modeling of compact binaries, and the increasing number of bridges across fields, the hope is that degeneracies, along with problems of theory-ladenness and circularities, which may look irresolvable viewed in isolation, might nonetheless be broken when viewed in their full astrophysical context.

---

limited basis for self-improvement on the basis of coherence with other measurements. Overall, on an abstract level, Chang's idea of epistemic iteration seems like a good fit for what I have described. However, given the two major differences between epistemic contexts of building a temperature scale and observing gravitational waves, I think that more work is needed to draw out the relationship between iterative progress in these two cases.

Acknowledgments I would like to thank two anonymous referees for their helpful comments on an earlier version of this paper.

My work on this project was undertaken during my employment at Harvard's Black Hole Initiative, which is funded in part by grants from the Gordon and Betty Moore Foundation and the John Templeton Foundation. The opinions expressed in this publication are those of the author and do not necessarily reflect the views of these Foundations.

References

Abbott, B.P., et al. 2016a. GW150914: First results from the search for binary black hole coalescence with advanced LIGO. *Physical Review D* 93: 122003. https://doi.org/10.1103/PhysRevD.93.122003

Abbott, B.P., et al. 2016b. Observation of gravitational waves from a binary black hole merger. *Physical Review Letters* 116: 061102. https://doi.org/10.1103/PhysRevLett.116.061102

Abbott, B.P., et al. 2016c. Observing gravitational-wave transient GW150914 with minimal assumptions. *Physical Review D* 93: 122004. https://doi.org/10.1103/PhysRevD.93.122004

Abbott, B.P., et al. 2016d. Properties of the binary black hole merger GW150914. *Physical Review Letters* 116: 241102. https://doi.org/10.1103/PhysRevLett.116.241102

Abbott, B.P., et al. 2016e. Tests of general relativity with GW150914. *Physical Review Letters,* 116, 221101. https://doi.org/10.1103/PhysRevLett.116.221101

Abbott, B.P., et al. 2017a. Exploring the sensitivity of next generation gravitational wave detectors. *Classical and Quantum Gravity* 34(4): 044001. https://doi.org/10.1088/1361-6382/aa51f4

Abbott, B.P., et al. 2017b. A gravitational-wave standard siren measurement of the Hubble constant. *Nature* 551: 85–88. https://doi.org/10.1038/nature24471

Abbott, B.P., et al. 2019a. A gravitational-wave measurement of the Hubble constant following the second observing run of advanced LIGO and Virgo. arXiv:1908.06060v2

Abbott, B.P., et al. 2019b. GWTC-1: A gravitational-wave transient catalog of compact binary mergers observed by LIGO and Virgo during the first and econd observing runs. *Physical Review X* 9: 031040. https://doi.org/10.1103/PhysRevX.9.031040

Abbott, B.P., et al. 2019c. Tests of general relativity with the binary black hole signals from the LIGO-Virgo catalog GWTC-1. *Physical Review D* 100: 104036. https://doi.org/10.1103/PhysRevD.100.104036

Abbott, R., et al. 2021a. Constraints on the cosmic expansion history fromGWTC-3. https://doi.org/10.48550/ARXIV.2111.03604

Abbott, R., et al. 2021b. GWTC-2: Compact binary coalescences observed by LIGO and Virgo during the first half of the third observing run. *Physical Review X* 11: 021053. https://doi.org/10.1103/PhysRevX.11.021053

Abbott, R., et al. 2021c. GWTC-3: Compact binary coalescences observed by LIGO and Virgo during the second part of the third observing run. https://doi.org/10.48550/ARXIV.2111.03606

Abbott, R., et al. 2021d. Tests of general relativity with GWTC-3. https://doi.org/10.48550/ARXIV.2112.06861

Anderl, S. 2016. Astronomy and astrophysics. In *Oxford handbook of philosophy of science,* ed. P. Humphreys. Oxford: Oxford University Press. https://doi.org/10.1093/oxfordhb/9780199368815.013.45

Belczynski, K., et al. 2020. Evolutionary roads leading to low effective spins, high black hole masses, and o1/o2 rates for LIGO/Virgo binary black holes. *Astronomy and Astrophysics* 636. https://doi.org/10.1051/0004-6361/201936528

Bokulich, A. 2018. Using models to correct data: Paleodiversity and the fossil record. *Synthese* 198(24): 5919–5940. https://doi.org/10.1007/s11229-018-1820-x

Bokulich, A. 2020. Towards a taxonomy of the model-ladenness of data. *Philosophy of Science* 87(5): 793–806. https://doi.org/10.1086/710516

Bokulich, A., W. Parker. 2021. Data models, representation and adequacy-for-purpose. *European Journal for Philosophy of Science* 11(1): 31–31. https://doi.org/10.1007/s13194-020-00345-2

Boyd, N.M. 2018. Evidence enriched. *Philosophy of Science* 85(3): 403–421. https://doi.org/10.1086/697747

Boyd, N.M., J. Bogen. 2021. Theory and observation in science. In *The Stanford encyclopedia of philosophy*, ed. E.N. Zalta, (Winter 2021). Stanford: Metaphysics Research Lab, Stanford University.

Chang, H. 2004. *Inventing temperature: Measurement and scientific progress*. Oxford: Oxford University Press.

Chen, H.-Y., M. Fishbach, D.E. Holz. 2018. A two per cent Hubble constant measurement from standard sirens within five years. *Nature* 562(7728), 545. https://doi.org/10.1038/s41586-018-0606-0

Elder, J. Forthcoming. Black hole coalescence: Observation and model validation. In *Working toward solutions in fluid dynamics and astrophysics: What the equations don't say*. Springer Briefs. ed. L. Patton, E. Curiel. Berlin: Springer

Elder, J. 2023a. Black hole coalescence: Observation and model validation. In *Working toward solutions in fluid dynamics and astrophysics: What the equations don't say*. Springer Briefs. ed. L. Patton, E. Curiel. Berlin: Springer.

Elder, J. In preparation. Independent evidence in multi-messenger astrophysics.

Elder, J. In preparation. On the 'direct detection' of gravitational waves.

Elder, J., Doboszewski, J. In preparation. How theory-laden are observations of black holes?

Ezquiaga, J.M., D.E. Holz. 2022. Spectral sirens: Cosmology from the full mass distribution of compact binaries. *Physical Review Letters* 129: 061102. https://doi.org/10.48550/ARXIV.2202.08240

Farr, B., D.E. Holz, W.M. Farr. 2018. Using spin to understand the formation of LIGO and Virgo's black holes. *The Astrophysical Journal* 854(1): L9. https://doi.org/10.3847/2041-8213/aaaa64

Hulse, R.A., J.H. Taylor. 1975. Discovery of a pulsar in a binary system. *Astrophysical Journal* 195: L51–L53. https://doi.org/10.1086/181708

Jacquart, M. 2020. Observations, simulations, and reasoning in astrophysics. *Philosophy of Science* 87(5): 1209–1220. https://doi.org/10.1086/710544

Leonelli, S. 2016. *Data-centric biology: A philosophical study*. Chicago: University of Chicago Press.

Maggiore, M. 2008. *Gravitational waves. Volume 1, theory and experiments*. Oxford: Oxford University Press.

Maggiore, M., et al. 2020. Science case for the Einstein telescope. *Journal of Cosmology and Astroparticle Physics* 2020(03): 050–050. https://doi.org/10.1088/1475-7516/2020/03/050

Matarese, V., C. McCoy. In preparation. When "replicability" is more than just "reliability": The Hubble constant controversy.

Parker, W.S. 2017. Computer simulation, measurement, and data assimilation. *British Journal for the Philosophy of Science* 68(1): 273–304. https://doi.org/https://doi.org/10.1093/bjps/axv037

Patton, L. 2020. Expanding theory testing in general relativity: LIGO and parametrized theories. *Studies in History and Philosophy of Science Part B: Studies in History and Philosophy of Modern Physics* 69: 142–153. https://doi.org/https://doi.org/10.1016/j.shpsb.2020.01.001

Shapere, D. 1982. The concept of observation in science and philosophy. *Philosophy of Science* 49(4): 485–525. https://doi.org/https://doi.org/10.1086/289075

Soares-Santos, M., et al. 2019. First measurement of the Hubble constant from a dark standard siren using the dark energy survey galaxies and the LIGO/Virgo binary–black-hole merger GW170814. *The Astrophysical Journal* 876(1): L7. https://doi.org/10.3847/2041-8213/ab14f1

Tal, E. 2012. The epistemology of measurement: A model-based account, PhD Dissertation, University of Toronto http://search.proquest.com/docview/1346194511/

Tal, E. 2013. Old and new problems in philosophy of measurement. *Philosophy Compass* 8(12): 1159–1173. https://doi.org/https://doi.org/10.1111/phc3.12089

Taylor, J.H., J.M. Weisberg. 1982. A new test of general relativity - gravitational radiation and the binary pulsar PSR 1913+16. *Astrophysical Journal* 253: 908–920. https://doi.org/10.1086/159690

van Son, L.A.C., et al. 2022. The redshift evolution of the binary black hole merger rate: A weighty matter. *The Astrophysical Journal* 931(1): 17. https://doi.org/10.3847/1538-4357/ac64a3

Vitale, S., R. Lynch, R. Sturani, P. Graff. 2017. Use of gravitational waves to probe the formation channels of compact binaries. *Classical and Quantum Gravity* 34(3): 03LT01. https://doi.org/10.1088/1361-6382/aa552e

Yunes, N., F. Pretorius. 2009. Fundamental theoretical bias in gravitational wave astrophysics and the parametrized post-Einsteinian framework. *Physical Review D* 80: 122003. https://doi.org/10.1103/PhysRevD.80.122003

Chapter 5
Hybrid Enrichment of Theory and Observation in Next-Generation Stellar Population Synthesis

Lydia Patton

Abstract Next-generation observational surveys in astronomy provide empirical data with increasingly high resolution and precision. After presenting the basic methods of population synthesis (via Conroy C, Ann Rev Astronom Astrophys 51:393–455, 2013; Maraston C, Mon Not Royal Astronom Soc 362:799–825, 2005), this paper argues for several related conclusions. The increased precision of the new methods requires the development of improved theoretical resources and models to provide the richest interpretation of the new data (as argued by Maraston C, Strömbäck G, Monthly Not Royal Astronom Soc 418:2785–2811, 2011). The measurement of physical variables and parameters in population synthesis is best understood using a model-based account along the lines of (Tal E, The epistemology of measurement: a model-based approach. Dissertation, The University of Toronto, 2012) and (Parker WS, Br J Philos Sci 68:273–304, 2017). Finally, in the case of population synthesis, improved empirical data does not dispense with the need for theoretical reasoning in post-data analysis. In fact, the high-resolution data used in next-generation population synthesis demands ever richer theories and models, a process that results in hybrid enrichment of theoretical and observational methods and results.

Keywords Population synthesis · Model-based measurement · Models · Simulations

5.1 Introduction

New precision observational surveys in astronomy provide high-resolution empirical data for population synthesis. After presenting the basic methods of population synthesis (following closely Conroy 2013 and Maraston 2005), this paper argues

L. Patton (✉)
Ona, WV, USA
e-mail: critique@vt.edu

© The Author(s) 2023
N. Mills Boyd et al. (eds.), *Philosophy of Astrophysics*, Synthese Library 472,
https://doi.org/10.1007/978-3-031-26618-8_5

for several related conclusions. The increased precision of the new methods requires the development of improved theoretical resources and models to provide the richest interpretation of the new data (as argued by Maraston and Strömbäck 2011). Further, the measurement of physical variables and parameters in population synthesis is best understood using a model-based account along the lines of Tal (2012) and Parker (2017). The paper concludes that, in the case of population synthesis, improved empirical data does not dispense with the need for theoretical reasoning in post-data analysis. The high-resolution data used in next-generation population synthesis demands ever richer theories and models, a process that results in hybrid enrichment of theoretical and observational methods and results.

5.2   Stellar Population Synthesis in Astrophysics

Stellar population synthesis involves generating simulations of the evolution and properties of an ensemble of stars. The method yields an overall picture of the stars as they age and interact with the interstellar medium. Astronomers refer to the evolutionary process of stars as the Main Sequence.[1] Stellar population synthesis maps not only the stars' properties at the present time, but also their position on the Main Sequence, their histories, and their predicted evolution, yielding a moving picture of how stars age and interact with the interstellar medium.[2] Astronomers determine a star's metallicity through diverse methods of spectroscopic classification (Heiter et al. 2014, Sect. 1). Knowledge of the age-metallicity relationship allows for inferences, not only about the properties of stars, but about their evolution and position on the Main Sequence.[3]

A spectral energy distribution (SED) is "light emitted over all or a portion of the [far ultraviolet to far infrared] spectral domain, including broadband data and/or moderate-resolution spectra" (Conroy 2013, 393). Measurements of the spectral

---

[1] For an explanation of the Main Sequence see Küpper et al. 2008.

[2] "As time elapses, this stellar population ages. According to their mass, the stars eventually leave the main-sequence . . . and soon after die (supernovae) or enter a quiescence stage (white dwarfs), injecting metals into the interstellar medium in form of supernova explosions or quiet winds. The interstellar medium becomes richer in metals . . . As a result of this activity, an age-metallicity relationship [. . .] is built up" (Pasetto et al. 2012, A14:3).

[3] Pasetto et al. (2012, Sect. 1) provide an overview of approaches to stellar population synthesis, identifying key elements: "Every real (or realistically simulated) stellar system is a set of stars born at different times and positions, and with different velocities, masses and chemical compositions". Ultimately, "to completely define the [. . .] parameters at a generic time t, we need to specify the distribution in space of the masses, M, and metallicities, Z" of the members of the population. Finally, "the stars evolve with time, i.e. they continuously move in space, lose mass, enrich in metals, and move in the phase-space" (2012, A14,2).

energy distribution are a key source of empirical data. Charlie Conroy sums up the foundation of population synthesis:

> The spectral energy distributions (SEDs) of galaxies are shaped by nearly every physical property of the system, including the star-formation history, metal content, abundance pattern, dust mass, grain size distribution, stardust geometry, and interstellar radiation field. The principal goal of stellar population synthesis (SPS) is to extract these variables from observed SEDs. (2013, 393).

To get from empirical data (spectral energy distributions) to physical variables, one must make backwards inferences. Spectral energy distributions are 'shaped' by physical properties of stellar systems, including their interactions with the interstellar medium. Stellar population synthesis uses computer models and simulations to retrace the processes involved.

For instance, a theoretical Hertzsprung-Russell diagram plots a star's effective temperature (T_{eff}) and its luminosity on perpendicular axes, including the evolution of that relationship over a star's life course. The H-R diagram reflects the classification of stars, including variables that are neither temperature nor luminosity but that have an effect on those variables: "Variations in composition can [. . .] affect the stellar evolution timescales as well as the appearance of the evolution on the [Hertzsprung-Russell diagram]" (Hurley et al. 2000, Sect. 2). A star's metal content, abundance patterns, and other physical properties affect that star's evolution. With an accurate model of stellar evolution and a spectral energy distribution, one can determine the physical properties of the star by reverse inference.

The theory of stellar evolution is the foundation stone of modern population synthesis. From the late 1960s to the 80 s, "synthesis models were being developed that relied on stellar evolution theory to constrain the range of possible stellar types at a given age and metallicity . . . The substantial progress made in stellar evolution theory in the 1980s and 1990s paved the way for [this] approach to become the de facto standard in modeling the SEDs of galaxies" (Conroy 2013, 394). Stellar or evolutionary population synthesis is the name given to methods for modelling "spectrophotometric properties of stellar populations", using "knowledge of stellar evolution" (Maraston 2005, 799). As Claudia Maraston notes, "This approach was pioneered by [Beatrice] Tinsley in a series of fundamental papers[4] that provide the basic concepts still used in present-day computations. The models are used to determine ages, element abundances, stellar masses, stellar mass functions, etc., of those stellar populations that are not resolvable in single stars, like galaxies and extragalactic globular clusters (GCs)" (ibid., 799).

Two features of stellar population synthesis are key. First, modern methods generate simulations of simple or complex stellar populations, not individual stars.[5] Second, the theory of stellar evolution is used to determine which types of stars

---

[4] See Tinsley and Larson 1977 for a survey of this fundamental work.

[5] Simple and complex populations will be defined below.

could be represented in a given spectral energy distribution. While the spectral energy distributions are the observable empirical data in play, that data does not add up to much without the theory of stellar evolution to determine the types of stars that make up the target population.

Once that has been determined, one can move on to find values for the physical variables of interest. The simplest method is generation of a Simple Stellar Population (SSP), which "describes the evolution in time of the SED of a single, coeval stellar population at a single metallicity and abundance pattern[6]" (Conroy 2013, 395). An SSP, Conroy notes, "requires three basic inputs: stellar evolution theory in the form of isochrones, stellar spectral libraries, and an IMF [Initial Mass Function], each of which may in principle be a function of metallicity and/or elemental abundance pattern" (Conroy 2013, 395). An isochrone is the location of a type of star in the Hertzsprung-Russell diagram, which specifies that it belongs to a group of stars with the same age and metallic composition.[7] Isochrones are found by stellar evolution theory, and they determine the basic properties of a stellar population.

To move from stellar evolution theory to predicted observable SEDs, astrophysicists use stellar spectral libraries (Conroy 2013, Sect. 2.1.3, Sordo et al. 2010). There are two types of libraries, theoretical and empirical. Theoretical stellar spectral libraries use atomic and molecular spectral line lists to generate predictions of observable SEDs for ensembles of stars. Then "observed stars are assigned physical parameters based on a comparison with models" (Conroy 2013, 401). Population synthesis using a theoretical library generates values for physical parameters like age, stellar mass, and elemental composition using atomic and molecular spectral emission lines that are reasonably assumed to be appropriate for that type of star. Simulations using theoretical libraries are only as good as the data that goes into them. Atomic and molecular emission lines used in theoretical libraries may be incomplete, uncertain, or derived by theoretical calculation instead of empirical observation (Conroy 2013, 400–1).

Empirical stellar spectral libraries have the advantage that they are based on observed data. They do not rely on hypothetical values for the emission lines, so they do not have the kind of uncertainty associated with theoretical libraries. On the other hand, empirical libraries have the usual limitations of empirical observations.[8] Moreover, as Conroy notes, "the empirical libraries are woefully incomplete in their coverage of parameter space" (2013, 402). Current instruments may allow only for

---

[6] An abundance pattern specifies the changing elemental composition of the star over time (e.g., metals, dust, gas).

[7] "An isochrone specifies the location in the Hertzsprung-Russell (HR) diagram of stars with a common age and metallicity" (Conroy 2013, 397).

[8] Empirical libraries "are plagued by standard observational constraints such as correction for atmospheric absorption, flux calibration, and limited wavelength coverage and spectral resolution" (Conroy 2013, 402).

investigation of stars in certain areas, or of certain kinds of stars, which introduces sampling and detection bias.[9]

A standard approach to population synthesis is to combine theoretical and empirical libraries. The combination allows the weaknesses and strengths of theoretical and empirical stellar libraries to complement each other. Theoretical libraries cover more of the parameter space, but are more uncertain. Empirical libraries are patchier in their coverage and display observational uncertainty, but provide robust data in certain domains.

5.3 Next-Generation Population Synthesis

The new era of precision astrophysics since the early 2000s has increased the quality of available empirical data significantly: "Galaxy evolution studies are reaching a high level of sophistication due to the very high quality of observational data permitted by modern technology, and the level of spectral details that such observations carry in" (Maraston and Strömbäck 2011, 2785–6).[10] The atomic and molecular emission and absorption lines that can be detected now can be pinpointed much more precisely and at higher resolution. The result has been a marked increase in the coverage and resolution of empirical spectral emission libraries, and of the surveys and maps of stars and galaxies that are available.

From what Elisabeth Lloyd has called a direct empiricist perspective, the new high-resolution empirical data would provide an improved perspective on the galaxy, independently of the theoretical libraries or models. The basic position of direct empiricism is that "data are treated as windows on the world, as reflections of reality, without any art, theory, or construction interfering with that reflection" (Lloyd 2012, 392). From this perspective, to deal with the challenges of population synthesis requires only improvements to empirical, observational methods, in order to gather better and better data. Over time, according to direct empiricism, the theoretical emissions libraries would wither away, replaced by robust empirical data that provides an independently convincing picture.

The development of population synthesis over time runs counter to the direct empiricist perspective. Population synthesis methods employ theories and models in analyzing data and in simulations. Stellar evolution theory is the foundation of population synthesis, and models recruit theoretical and empirical stellar libraries to

---

[9] "This is a long-standing issue that is difficult to address owing to the fact that empirical libraries are drawn from samples of stars in the Solar Neighborhood. For example, hot main sequence stars at low metallicity are very rare, as are stars in rapid phases of stellar evolution such as WR and TP-AGB stars" (Conroy 2013, 402).

[10] Maraston and Strömback mention "the Sloan Digital Sky Survey – SDSS, the Galaxy Mass Assembly – GAMA, RESOLVE 1, GMASS2, the SINS survey". The GAIA project carried out by the European Space Agency is a significant recent survey. See Perryman et al. 1997 for the HIPPARCOS Catalogue and Cappelluti et al. 2017 for the Chandra COSMOS survey.

generate simulations. Thus, the analysis that follows will employ the tools of model-based philosophy of science (Suppes 1962; Giere 2006; van Fraassen 2008; Lloyd 2012; Parker 2017).[11] The approach is based on the perspective that understanding a complex system "require[s] a combination of tools, including models, theory, the taking of measurements, and manipulations of raw data" (Lloyd 2012, 392). The analysis that follows will not defend any particular account of model-based reasoning, but it will assume the use of theories and models in the measurement and analysis of physical variables, in keeping with contemporary methods in population synthesis.

The two sections that follow will argue that model-based methods in next-generation population synthesis reinforce the following conclusions:

1. New high-resolution empirical data does not necessarily provide an improved scientific outlook independently. In fact, higher spectral resolution demands concomitant improvements in theoretical models or methods to provide a better overall perspective (Sect. 5.3.1).
2. The physical variables that are the target of population synthesis cannot be measured without models that employ significant theoretical resources,[12] at least, they cannot be so measured using current methods (Sect. 5.3.2).[13]

5.3.1 High-Resolution Surveys and Theoretical Reasoning

The past few decades have seen exciting developments in new precision instruments that allow for a significant increase in the spectral resolution achievable. Higher resolution allows for the determination of narrower wavelength bands. The instruments of precision astrophysics provoke a tradeoff between the virtues of theoretical and empirical approaches to modeling stellar populations.

---

[11] As Lloyd summarizes Giere's view, "the fit of a model to a real system [. . .] is never a direct comparison of model to reality, but rather a fit between a data model and a model derived from theories. The data model (from the observation side) and the model with which it is compared (from the theory side [. . .]) are gradually built up toward one another, eventually converging toward structures that can be directly compared or matched. Thus, a great deal depends on how the data model (dataset) is derived from the raw data" (Lloyd 2012, 393; see Giere 2006, 68–9).

[12] One might ask what is meant by a 'theoretical' model or resource, as opposed to an empirical one. In the context of population synthesis, scientists refer to stellar evolution theory and hypothetical models of types of stars as 'theoretical' reasoning, and to emissions libraries based on observations with instruments as 'empirical'. I will follow this usage, without making any particular assumptions about a neat cleavage between theories and data.

[13] This is in line with recent work by Nora Boyd, who notes that "empirical results are typically generated and interpreted by recruiting significant theoretical resources" (2018, 404). Nonetheless, empirical constraints on "theorizing about nature" are desirable, "such that some theories are empirically viable and some are not" (Boyd 2018, 404). The methods of population synthesis exemplify this hybridization of theoretical and empirical reasoning.

Astronomers Claudia Maraston and Gustav Strömbäck describe a trade-off between theoretical and empirical population synthesis methods occasioned by next-generation precision astronomy. It is clear that higher resolution data and images are preferable. Higher spectral resolution "is required for a detailed modelling of emission and absorption lines" (Maraston and Strömbäck 2011, 2786). While they note that "a high spectral resolution SED can be obtained either with theoretical or empirical stellar spectra" (ibid., 2786), Maraston and Strömbäck concur with Conroy that theoretical and empirical spectra have comparative advantages and limitations.[14] "Hence," they conclude, "the approach of combining empirical and theoretical spectra is the most convenient one" (ibid.).

The new era of precision, high-resolution empirical data calls on significant theoretical and modeling resources, not fewer. "In this era of precision astrophysics," Maraston and Strömbäck argue, "interpretative models, such as stellar population and galaxy models, need to keep pace with the fast observational development" (2011, 2785–6). Theories and models are used to interpret and analyze the data, and are even needed to measure physical parameters, as discussed in Sect. 5.3.2 below. Theoretical improvements must keep pace with the improvements in observational methods.

5.3.2   Model-Based Measurement of Physical Parameters

Determining the physical variables of a stellar system using population synthesis is the endpoint of a process that involves, not only reading off the state of a measuring instrument, but also two further steps: first employing the theory of stellar evolution as a constraint, and then using simulation to determine the value of the variables. According to some philosophical understandings of measurement, the second two steps would be considered inferences from measurement, rather than measurements.

Model-based theories include models and simulations in the process of measurement. Eran Tal's theory of measurement argues that "a necessary precondition for the possibility of measuring is the specification of an *abstract and idealized model of the measurement process*" (Tal 2012, 17). For instance, idealizing assumptions may be employed to link a background theory to the experimental setup necessary to make a given measurement. Parker argues that, given a robust theory of measurement like Tal's, "it is possible for computer simulations to be embedded in measurement practices and, indeed, for them to be embedded in measurement

---

[14] "While theoretical spectra give in principle high flexibility on the stellar parameter coverage (temperature, gravity, metallicity) and in the wavelength extension, known problems exist in the modelling of specific lines [...] as well as continuum regions [...]. Empirical stellar spectra have all lines that are observed in stars, but their intensity and flux ratios depend on the chemical enrichment history of the site where the star is found. Moreover, empirical libraries usually cover a limited region of stellar parameter space and a limited λ range" (Maraston and Strömbäck 2011, 2786).

practices in such a way that simulation results constitute raw instrument readings or even measurement outcomes" (2017, 285). It may seem implausible that this would be the case. How can a simulation yield a more accurate measurement than a direct observation? Building on an example from Bas van Fraassen (2008, p. 146), Parker explains:

> Suppose we are interested in measuring the temperature of a very small cup of hot tea at time t_0, and we insert a mercury thermometer at that time; we wait a short while for the mercury to stop rising in the tube and take a reading. But thermodynamic theory tells us that the thermometer itself will affect the temperature of the tea and hence the reading obtained. To arrive at a more accurate temperature estimate for t_0, our measurement process will need to include a step that corrects the thermometer reading for this interference. This might involve calculating the earlier temperature of the tea using the thermometer reading, thermodynamic theory, and our knowledge of the initial temperature of the thermometer. In this example, the equation that needs to be solved to obtain the corrected value might be solved directly, but in other cases corrected values might be obtained with the help of computer simulation. In those cases, simulation results can be direct measurement outcomes. (2017, 285)

Analogous reasoning applies, with some interesting changes, for population synthesis methods. The direct measurement in the astronomical case is the set of observed spectral energy distributions or SEDs. But SEDs are not as informative in the absence of population synthesis. Using the methods developed in the later twentieth century and described in Sect. 5.2 above, direct measurements can be combined with a theoretical analysis that identifies the kind of stars that are emanating the energy, and determine the type of stellar evolution in play. That analysis is then the source of simulations (population synthesis) that provide values for the target physical variables. These variables are measured, but not directly. Both the theoretical determination of the kinds of stars involved, and the synthesis of stellar populations on that basis, are necessary to measuring the physical properties of the target systems.

Population synthesis is an excellent case in support of Tal's and Parker's account of simulation-based measurement. Improving theories and models might expand the class of measurable phenomena equally as well as improving empirical instruments and data. Moreover, theory and observation work together in many cases. Improvements to empirical instruments can require enrichment of the theoretical resources that can be employed (Sect. 5.3.1). But by the same token, development of theory and models can allow for better interpretation of the data, better post-data analysis, and even better methods of instrumentation and calibration.

5.4 Conclusion

Next-generation methods of astronomical observation increase the precision and resolution of sky surveys, which in turn enriches the resources available for population synthesis via empirical stellar libraries. The improved observational methods in turn demand enriched theoretical and modeling resources, which ideally develop in tandem with novel data. The process of hybrid enrichment between theoretical and observational methods reinforces theory- and model-based philosophy of science. Population synthesis displays hybrid enrichment in two ways: the combined theoretical and observational methods complement each other and develop in tandem, and the process of measuring physical variables requires theories and models for robust post-data analysis.

References

Boyd, Nora. 2018. Evidence Enriched. *Philosophy of Science* 85: 403–421.
Cappelluti, Nico, Yanxia Li, Angelo Ricarte, Bhaskar Agarwal, Viola Allevato, Tonima Tasnim Ananna, Marco Ajello, Francesca Civano, Andrea Comastri, Martin Elvis, Alexis Finoguenov, Roberto Gilli, Günther Hasinger, Stefano Marchesi, Priyamvada Natarajan, Fabio Pacucci, E. Treister, and C. Megan Urry. 2017. The Chandra COSMOS Legacy Survey: Energy Spectrum of the Cosmic X-Ray Background and Constraints on Undetected Populations. *The Astrophysical Journal* 837 (1): 19.
Conroy, Charlie. 2013. Modeling the Panchromatic Spectral Energy Distributions of Galaxies. *Annual Review of Astronomy and Astrophysics* 51: 393–455.
Giere, Ron. 2006. *Scientific Perspectivism*. Chicago: University of Chicago Press.
Heiter, Ulrike, Caroline Soubiran, Martin Netopil, and Ernst Paunzen. 2014. On the Metallicity of Open Clusters. II. Spectroscopy. *Astronomy and Astrophysics* 561: A93.
Hurley, Jarrod, Onno Pols, and Christopher Tout. 2000. Comprehensive Analytic Formulae for Stellar Evolution as a Function of Mass and Metallicity. *Monthly Notices of the Royal Astronomical Society* 315 (3): 543–569.
Küpper, Andreas, Pavel Kroupa, and Holger Baumgardt. 2008. The Main Sequence of Star Clusters. *Monthly Notices of the Royal Astronomical Society* 389 (2): 889–902.
Lloyd, Elisabeth. 2012. The Role of 'Complex' Empiricism in the Debates about Satellite Data and Climate Models. *Studies in History and Philosophy of Science* 43: 390–401.
Maraston, Claudia. 2005. Evolutionary Population Synthesis: Models, analysis of the ingredients and application to high-z galaxies. *Monthly Notices of the Royal Astronomical Society* 362 (3): 799–825.
Maraston, Claudia, and Gustav Strömbäck. 2011. Stellar Population Models at High Spectral Resolution. *Monthly Notices of the Royal Astronomical Society* 418: 2785–2811.
Parker, Wendy S. 2017. Computer Simulation, Measurement, and Data Assimilation. *British Journal for the Philosophy of Science* 68 (1): 273–304.
Pasetto, Stefano, Cesare Chiosi, and Daisuke Kawata. 2012. Theory of Stellar Population Synthesis with an Application to N-Body Simulations. *Astronomy and Astrophysics* 545: A14.
Perryman, M., et al. 1997. The HIPPARCOS Catalogue. *Astronomy and Astrophysics* 323: L49–L52.
Sordo, Rosanna, et al. 2010. Synthetic Stellar and SSP Libraries as Templates for Gaia Simulations. *Astrophysics and Space Science* 328: 331–335.
Suppes, Patrick. 1962. Models of Data. In *Logic, Methodology, and Philosophy of Science*, 252–261. Stanford: Stanford University Press.
Tal, Eran. 2012. *The Epistemology of Measurement: A Model-Based Approach*. Dissertation, The University of Toronto.
Tinsley, Beatrice, and Richard Larson, eds. 1977. *The Evolution of Galaxies and Stellar Populations: Conference at Yale University, May 19–21, 1977*. New Haven: Yale University Observatory.
van Fraassen, Bas. 2008. *Scientific Representation*. Oxford: Oxford University Press.

Chapter 6
Doing More with Less: Dark Matter & Modified Gravity

Niels C. M. Martens ⓘ **and Martin King** ⓘ

Abstract Two approaches have emerged to resolve discrepancies between predictions and observations at galactic and cosmological scales: introducing dark matter or modifying the laws of gravity. Practitioners of each approach claim to better satisfy a different explanatory ideal, either unification or simplicity. In this chapter, we take a closer look at the ideals and at the successes of these approaches in achieving them. Not only are these ideals less divisive than assumed, but moreover we argue that the approaches are focusing on different aspects of the same ideal. This realisation opens up the possibility of a more fruitful trading zone between dark matter and modified gravity communities.

6.1 Introduction

One of the most startling discoveries of twentieth century physics is that applying the gravitational theories of Newton and Einstein to the visible matter of the universe fails strikingly to account for the astrophysical and cosmological behaviour of that matter. The discrepancies with observations appear at many different scales: at the cosmological scale, in galaxy clusters, and in individual galaxies. In order to match observations, some new component must be introduced: either one postulates a significant amount of additional dark matter, or one modifies the laws of gravity, or perhaps both. Dark matter, as it is encapsulated in what is by now the standard

N. C. M. Martens (✉)
Freudenthal Institute and Descartes Centre for the History and Philosophy of the Sciences and the Humanities, Utrecht University, Utrecht, The Netherlands

Lichtenberg Group for History and Philosophy of Physics, University of Bonn, Bonn, Germany
e-mail: n.c.m.martens@uu.nl

M. King
Munich Center for Mathematical Philosophy, Ludwig Maximilian University of Munich, Munich, Germany

Lichtenberg Group for History and Philosophy of Physics, University of Bonn, Bonn, Germany

N. Mills Boyd et al. (eds.), *Philosophy of Astrophysics*, Synthese Library 472,
https://doi.org/10.1007/978-3-031-26618-8_6

model of cosmology, ΛCDM, has been heralded as the clear winner at the scales of cosmology and galaxy clusters, whereas modified gravity excels at the level of individual galaxies. Dark matter simulations of structure formation still suffer from several well-known "small-scale problems" (De Baerdemaeker and Boyd 2020) but are making progress in fitting some of the empirical correlations within and between individual galaxies which were once only accounted for by modified gravity. Dark matter and modified gravity are often seen as incompatible communities, as "two paradigms locked in mortal combat" (Milgrom 2012). Whereas Ryle (1954) welcomes contests such as those between ΛCDM and modified gravity, as they help to test and develop the power of the arguments in favour of the survivor, Galison (1997) is careful to add that, for the progression of science to be strong and stable, 'trading zones' between the various communities are required, i.e. local coordination of tools, problems, solutions, etc. via a local contact language. However, the relationship between the ΛCDM and modified gravity communities is notoriously polemical, with barely any trading zone existing (Martens et al. 2022).

We contend that there are at least four key aspects to understanding and thereby potentially alleviating this feud: (i) it cannot be won merely by pointing to which data is covered or not; (ii) there are sociological (or non-physics-based) reasons for the divide; (iii) against common lore, it is in fact possible to construct hybrid theories that do not exclusively take one approach; and (iv) lastly, even though proponents of the two approaches tout different aims, successes, and explanatory ideals, these can in fact be brought into a discussion together.[1] The latter of these will be the focus of this chapter. We find that one of the more significant reasons for the dispute—that the communities simply have different explanatory goals—is not such a good reason after all, since the goal is in fact shared. Understanding this may remove one obstacle between these communities and helps us to show that this divide is not unbridgeable. We will assume that each research programme has something of value to offer that the other does not—see also Sect. 6.5—and that a trading zone would therefore be mutually beneficial.

We briefly discuss each of the above four aspects in a bit more detail. The first key aspect is that current appeals to empirical adequacy will not by themselves resolve the debate. The presumption that solving the debate is a simple matter of comparing the data against the predictions of each research programme is not fruitful for reconciling the two research programmes. Neither dark matter nor modified gravity is fully empirically adequate as it stands: small-scale problems remain for ΛCDM and accurate descriptions of galaxy clusters and cosmological observables still plague modified gravity. Both communities understand and approach the data in different ways. It is common to hear that one of the research programmes accounts for "90% of the data" or for the "most important data". However, it is of course unclear how one would quantify the fraction of the data that has been accounted for or how to establish that certain data is more important than some other data. And

---

[1] This coheres with Vanderburgh's (2014, Sect. 6) insight that we should not artificially separate out different methodological aspects/theoretical virtues, but consider them in a holistic fashion.

although it may well be fair to prioritise certain explananda for now, the eventual aim is for an empirically adequate theory to account for all the data.

The second aspect, which is worth mentioning but will not be discussed much here, is that there are sociological factors that influence which research programme one adopts and which explananda are targeted as salient. In cosmology or relativity departments, institutes, and research groups, the focus is obviously on cosmological observables and it seems that dark matter is by far the favoured approach. Particle physicists follow suit in focusing on models of particle dark matter. Modified gravity approaches seem to gain their followers within communities of observational (galactic) astronomers.[2]

Third, the debate is often cast as a battle of incompatible paradigms with modified gravity and dark matter being mutually exclusive concepts—a newly postulated field can only be pure matter or a pure modification of the gravitational field. It has been argued by Martens and Lehmkuhl (2020a,b) that this is contested by a recent trend of hybrid theories that postulate a single novel entity that, in one of several possible ways, is both a dark matter field and an aspect of gravity. Such hybrid theories could thereby play an important role as boundary objects (Star and Griesemer 1989) or aspects of a trading zone.

The fourth aspect is that a large part of the stalemate is due to the fact that practitioners of each of these competing research programmes focus on distinct explanatory ideals and furthermore believe that by their own standards their own research programme is clearly favoured. We first establish that proponents of ΛCDM employ notions of explanation that draw on aspects of unification and that proponents of modified gravity employ those notions that focus on (parametric) simplicity (Sect. 6.3).[3] We then critically evaluate each approach according to both its own explanatory standard and that of the other approach. We argue in Sect. 6.4 that ΛCDM is less unifying than often assumed, but at the same time scores better with respect to simplicity/lack of fine-tuning or curve-fitting than its critics maintain.

---

[2] In future work, we intend to systematically quantify and further explore these suspected trends, using tools from the digital humanities.

[3] Compare this to Massimi's (2018) analysis of the debate between ΛCDM and MOND, where she identifies a "downscaling problem" for ΛCDM—going from the large scale of structure formation to the galactic scale—and an "upscaling problem" for MOND—going in the other direction. We agree with her claims (i), (ii) and (v) (ibid., 27). However, we would nuance her third claim, that the upscaling and downsaling problems are different in nature and that different physical solutions to them have been given. Massimi is right that different solutions are required for each problem, and also correct that the upscaling but not the downscaling problem is explicitly about consistency. However, we disagree that (therefore) only the downscaling problem is an issue of explanation. As explained in our Sect. 6.4, a central notion is that of unification, which goes beyond mere consistency and scope, in a way that renders it a form of explanation as well. For instance, some MOND advocates combine MOND with some neutrino dark matter to account for galaxy cluster phenomenology; while this may well increase consistency, it is not therefore an instance of (substantial) unification (when compared to the ΛCDM-only alternative). Thus, although there are important differences between the upscaling and downscaling problems, they are both best understood as problems (that are partially) about explanation.

Similarly, we find that modified gravity is less simple than often claimed, but also more unifying than often presupposed. Tackling this problem from an explanation viewpoint allows us to distil three important philosophical lessons in Sect. 6.5.

6.2 Astronomical and Cosmological Explananda

Let us begin by briefly describing the two research programmes that are at the centre of our analysis and the distinct explananda that they cover.

ΛCDM is the standard model of cosmology. It describes the universe's space-time geometry, its matter, and its dynamical evolution. It is developed around the Friedman-Lemaître-Robertson-Walker (FLRW) metric, which makes use of symmetry assumptions (that the universe is homogeneous and isotropic) in order to reduce the Einstein Field Equations to just two equations governing the scale factor. In this picture, the standard model of particle physics combined with general relativity lacks the resources for 'seeds' to develop the observed structures of the universe, such as galaxies. Measurements of the cosmic background radiation left over from atomic recombination in the early universe indicate that the majority of the matter and energy of the universe must be dark. After recombination, dark matter, in particular cold dark matter (CDM), would help provide the perturbations that seed structure formation. The model also includes a dark energy component that is, or mimics, a cosmological constant Λ to account for accelerated expansion of the universe.

A non-exhaustive list of some of the main quantitative and qualitative large-scale, cosmological explananda emphasized by ΛCDM advocates and typically best accounted for by the current state of their research programme is as follows:

1. The relative height of the second and third peak in the angular power spectrum of the anisotropies of the cosmic microwave background (CMB);
2. The velocity dispersion of galaxies within galaxy clusters, assumed to obey the virial theorem;
3. The strength of gravitational lensing around galaxies and galaxy clusters;
4. The displacement between the centers of mass of baryonic matter and of dark matter in clusters such as the Bullet cluster and El Gordo.

In a similar spirit to Le Verrier's hypothesis of the planet Neptune, here, more matter (though a different kind of matter) is proposed to account for deviations from theoretical predictions. Some find this addition of different matter ad hoc. A different approach is to liken the situation to the solution of the anomalous perihelion precession of Mercury, where not more matter, but a new theory of gravity was ultimately needed. This is the Modified Newtonian Dynamics (MOND) approach introduced by Milgrom in 1983 (Milgrom 1983). Milgrom proposed that the reason that there is a mass discrepancy is that we are attempting to apply a gravitational theory well beyond its well-tested domain of applicability, viz., solar systems. One version of MOND is a modification of Newton's inverse square law of gravity, which is replaced by:

$$F_G = G \frac{Mm}{\mu(\frac{a}{a_0})r^2} \quad \begin{cases} \mu \approx 1, & \text{if } a \gg a_0 \\ \mu \approx a/a_0, & \text{if } a \ll a_0 \end{cases} \tag{6.1}$$

where a_0 is a new constant of nature with the dimensions of acceleration, which has been empirically determined to be $1.2 \times 10^{-10} \frac{m}{s^2}$ (Li et al. 2018; McGaugh et al. 2018).

A non-exhaustive list of some of the main quantitative and qualitative small-scale, galactic explananda emphasized by MOND advocates and typically best accounted for by MOND rather than dark matter, is as follows. Each item on this list comprises a different aspect of the tight connection between baryonic matter and the mass discrepancy. (From the perspective of ΛCDM this would correspond to a surprisingly tight connection between baryonic matter and dark matter.)

1. The baryonic distribution suffices (in combination with a single fundamental parameter a_0) to determine the full galaxy rotation curves. Moreover, this deterministic algorithm is in accord with Renzo's rule: qualitative features in the baryonic galaxy rotation curve are mimicked in the total galaxy rotation curve;
2. The baryonic Tully-Fisher relation, i.e. the relation between baryonic mass and rotation velocity in galaxies ($M \propto V_{rot}^4$);
3. The mass-discrepancy acceleration relation (MDAR), i.e. the anti-correlation of the mass discrepancy with the baryonic acceleration within galaxies;
4. The small scatter of the observed MDAR, consistent with zero intrinsic scatter;

Going beyond galactic phenomena in an attempt to compete with the full empirical scope of ΛCDM requires a relativistic extension of MOND. The modified gravity research programme thus consists of MOND plus a plethora of relativistic theories which each have MOND rather than standard Newtonian gravity as the appropriate limit, such as Tensor-Vector-Scalar Theory (TeVeS) and Relativistic MOND (RMOND).

6.3 Unification and Simplicity

This section elaborates upon the two differing explanatory ideals that practitioners of each of the two research programmes emphasize when motivating their own approach. Dark matter advocates tend to focus on the explanatory ideal of unification—the characteristic virtue of ΛCDM, the 'concordance model', is that it can bring together so many different kinds of phenomena. Modified gravity advocates focus on the benefits of simplicity in number of parameters and in avoiding problems of falsifiability.

Although the dark matter story is typically told by starting with Zwicky's work in the 1930s, it was not until the 1970s that the dark matter concept was taken seriously (de Swart et al. 2017), when it was realised that it provided a *single solution to multiple problems*: velocity dispersions in clusters, flat rotation curves in galaxies, instabilities of simulated disk galaxies, and the cosmologist's need for extra massive matter given their *a priori*, philosophical, Machian desire to close the universe (de Swart et al. 2017; de Swart 2019; Sanders 2010). The solutions to these problems could have a single, common origin: dark matter.

From that point onward, one cannot discuss the motivations and justifications for dark matter without the broader context of the cosmological model in which it became embedded, ΛCDM, as well as the various more specific accounts, often from particle physics, for filling in the titular CDM slot. ΛCDM is often referred to as the concordance model, as it manages to incorporate, to unify, a large swath of cosmological and astrophysical phenomena from all epochs from the very early universe up till now. It turns out to be the case that there exists a choice of values for the six or seven parameters of this single model, such that it is consistent with most of the 'relevant' data (Hawley et al. 2005; Olive 2014; Merritt 2017).[4]

Importantly, unification is more than merely scope. When Kitcher (1981, 1989) argued for the explanatory power of unification, for example, he emphasised the need for particular derivations of phenomena to be part of a theory—a set of consistent argument patterns used to derive different kinds of phenomena. The explanatory power of unification does not come from the logical structure of the derivation of the phenomenon alone—it involves more than merely providing a potential common origin (as will be discussed further in Sect. 6.4 when discussing the link between unification and simplicity)—but stems from its bringing new phenomena into a broader theory or set of laws that provide an explanatory structure. This is an aspect of explanations in physics that has been recently highlighted by Wayne (2017) and King (2020).

This is the condition that aims to prevent merely tacking one theory onto another and considering it as one theory with increased scope—so called 'spurious unification'. This condition for unification may indeed be satisfied for a particle model of dark matter, which may embed a local description of a phenomenon into a global theory like supersymmetry.

Indeed, the strongest emphasis on unification appears when going beyond pure ΛCDM by considering various popular particle physics candidates precisifying the rather high-level dark matter concept as it features in ΛCDM. They are typically motivated in terms of solving several independent problems, while also solving the dark matter problem 'for free': " [multiple] birds with one stone' theor[ies]". "Theoretical constructions that extend the [Standard Model of particle physics] are clearly more appealing when they are able to solve more than one [...] issue [...] with the same amount of theoretical input" (Di Luzio et al. 2020, Sect. 1). For instance, supersymmetric WIMPs, a popular dark matter candidate, provide a solution to the

---

[4] See Liddle (2004) for a discussion of the number of parameters.

hierarchy problem, they can unify the coupling constants of the three interactions of the standard model of particle physics, and supersymmetry is claimed to play a role in solving the matter-antimatter asymmetry problem and the problem of quantum gravity.

An additional important fact is that supersymmetry was not introduced in order to solve the dark matter problem. Many take it as a sign that a theory might be true, or viable, if it can solve a problem it was not introduced to solve (Dawid 2019). This notion is called 'unexpected explanatory interconnections' by Dawid. For example, axions, which were introduced to solve the strong CP problem,[5] are also a dark matter candidate—a stable field that interacts gravitationally and at most very weakly electromagnetically. Sterile neutrinos, introduced to account for the suppression of the mass scale of the standard model neutrinos, also naturally solve the matter-antimatter asymmetry problem and are dark matter candidates. All of these mainstream dark matter candidates are motivated by solving several independent problems at once, thereby allegedly unifying the associated phenomena. The real benefit comes not from simply including more phenomena, but bringing more different classes of phenomena that would otherwise be unrelated together in the same theory.

Modified gravity advocates on the other hand emphasise simplicity, in particular the parametric simplicity of MOND with its single parameter a_0.[6] Once one fixes the stellar mass-to-light ratio of each galaxy and the acceleration parameter a_0 that applies universally, the MOND formalism serves as an "algorithm" that spits out galaxy rotation curves from the distribution of baryonic matter in each galaxy (Sanders and McGaugh 2002; Sanders 2019). Moreover, it uniquely predicts the correlations mentioned in Sect. 6.2. If we were to allow a_0 to vary across galaxies, this would not even improve the fit to, for instance, the radial acceleration relation (which is equivalent to the MDAR) (Li et al. 2018).

Simplicity is not desired merely for simplicity's sake or for tractability. The appeal to simplicity by modified gravity sympathisers is typically motivated in terms of avoiding two related negative features attributed to dark matter approaches to galactic data: curve-fitting/fine-tuning and unfalsifiability (or being less falsifiable than its competitor). Manually fitting dark matter halos to galaxies typically

---

[5] This problem refers to the non-observation of Charge+Parity symmetry violation in the context of quantum chromodynamics, although violation of this symmetry is allowed in the most general Lagrangian.

[6] Modified gravity advocates typically give no explicit justification for only focusing on a specific type of syntactic simplicity (i.e. parametric simplicity) and not (also) on ontological simplicity (i.e. either qualitative parsimony—minimising the number of types of entities postulated by the theory—or quantitative parsimony—minimising the number of (token) entities (of a given type) postulated by the theory). It could be that it is simply too difficult to measure and compare the ontological simplicity of dark matter vs. modified gravity (especially without having quantised the modified gravitational field), and/or that these other notions of simplicity are not (obviously) connected to explanatory power (and falsification). See Vanderburgh (2014) for a discussion of both these considerations—particularly interesting is his footnote 3. See also (Vanderburgh 2001, Sect. 6.4.1).

includes, besides the stellar mass-to-light ratio, at least two parameters describing the halo, which are barely constrained and can take on different values for each galaxy. Simulations take into account the common origin story of all these dark matter halos, but they still require many parameters, for instance to describe the astrophysical contribution of gas to galaxy formation and evolution—sometimes called 'gastrophysics'—including important feedback processes (i.e. relatively small processes that have large effects), such as the reheating of gas via supernovae feedback. MOND advocates disapprove, referring to this approach as a mere exercise in curve-fitting or fine-tuning. Not only does MOND have only one parameter (besides having to fix the stellar mass-to-light ratio), this parameter could in principle have differed between galaxies which, one might have expected, could make it much easier to fit multiple galaxy rotation curves, but (as indicated above) it turns out that a universal value of a_0 suffices. The reason curve-fitting/fine-tuning is considered undesirable is not so much that a probability distribution over parameter values is presumed under which the fine-tuned values are highly improbable, as is the case in other fine-tuning worries, but because the large freedom to curve-fit makes it (more) difficult for the dark matter approach to fail. Whatever observation there is for a given galaxy, the right amount of dark matter can be postulated. Observing more galaxies, therefore, does not make a strong test of the hypothesis that there is dark matter and would not confer much confirmation. This is an explicit result, e.g. on the error statistics approach (e.g. Mayo 1996), where a hypothesis can only be confirmed by a test if that test could reasonably show that the hypothesis is false if it in fact is. MONDians accuse the dark matter approach of being difficult to falsify because it makes no unique predictions, as varying the parameters would result in substantially distinct values of the observables. Dark matter, as a class of models, thus does not go out on a limb as much as MOND when it comes to matching observations to theoretical predictions.

6.4 Assessment

In this section we provide a brief evaluation of the extent to which each research programme is unifying and the extent to which it is simple. Since unification and simplicity are the two explanatory ideals that dominate the discussion, we will avoid discussing other accounts of explanation, or assessing the theories' empirical confirmation, pursuitworthiness, etc.[7]

---

[7] Vanderburgh (2001, Sect. 6.4.1) argues that in such situations—that is, when we do not have any theory on the table that is fully empirically adequate, nor are we comparing two theories that are completely empirically equivalent—it is premature to use theoretical virtues such as simplicity to attempt to break a non-existent tie. We take our paper to be complementary to this point: given that both communities do in fact invoke different virtues, we argue that this does not justify the divide that currently exists between the communities (since these virtues are misunderstood and misapplied), not even in this current, premature, empirical situation.

Consider first the extent to which ΛCDM is unifying. That there exists a set of values of the six or seven parameters of this concordance model such that it is consistent with the data is just to say that it is empirically adequate. This is an issue merely of empirical scope; the model (supposedly) can account for the data but this does not by itself imply that the stronger notion of 'explaining the data by unifying it' is appropriate. Without theoretical reasons for the values of the parameters and/or *convergence* of independent lines of empirical evidence for the values of each parameter, concordance boils down to curve-fitting (which is being condemned by MOND advocates). The ability of the model to accommodate is due to its flexibility, the variability of its parameters. Merritt (2017) argues that such convergence of independent lines of evidence does not obtain. On the one hand, there are various degeneracies between some of these parameters, e.g. between the matter density and the dark energy density when applying the Hubble diagram test. On the other hand, in some of the scenarios where we do have independent determinations of a single parameter, these do not converge: the infamous and controversial Hubble tension, and the Lithium problem. Add to this that ΛCDM is not even empirically adequate in that it cannot account for various small-scale problems and it becomes quite clear that while it has a broad empirical scope, it does not unify everything.

As mentioned in Sect. 6.3, the strongest claim of dark matter's explanatory power in terms of unification comes from various particle models to be paired with ΛCDM, such as WIMPs, axions, or sterile neutrinos. The problem here is that colliders and direct and indirect detection methods have failed, as of yet, to detect any such particle. Their explanatory power is thus currently only a promise. However, as the parameter space gets constrained further and further, these promises not only become less likely to be true, but are also being watered down. For instance, attention is redirected towards axion-*like* particles, which do not solve the strong CP problem as is the case for the original axion—which has been severely constrained. This reduces the explanatory power (in terms of unification) of such dark matter approaches.

Myrvold (2003) looks at the evidential import of unification. He finds that, on a Bayesian scheme, the evidential benefit of unification is only found for explanations where distinct phenomena provide additional information about each other, by for example providing constraints on the values of parameters, but not in the general case where an explanation accommodates different phenomena by providing a common origin. He finds that the common origin unification is at best heuristic and he poses a challenge for defenders of this kind of unification to demonstrate the epistemic value of these common origin cases. As we saw above, one of the key reasons for thinking that dark matter is unifying is that the theory or model that contains a dark matter candidate has the potential to explain many cosmological puzzles as well as the entirety of Standard Model physics and account for some of its explanatory deficiencies. This is a fairly dramatic increase in scope, but this unification would only be explanatory to the extent that these distinct phenomena provide mutual information about each other. This may well be the case as the dark sector is likely to couple to the SM and thus provide some constraints that may help explain the values of certain parameters. Of course, no such theory has been confirmed and the parameter spaces where many such theories may dwell has been greatly reduced (e.g. Bechtle et al. 2016).

It is thus fair to say that the typical claims of the explanatory power of ΛCDM in virtue of its unifying nature are somewhat exaggerated. At the same time, there is a sense in which dark matter does score some points with respect to avoiding fine-tuning—a desideratum associated with the simplicity ideal of the modified gravity advocates. Both the hierarchy problem—addressed by supersymmetric WIMPs— and the strong CP problem—addressed by the original axion—can be construed as fine-tuning problems (of course, as described above, these explanatory powers are currently only a promise). Additionally, where MONDians accuse ΛCDM of curve-fitting to the extent that it could fit any possible data and thereby be vacuous, we have seen that ΛCDM is currently not empirically adequate and thus to some extent falsifiable after all.

We now consider the extent to which modified gravity approaches are as simple as claimed. MOND is indeed parametrically simple in that it only contains a single parameter, a_0. Applying it to a galaxy further requires fixing the stellar mass-to-light ratio, but that is all. However, we not only know theoretically that MOND must be embedded in some relativistic theory, but such relativistic extensions are indispensable when accounting for observables at scales larger than galaxies. Perhaps unsurprisingly, the relativistic theories that stand a chance at accounting for this larger scope of empirical phenomena tend to be much less simple than MOND. Take for instance TeVeS, Bekenstein's 2005 tensor-vector-scalar theory, which was the flagship theory of the modified gravity research programme until the disastrous constraints arising from LIGO's 2017 detection of gravitational waves from the coalescence of binary neutron stars. The action describing it is rather elaborate, and includes besides a dynamical ('Einsteinian') metric field a dynamical, timelike unit vector field, a dynamical scalar field, a non-dynamical scalar field, a free dimensionless function, two dimensionless parameters and a parameter/constant with the units of length (as well as Newton's gravitational constant). Moreover, matter is coupled to a physical metric that is determined in terms of the dynamical Einstein metric, vector and scalar fields, rather than being coupled directly to the Einstein metric. This is a substantial reduction in simplicity compared to MOND (Abelson 2022, 31), construed narrowly in terms of parametric simplicity as well as when one considers a broader notion that also takes into the account the additional fields being postulated. The new, TeVeS-inspired flagship theory of MOND sympathisers, a version of RMOND by Skordis and Złośnik (2021), which improves upon TeVeS by dealing with the constraints arising from gravitational waves detected by LIGO and by accounting for the cosmic microwave background and matter power spectra, only manages to do so by being yet more complex than TeVeS was. According to Spergel, such models only work by "effectively positing a complex form of dark matter"—they are "baroque" dark matter (Schirber 2021).

On the other hand, MOND arguably does score some points with respect to unification. Each new galaxy rotation curve is an independent test of MOND. The same is true for the (otherwise) independent correlations mentioned in Sect. 6.2.[8] Moreover, the rotation curves and correlations all constitute independent determinations of the acceleration scale—and they converge on the same value.

If ΛCDM would score very high in terms of unification and very low in terms of simplicity, with the opposite being the case for MOND, this would provide some justification for a divide between the communities associated with each research programme. However, it has been argued that each programme is somewhat less successful with regards to their own favoured standard of explanation than is typically claimed to be the case, and somewhat more successful with regards to the standard of explanation typically favoured by the other community. This reduces one further obstacle in bringing these communities together.

6.5 Philosophical Lessons

In this section, we wish to distil three philosophical lessons from the foregoing discussion. Firstly, it is interesting to distinguish between explanations that arise from the common core concept of a research programme, and those that arise from specific models/theories. The common core concept of all dark matter models (De Baerdemaeker 2021; Martens 2021) is rather thin, both semantically and explanatorily speaking; the unificatory promises arise predominantly from specific dark matter models. On the other hand, the common core of most modified gravity models, i.e. its MOND-limit, contains most of its explanatory power, with its relativistic extensions typically reducing the simplicity and hence the explanatory power of modified gravity. This adds an extra dimension to the way in which the two communities talk past each other: in a sense, dark matter advocates focus on a promising-but-not-guaranteed future whereas MOND advocates focus on a somewhat outdated past.

Secondly, in contrast to what sometimes seems to be implicitly assumed by both communities, it is far from clear a) why unification and simplicity would be mutually exclusive explanatory ideals, and b) that there is a research programme-independent way of privileging one of the two explanatory ideals over the other. More importantly, not only does it seem to be false that these explanatory ideals would be mutually exclusive, they are not conceptually independent in the first place. For many, the core of a good explanation is: doing more with less. Unifying more phenomena (or, as Thagard 1978 and Whewell 1840 call it, 'consilience') focuses on the 'doing more' part of this slogan, with simplicity focusing on the 'with less' part. Simplicity and unification are thus best understood as being two sides of the same coin, rather than competing or even mutually exclusive ideals. Unification is, as we stressed, not merely a matter of scope or coherence. A good explanation

---

[8] Indeed, Milgrom has quite recently used the terms "convergence" and "unifying" when referring to MOND (Milgrom 2020).

is one that maximises coherence while minimising flexibility. For example, in Kitcher's (1981; 1989) formulation of unification, the stringency of the argument pattern used also plays an important role. Kitcher notoriously never provides a method for quantifying or trading off stringency against scope, but both are in tension with each other.[9] One can derive just about everything from the simple argument pattern 'God wills X, therefore X', but because this pattern is so flexible that any phenomena can be fit, it fails to be explanatory. This is not a novel point, but something important to keep in mind, as dark matter and modified gravity camps that seem to favour different explanatory ideals, often imagined to be exclusive, are in fact focusing on different aspects of the same explanatory ideal. This is not as strong of a division as one group aiming at a goal the other group will not or cannot accomplish—they in fact share a goal with different emphases.

This is good news for the viability of a trading zone. It had already been pointed out by Galison (2010) that trading between enemies is generally possible, since trading does not require that both groups of merchants share the same understanding and *value* of the goods that are to be traded—local coordination is all that is required. We have argued that the common ground for trading between dark matter and modified gravity sympathisers is more fertile than this minimal requirement, as their explanatory values are in fact much closer to one another than is usually assumed. This should reduce to some extent Galison's worry that the concept of a trading zone loses its applicability in the limit of an asymptotically large power difference, with the asymmetry in size and popularity between the dark matter and modified gravity communities indeed being rather pronounced.

Thirdly, it is important to determine whether the ideals of explanation in terms of unification and simplicity are epistemic or non-epistemic: are they 'merely' aesthetic, heuristic or fruitful; or are they a sign of a theory latching on to the truth? If they are non-epistemic, this strengthens our case against an inevitable divide between the dark matter and modified gravity communities. Their differing ideals would just resemble a difference in preference, not a more fundamental disagreement about essential characteristics of a theory for it to be correct.

So can these explanatory ideals be understood as having epistemic value? Some have argued for the epistemic benefits of unification (see Myrvold 2003). We have already covered this so here we focus on simplicity. There is a strand of literature on (the philosophy of) statistics that motivates the epistemic relevance of parametric simplicity as follows (Forster and Sober 1994; Myrvold and Harper 2002; Sober 2002). Assume the true curve describing some system, say the rotation curve of a galaxy, is an nth order polynomial, i.e. it contains n parameters/coefficients. Given that data always exhibits some measurement error, i.e. the data points are scattered around the true curve, one would always obtain a better fit—for instance in terms of the least sum of squares—by fitting the data to a polynomial of a higher order than n. "Curves that fit a given data set perfectly will usually be false; they will perform

---

[9] We do not endorse Kitcher's particular account of explanation but use this description to highlight that unification is more than scope and itself involves simplicity.

poorly when they are asked to make predictions about *new* data sets" (italics in original) (Forster and Sober 1994, 8).

Various so-called information criteria/ theorems—most notably the Akaike and Bayesian information criteria—aim to address this overfitting to the noise in the data. Such information criteria subtract a penalty from the log-likelihood of a (polynomial) curve that is fitted to the data, where that penalty is a function of the number of parameters of the polynomial that is used. Without this penalty, more parameters would always reduce the sum of least squares and thereby increase the log-likelihood; with the appropriate penalty the sweet spot will occur when the 'true' number of parameters is being used. Now, if we keep observing more and more galaxies, it may seem to be the case that although the dark matter and MOND fits will incur an equal penalty for the stellar mass-to-light ratio, MOND will only receive a one-off penalty for the universal parameter a_0 whereas the penalty for the dark matter approach will keep increasing with every two new parameters introduced per new galaxy. It thus seems to be the case that one can make the penalty for a dark matter fit arbitrarily large by observing sufficiently many galaxies, such that the parametrically simpler MOND fit will always win. Difficulties with this type of argument are that the information criteria apply to a single curve, e.g. fitting a single galaxy rotation curve, and it is not obvious how to apply them to fitting a collection of data sets, each with a member of dark matter or of a MOND family of curves. Even if this difficulty is overcome, the burden is on MOND advocates to perform this quantitative analysis, and to show that the increasing penalty indeed disfavours dark matter, i.e. to show that the bare log-likelihoods (without the penalty) of the dark matter fit are not so much better than those of the MOND fits that they can overcome the subsequent penalty.

Perhaps then the best shot at justifying the epistemic nature of simplicity would be as follows. MOND advocates do not emphasise simplicity (just) because of the aesthetically pleasing elegance of a simpler theory. They claim that it is in virtue of the relative simplicity of MOND compared to dark matter fits to galactic data that MOND is more falsifiable. Due to its smaller number of parameters, it could account for a smaller fraction of all the imaginable observations of galaxies, and thereby makes stronger predictions. Even if this is true in some restricted sense (which, Vanderburgh 2001, Sect. 6.4.2. argues, is not the case)—e.g. when comparing only galactic dynamics, and only the manual fitting of dark matter halos without taking into account their common origin by simulating structure formation from the early universe until now—we have already seen that the complete situation is more intricate. When simulating structure formation, the resulting distribution of various types of halos is indeed inconsistent with observations in a variety of ways—the so-called small-scale problems. This is an example of dark matter being falsifiable. MONDians are right though in pointing out that the typical response by dark matter advocates is that these will be solved when the messy gastrophysics is taken into account, i.e. when more parameters are added (De Baerdemaeker and Boyd 2020) and parametric simplicity is thus further reduced. However, modified gravity theories tend to predict the wrong answers at the level of galaxy clusters and the CMB (if they say anything about the latter at all). In order to avoid falsification of their research programme, the response is to add dark matter in galaxy clusters (for instance in the form of neutrinos) and/or to design ever more complex relativistic extensions of MOND. Both of these options tend to reduce the simplicity and falsifiability of the theory, in order to remain epistemically viable.

6.6 Conclusion

We contend that there are at least four important aspects to understanding and perhaps resolving some of the tensions in the dark matter/modified gravity debate. Here in this chapter, we shed light on the role that different explanatory ideals play in the assessment of these theories. We find that a careful look at the explanation literature, in particular that involving unification and simplicity, shows that these two approaches are in fact focusing on different aspects of the same explanatory ideal: to explain more with less. The chapter concludes that, although part of the divide between the dark matter and modified gravity communities may have arisen in a self-reinforcing way, i.e. from each community believing that different explanatory measures are important and that by their own favoured measure only their own approach is satisfactory, the actual explanatory structure of both approaches is much more complex and does not justify a strong divide between the two communities. This realisation opens the door towards a dark matter/modified gravity trading zone.

Earlier work on the conceptual interpretation of hybrid dark matter/modified gravity theories (Sect. 6.1) (Martens and Lehmkuhl 2020b,a) pushed back against an abstract obstacle that stood in the way of a trading zone, i.e. the idea that both camps are enemies in the sense that their approaches are conceptually exclusive of one another, and that it was therefore impossible for both camps to be (partially) 'right' at the same time. The positive, more concrete upside to the removal of this abstract obstacle is that the hybrid theories themselves, by providing a natural (i.e. non-ad-hoc) physical mechanism for combining the strengths of both camps, could provide the required common ground for both communities to come together and trade ideas, solutions, methods and tools. Similarly, this chapter has pushed back against another obstacle, the idea that each camp has diametrically opposed aims, in terms of explanatory ideals. (Even if these aims were diametrically opposed, we have argued that they do not favour their associated research programmes as straightforwardly as is usually being assumed.) On top of this, the positive message is that, rather than it being inevitable that both camps talk past each other, there turns out to be a point of contact. Unification and simplicity are different nuances within the 'doing more with less' language of explanation. Although it is well known that trading between different communities is possible even if there is no common currency—recall in this regard that anthropological work on trading of material goods between communities is one of the motivations for Galison's concept of a trading zone between scientific communities—there is to some extent a single explanatory currency in use in the context of dark matter and modified gravity, with explanation in terms of unification and simplicity being two sides of the same coin.

Acknowledgments We would like to acknowledge support from the Deutsche Forschungsgemeinschaft (DFG) Research Unit "The Epistemology of the Large Hadron Collider" (grant FOR 2063). We are grateful to Miguel Ángel Carretero Sahuquillo, Sophia Haude, Michael Krämer, Dennis Lehmkuhl, Stacy McGaugh, Erhard Scholz and the Lichtenberg Group for History and Philosophy of Physics (University of Bonn) for valuable discussions and feedback.

References

Abelson, S.S. 2022. The fate of tensor-vector-scalar modified gravity. *Foundations of Physics* 52(1): 31.

Bechtle, P., et al. 2016. Killing the cMSSM softly. *European Physical Journal* C76(2): 96.

Bekenstein, J.D. 2005. Modified gravity vs dark matter: Relativistic theory for MOND. *PoS* JHW2004: 012.

Dawid, R. 2019. The significance of non-empirical confirmation in fundamental physics. In *Why trust a theory? Epistemology of modern physics*, ed. Dardashti, R., Dawid, R., and Thebault, K., pp. 99–119. Cambridge: Cambridge University Press.

De Baerdemaeker, S. 2021. Method-driven experiments and the search for dark matter. *Philosophy of Science* 88(1): 124–144.

De Baerdemaeker, S., and N.M. Boyd. 2020. Jump ship, shift gears, or just keep on chugging: Assessing the responses to tensions between theory and evidence in contemporary cosmology. *Studies in History and Philosophy of Science Part B: Studies in History and Philosophy of Modern Physics* 72: 205–216.

de Swart, J. 2019. Closing in on the cosmos: Cosmology's rebirth and the rise of the dark matter problem. *Einstein Studies* 16: 257–284.

de Swart, J., G. Bertone, and J. van Dongen. 2017. How dark matter came to matter. *Nature Astronomy* 1(3): 59.

Di Luzio, L., M. Giannotti, E. Nardi, and L. Visinelli. 2020. The landscape of QCD axion models. *Physics Reports* 870: 1–117. The landscape of QCD axion models.

Forster, M.R., and E. Sober. 1994. How to tell when simpler, more unified, or less ad hoc theories will provide more accurate predictions. *British Journal for the Philosophy of Science* 45(1): 1–35.

Galison, P. 1997. *Image and logic: A material culture of microphysics.* Chicago: University of Chicago Press.

Galison, P. 2010. Trading with the enemy. In *Trading Zones and Interactional Expertise*, ed. Gorman, M.E., chap. 2, pp. 25–52. Cambridge: The MIT Press.

Hawley, J.F., J.F. Hawley, and K.A. Holcomb. 2005. *Foundations of modern cosmology.* Oxford: Oxford University Press.

King, M. 2020. Explanations and candidate explanations in physics. *European Journal for Philosophy of Science* 10(1): 1–17.

Kitcher, P. 1981. Explanatory unification. *Philosophy of Science* 48(4): 507–531.

Kitcher, P. 1989. Explanatory unification and the causal structure of the world. In *Scientific explanation*, ed. Kitcher, P., and Salmon, W. Minneapolis: University of Minnesota Press.

Li, P., F. Lelli, S. McGaugh, and J. Schombert. 2018. Fitting the radial acceleration relation to individual SPARC galaxies. *Astronomy & Astrophysics* 615: A3.

Liddle, A.R. 2004. How many cosmological parameters? *Monthly Notices of the Royal Astronomical Society* 351(3): L49–L53.

Martens, N.C.M. 2021. Dark matter realism. *Foundations of Physics* 52(1): 1–19.

Martens, N.C.M., M.Á. Carretero Sahuquillo, E. Scholz, D. Lehmkuhl, and M. Krämer. 2022. Integrating dark matter, modified gravity, and the humanities. *Studies in History and Philosophy of Science Part A* 91: 1–5.

Martens, N.C.M., and D. Lehmkuhl. 2020a. Dark matter = modified gravity? Scrutinising the spacetime–matter distinction through the modified gravity/dark matter lens. *Studies in History and Philosophy of Science Part B: Studies in History and Philosophy of Modern Physics* 72: 237–250.

Martens, N.C.M., and D. Lehmkuhl. 2020b. Cartography of the space of theories: An interpretational chart for fields that are both (dark) matter and spacetime. *Studies in History and Philosophy of Science Part B: Studies in History and Philosophy of Modern Physics* 72: 217–236.

Massimi, M. 2018. Three problems about multi-scale modelling in cosmology. *Studies in History and Philosophy of Science Part B: Studies in History and Philosophy of Modern Physics* 64: 26–38.

Mayo, D.G. 1996. *Error and the growth of experimental knowledge.* Chicago: Chicago Univiersity Press.

McGaugh, S.S., P. Li, F. Lelli, and J.M. Schombert. 2018. Presence of a fundamental acceleration scale in galaxies. *Nature Astronomy* 2: 924.

Merritt, D. 2017. Cosmology and convention. *Studies in History and Philosophy of Science Part B: Studies in History and Philosophy of Modern Physics* 57: 41–52.

Milgrom, M. 1983. A modification of the Newtonian dynamics as a possible alternative to the hidden mass hypothesis. *The Astrophysical Journal* 270: 365–370.

Milgrom, M. 2012. Light and dark in the universe. arXiv:1203.0954 [physics.pop-ph].

Milgrom, M. 2020. MOND vs. dark matter in light of historical parallels. *Studies in History and Philosophy of Science Part B: Studies in History and Philosophy of Modern Physics* 71: 170–195.

Myrvold, W.C. 2003. A Bayesian account of the virtue of unification. *Philosophy of Science* 70(2): 399–423.

Myrvold, W.C., and W.L. Harper. 2002. Model selection, simplicity, and scientific inference. *Philosophy of Science* 69(S3): 135–149.

Olive, K. 2014. Review of particle physics. *Chinese Physics C* 38(9): 090001.

Ryle, G. 1954. Dilemmas. *Philosophy* 69(269): 378–380.

Sanders, R.H. 2010. *The dark matter problem: A historical perspective.* Cambridge: Cambridge University Press.

Sanders, R.H. 2019. Dark matter – modified dynamics: Reaction vs. prediction. https://doi.org/10.48550/arXiv.1912.00716

Sanders, R.H., and S.S. McGaugh. 2002. Modified Newtonian dynamics as an alternative to dark matter. *Annual Review of Astronomy and Astrophysics* 40(1): 263–317.

Schirber, M. 2021. Dark matter alternative passes big test. *Physics Magazine* 14. https://physics.aps.org/articles/v14/143

Skordis, C., and T. Złośnik. 2021. New relativistic theory for modified Newtonian dynamics. *Physical Review Letters* 127: 161302.

Sober, E. 2002. Instrumentalism, parsimony, and the Akaike framework. *Philosophy of Science* 69(S3): 112–123.

Star, S.L., and J.R. Griesemer. 1989. Institutional ecology, 'translations' and boundary objects: Amateurs and professionals in berkeley's museum of vertebrate zoology, 1907-39. *Social Studies of Science* 19(3): 387–420.

Thagard, P.R. 1978. The best explanation: Criteria for theory choice. *Journal of Philosophy* 75(2): 76–92.

Vanderburgh, W. 2001. Dark matters in contemporary astrophysics: A case study in theory choice and evidential reasoning. Ph.D. Thesis.

Vanderburgh, W.L. 2014. Quantitative parsimony, explanatory power and dark matter. *Journal for General Philosophy of Science/Zeitschrift für Allgemeine Wissenschaftstheorie* 45(2): 317–327.

Wayne, A. 2017. Explanatory integration. *European Journal of Philosophy of Science* 8: 347–365.

Whewell, W. 1840. *The philosophy of the inductive sciences, founded upon their history.* New York: Johnson Reprint.

Part II
Models and Simulations

Chapter 7
Stellar Structure Models Revisited: Evidence and Data in Asteroseismology

Mauricio Suárez

Abstract This paper advances further an ongoing project to understand the history of stellar structure modelling and its inferential practice. It does so by taking a harder look at the data: how it is collected, analysed statistically, and represented in HR diagrams and stellar structure models alike. The focus is ultimately on the sorts of strong observational constraints revealed in the last two decades within the new and expanding field of asteroseismology. It is argued that the typical inferential practices in asteroseismology, while richly loaded with modelling assumptions of their own, do not raise any circularity worries that may compromise the quality or value of the data.

7.1 Three Aims in the Philosophy of Stellar Astrophysics

There are three aims to this chapter. First of all, I aim to provide a succinct introduction to stellar astrophysics – particularly as regards stellar structure modelling, with a focus on the sorts of observational data and constraints that are nowadays operative within the field. The chapter may thus provide an informative first point of contact with a fascinating scientific field. The hope is that this will be of use to philosophers and other humanists not specialised in the philosophy of astrophysics – or, for that matter, not even specialised in the philosophy of science.

The second aim is most distinctly philosophical. I claim that there is one lesson regarding scientific modelling at large that becomes very apparent in stellar structure modelling. It concerns the nested nature of modelling; the fact that most models operate against a background that incorporates further models, and where data is routinely tested backwards, as it were – by deducing the values of theoretical parameters from the data given some background assumptions, and where those parameters are then fed into new models that can appropriately generate the data.

M. Suárez (✉)
Complutense University of Madrid, Clare Hall, Cambridge, UK
e-mail: msuareza@ucm.es; ms2949@cam.ac.uk

© The Author(s) 2023
N. Mills Boyd et al. (eds.), *Philosophy of Astrophysics*, Synthese Library 472,
https://doi.org/10.1007/978-3-031-26618-8_7

While this method seems amenable to a bootstrapping model of confirmation (Glymour 1980), I do not here consider in detail issues of confirmation. Instead, I look at how, in the nowadays booming field of observational asteroseismology, our models of stars and their evolution are sensitive to further models of layered stellar interiors, to models of the physics of radiative materials, to models of stellar atmospheres in coronal astrophysics, and to models of the vibration modes that yield the astrophysical data that in turn supports those models.

Let me emphasise from the start that none of this 'nesting' of models within models can serve to deny that we have by now robust knowledge of the physics of stellar interiors, and that the accepted typologies of stars, and their evolutionary phases, are a very secure part of our stock of scientific knowledge. As the slogan goes: we do know the interior of our local star, the Sun, rather better than we know the interior of our own planet, the Earth. Yet, it is noteworthy (certainly to an epistemologist) that none of this knowledge is supported by any sort of active laboratory intervention on the system that is the object of our knowledge. We certainly don't experiment on stars the way we experiment on the objects of our laboratory studies. In stellar astrophysics, we can't actively intervene on the conditions of the production of the phenomena we study, and we certainly cannot have the sort of experimental warrant that is acquired in the actual causal manipulation of the system of interest in most laboratory experiments.[1] In other words, we don't materially probe into stars' interiors in the way we can probe materially into (the surface layers of) our Earth. It is then at least paradoxical that it should be common lore that we *know* the interior of stars. My second aim is to provide some philosophical insight into why this is so.

As a word of warning, the idea that models are 'nested' in astrophysics, which I defend, must be distinguished carefully from Harry Collin's (1992) much discussed 'experimenter's regress'. I do not believe that there is a pernicious circularity in these models of the sort Collins denounces for experiments on gravitational waves.[2] Nonetheless, there is an obvious circularity in the fact that all models contain assumptions that involve or result from further models. However, I shall argue that, at least as regards stellar astrophysics, such 'nesting' of models within models is innocuous to the justification of the models to the extent that the supporting models do not themselves employ the same questionable idealisations that appear downstream the modelling chain. This view gains support from a consideration of the 'inverse' or 'forward' modelling practices typically engaged in by astrophysicists, which is the focus of much of the discussion throughout.

---

[1] This point has often been made in connection with the application to astrophysics of Ian Hacking (1983)'s 'experimental realism' (See also Jacquart 2020). In the version of experimental realism that I find defensible (Suárez 2024), the sort of warrant acquired in laboratory manipulations is not different in kind but only in degree from the sort of warrant provided by theoretical inferences based on observational data. This explains why, unlike Hacking (1989), I am not tempted by antirealism concerning the objects of astrophysics.

[2] Not, at any rate, any circularity that would require social, or even generally extra-empirical, considerations to determine the models that we do have of stellar structure and evolution.

The final third goal to this chapter involves updating previous work of mine in this area, and in particular a series of papers produced over a decade ago on the fictional nature of the conditional modelling assumptions in stellar structure modelling. Those papers argued that stellar structure inferences are supported by 'fictional conditionals'.[3] The new and exciting observational science of asteroseismology, and the remarkable data thrown out by the CoRot and Kepler missions, certainly force a revision of some of those claims. At the very least, it is necessary to finess some of the claims regarding the fictional nature of the antecedents/background conditions. Many of those fictional assumptions turn in the light of recent evidence to have been false idealisations.[4] Yet, it all remains of a piece with the idea that stellar structure models involve indicative conditionals with hypothetical assumptions in their antecedents from which the observational consequences that appear in their consequents can be derived. It may be just as well to start at this point, recapitulating some of those claims before launching into a discussion of how the new observational science of asteroseismology changes, corrects, and enhances the picture.

7.2 A Very Brief History of Stellar Astrophysics

Arthur Eddington's *The Internal Constitution of the Stars* (Eddington 1926) is regarded as a milestone in the history of stellar astrophysics, and as having set the foundations of the discipline for years to come. By the time of its publication, Arthur Eddington (1882–1944) had already been a very young Director of the Cambridge Observatory, secretary and then president of the Royal Astronomical Society, and a fellow of the Royal Society for over a decade. He had played a major role in the 1918 expedition (sometimes named the 'Eddington expedition' in his honour) that collected the data that served to confirm Einstein's theory of general relativity. Eddington is widely credited for the discovery that the energy produced in stars is generated by hydrogen fusion into helium. His 1920 paper with the same title (Eddington 1920) already advanced the hypothesis that Einstein's postulate of the

---

[3] See particularly Suárez (2013) which summarises the results of the earlier papers and culminates research over the previous five-year period.

[4] This is what Hans Vaihinger (1924) took to be the inevitable and rather honourable fate of most fictions in the long-term course of scientific research, anyway. See the essays in Suárez, ed. (2009), including the introduction, for a review. Throughout the essay, as applied to model assumptions, I use the terms 'fictional' and 'idealised' with the meaning ascribed in that volume. Roughly: a fictional assumption may be truth-apt, but it is not in practice even a candidate for truth – its purpose and cognitive value lies elsewhere entirely. An idealised assumption, by contrast, is not only truth-apt, but it 'hankers after' the truth. Yet, an idealisation typically falls short of its goal, i.e., it fails to be true. There may nevertheless be ways to figure out, at least in principle, its manner of departure from truth. I realise this is not the universally agreed usage of the terms amongst philosophers of science these days, even though they correspond naturally to those in Vaihinger's seminal (1924).

equivalence of mass and energy, encapsulated in the famous $E = mc^2$ equation, allowed for the conversion of mass into energy at rates that would explain the range of luminosity outputs observed for most stars, and he consequently predicted that the temperatures in the interior of stars could reach millions of degrees Celsius. These hypotheses were all confirmed in due course, cementing Eddington's reputation as the first and foremost expert in the theory of stellar evolution, and hence helping to launch the discipline of stellar astrophysics.

Eddington's other great contribution to stellar evolution is his re-interpretation of the central or most important regularity in observational astrophysics, the so-called Hertzsprung-Russell (HR) law, as a 'mass-luminosity relation' (Eddington 1926, chapter VII). The HR law had been independently discovered in the years 1911–13 by the Danish astronomer Ejnar Hertzsprung (1873–1967) and the American Henry Norris Russell (1877–1957). On analysing the spectral lines of the stars surveyed at the Harvard College Observatory for the Henry Draper catalogue, Antonia Maury (1866–1952) had found a way to group them neatly into spectral classes while retaining information regarding their temperature.[5] This gave rise to the identification of the spectral classes of stars (O, B, A, F, G, K, M) in the Harvard Classification Scheme, with specific ranges of surface temperatures associated to each class. It was not difficult then to plot in a diagram increasing luminosity versus decreasing surface temperature (spectral class). This soon was noted by Hertzsprung, Russell and others to express a clear correlation between a star's luminosity (L), a measure of the star's 'energy power', on the one hand, and its effective surface temperature (T_{eff}), the temperature of the outer layer or photosphere, on the other. In a typical Hertzsprung-Russell diagram (such as the one reproduced in Fig. 7.1), this correlation is represented as a descending sequence from the hottest and brightest stars in the top left-hand corner to the coolest and dimmer ones in the bottom right-hand corner. This is the notorious *main sequence*, which Eddington was amongst the first to understand, via his postulate of a mass-luminosity relation, as the central sequence of temporal evolution in the lifecycle of stars.

The HR correlation expressed in this diagram is sometimes referred to as the main 'empirical law' of stellar astrophysics, but this is a misnomer for a couple of reasons. First, the quantities plotted in an HR diagram are not directly observed in any meaningful sense of the term. They are rather inferred, by means of some simple phenomenological laws and extrapolations, from those quantities that can in fact be observed. The two quantities that are directly observable are the incident radiation flux from a stellar source into a telescope on earth, also known as the star's apparent brightness (I_{obs}), and the characteristic set of spectral lines of the

---

[5] The work is collected in Maury (1897), and Hertzsprung was explicit (in correspondence in 1908 with the Director of Harvard College, Edward Pickering) that Maury's classification was critical in the development of what became known as the Hertzsprung-Russell law: "In my opinion the separation by Antonia Maury of the c-and ac stars is the most important advancement in stellar classification since the trials by Vogel and Secchi ...". The history is recounted succinctly in Gingerich (2013).

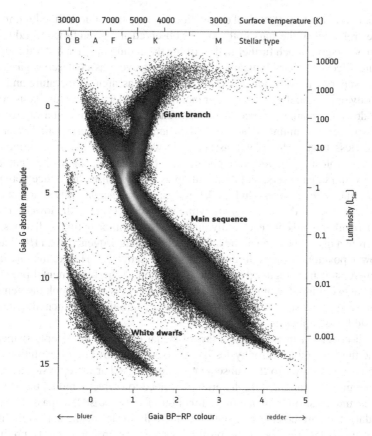

Fig. 7.1 A Hertzsprung-Russell diagram displaying the HR law for GAIA data. (© ESA public domain)

radiation received on earth, or distribution of radiation intensity per wavelength, also known as the electromagnetic spectrum (I_λ) of the source. Add a third, the star's distance from earth (d), which is not strictly speaking observable but can be calculated by geometric means independent of any consideration of stellar structure, namely 'parallax', or cluster analysis. Given precise values for these three quantities (I_{obs}, I_λ, d) for any given stellar source, it is possible to derive the luminosity (L), effective temperature (T_{eff}), and chemical composition of the star as follows (see Suárez 2013, 239–240; or Tayler 1994, chapter 2). The luminosity (L) of a star is its 'energy power' or the amount of energy radiated per unit time – a simple function of distance and observed incident radiation flux: $L = 4\pi d^2 I_{obs}$. The effective surface temperature can then be derived from the luminosity under the assumption that the star is a *blackbody spectrum* as follows: $T_{eff}^4 = L/4\pi R^2 \sigma$, where σ is the Steffan-Boltzmann constant and R is the radius of the star.

The second and most important reason why the HR law is not 'empirical', though, is that the 'main sequence' in an HR diagram is not merely the diagrammatic representation of a bare statistical correlation between the values of the luminosities and spectral types of the stars in the sky, as observed at a given time. As Eddington pointed out, it goes much further in describing the evolution of each star throughout its entire lifecycle. Stars typically begin their lifecycle as coalescing gas in the wake of supernova explosions. They thus start out as high temperature and great luminosity objects, and they reach their maturity as colder and less luminous objects further down the main sequence. All stars eventually meet their end away from the main sequence, as luminous but cool 'red giants' if they were initially not very massive (less than eight solar masses); otherwise, the star's core explodes in a supernova explosion – a very rare occurrence. Eventually a star will radiate away its energy and outer layers, and then end up as a very dense, and hence hotter, but much dimmer 'white dwarf' (In Fig. 7.1, red giants correspondingly sit to the upper right-hand corner of the HR diagram, while white dwarfs are in the lower left-hand corner). In other words, for each individual star, its evolution in time will necessarily take it from a higher position at the left side of the main sequence in an HR diagram to a lower position at the right end of this same sequence. Hence, the HR law is not a mere statistical regularity but the expression of the fundamental process of dynamical evolution that applies to the lifecycle of every star. It took the genius of Eddington to turn a statistical empirical regularity into the fundamental dynamical law of stellar astrophysics.

The discovery of this dynamical law of stellar evolution has deeply shaped the way the field of stellar astrophysics has evolved. The HR law of evolution down the main sequence led to the subsequent history of stellar astrophysics as a field overwhelmingly dominated by the effort and need to correctly model the structure of stellar interiors, often as a simple function of only the stellar mass. For if we can reduce the parameters describing the internal physics of the star to the initial mass of its coalescing gas, at the birth of the star, we are then able to predict its entire subsequent lifecycle, as it moves down the HR diagram's main sequence. So, we are in an optimal position to generate accurate predictions for what are known as the 'observable quantities' of stellar astrophysics, namely the luminosity and spectral type (or effective temperature of the photosphere), precisely the quantities that get plotted in an HR diagram. This mode of inference from model parameters to observable quantities is what is known in the field as 'forward' or 'inverse' modelling. It does not allow us to infer the physical conditions prevalent in the star from its observable conditions at the photosphere, but rather the other way round: One must start hypothesising some values of the parameters (some description of the internal structure of the star) to derive the observable consequences that can then be tested against the empirical data.

This inferential procedure can evidently lead to suitably modify some of these parameters retrospectively, by fine-tuning the initial description within some margins both for the parametrization chosen, and the initial values for some of those parameters. It is not a procedure that can ever settle, or in any way determine, the value of those parameters. On the contrary, underdetermination is rife here since we can produce models with critically different hypothetical descriptions of the (presumably unobservable) properties of the interior of a star yielding nonetheless approximately equally correct values for its observed quantities.[6] That is, nothing in an HR diagram can determine precisely what goes on in the actual interior of any star. We can merely postulate some description of a star's age, chemical composition and initial mass at birth, shape and layered structure, energy transfer mechanism and so on, and then use the models to appropriately deduce from these hypothetical physical processes and parameters within the star some of the star's observed quantities. In other words, the 'forward' inferences that take us from the values of the central parameters in stellar structure models describing the (unobservable, hypothetical) stellar interior, to the (observable, actual) photosphere, cannot ever hope to settle what goes on inside a star, for any star, whatever its position in the main sequence. Thus, either the interior of a star remains a useful fiction (as useful as many other fictions that have routinely been employed to great benefit for inferential purposes throughout the history of science); or there is another set of 'observable' quantities beyond those plotted in HR diagrams, one that can provide us with information regarding those elusive interiors of stars.

7.3 'Fictional Conditionals' in Stellar Structure Modelling

Let us consider the first option first: the modelling descriptions of stellar interiors within stellar structure models are essentially fictions, adequate only for the purpose of expedient inference to the star's observable quantities, namely its luminosity (L), and the effective temperature of its photosphere (T_{eff}).[7] Such fictional assumptions have no further cognitive value beyond the convenience of their expediency. At the most abstract level, stellar structure modellers assume that a star is defined as a cloud of gas uniformly constituted by a mixture of hydrogen and helium, bound together by self-gravitation, and radiating energy from an internal source at its core (Prialnik 2000, 1). It then seems extraordinary that the star would maintain itself in equilibrium for vast periods of its lifetime (for millions of years). This is due to the exquisite way in which the forces balance themselves out in the cloud of gas. The inward gravitational force is perfectly balanced by the outward radiation pressure; otherwise, the star would collapse under its own weight. Conversely, gravity prevents the star's matter from blowing away under the outwards pressure exerted by the radiation. As Eddington himself put it almost a century ago (1926, 20): "We may think of a star as two bodies superposed, a material body (atoms and electrons) and an aethereal body (radiation). The material body is in dynamical equilibrium but the aethereal body is not; gravitation takes care that there is no outward flow of matter, but there is an outward flow of radiation". What prevents all

---

[6] See Belot (2015) and Miyake (2015) for a related discussion of underdetermination in forwards modelling in geophysics (I thank the editors for pointing me towards their works).

[7] This is, with caveats, the view defended in Suárez (2009, 2013), and this section both draws on, and builds upon, those ideas.

radiation from diffusing away instantaneously is the *opacity* of the material of the star – which we represent by means of an absorption coefficient.

Nevertheless, this abstract description of a star is clearly a convenient fiction. A star's boundaries are rather imprecise, and the extensive area where the surrounding atmosphere interacts with the inter-stellar gas is the locus of extraordinary physical processes and events which are interesting in themselves, and which affect the radiation passage from emission to reception on earth. In other words, a star is not a closed system, but is in constant contact with its environment, the interstellar medium. The study of this interaction is now the remit of an expanding field known as 'coronal astrophysics', which has nonetheless traditionally been ignored for the purposes of modelling stellar structure and stellar evolution. As regards the forces acting on the constitutive gas, although self-gravity dominates it clearly is not the only force. Besides the radiation forces, there are also magnetic forces at play, which can occasionally have dramatic effects on the shape of the star and the ensuing surface temperature distribution over the photosphere. Finally, while young stars tend to be composed mainly of hydrogen and helium, as they move down the main sequence, they will generate elements with heavier atomic numbers such as oxygen, carbon, and nitrogen, which they then eject into the interstellar medium (and which can thereby be present in still younger stars formed in the vicinity, particularly in the wake of supernova explosions). This is the sense in which stars are popularly said to be the 'kitchen' of the universe, where the heavy elements that make life possible are 'cooked up'.

Thus, a real star will typically not look much like the perfectly symmetrical ball of uniformly distributed hydrogen and helium gas mixture in perfect equilibrium described in stellar structure textbooks. A real star like our Sun looks a bit more like the object depicted in Fig. 7.2. We suppress a great deal of the physical detail that we know to be present in a star when we model it in accordance to the four ubiquitous assumptions in stellar structure modelling: (i) isolation (IA) from interstellar medium, (ii) blackbody radiative equilibrium (EA), (iii) uniform composition (UCA) of hydrogen and helium (roughly at 70–30% respectively), (iv) gravity as the only self-bounding force, which yields the assumption of perfect spherical symmetry (SSA) of the star's layers, including the photosphere. These assumptions together combine to great effect in the building of concrete stellar structure models that take in as initial conditions the description of the internal state of the star at each of its layers, as parametrized solely by the mass of the star (the ability to parametrize singly by mass is a consequence of the UA and SSA assumptions which together entail that the mass of the star grows linearly and monotonically with radial distance from the centre of the sphere).

These assumptions yield the four equations of stellar structure: hydrostatic equilibrium, continuity, radiative transfer, and thermal equilibrium (Prialnik 2000, Ch. 5; Tayler 1994, Ch. 3), and it is these four equations jointly that swiftly yield values for the 'observable' properties that are plotted in an HR diagram: luminosity (L), effective temperature (T_{eff}), and mass fraction (I_λ) at the photosphere. In other words, the assumptions are in place not because they are idealised approximations to the nature of stars. Barring uniform chemical composition (UCA), which clearly is

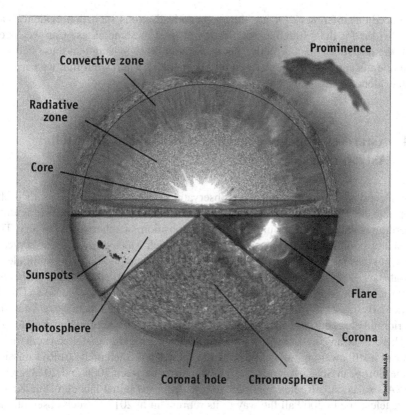

Fig. 7.2 The internal structure of a typical average main sequence star. (© ESA/NASA public domain)

an idealisation, the others would seem, at this point, to be working rather as fictional posits. They are employed because they are effective in allowing modellers to derive expedient predictions for the observable quantities in an HR diagram from mere estimates of radius, or overall mass.

Suárez (2013) argued that the inferences that these four equations licence towards the 'observable' HR quantities can be formally represented by means of *fictional conditionals*: Indicative conditionals that operate against the background of the four fictional assumptions (IA, EA, UCA, SSA), or have those assumptions appear in their antecedent.[8] If so, such fictional conditionals allow for expedient inference without necessarily requiring the truth, or even truth-aptness, of the assumptions – just figuring as presuppositions or as part of the antecedents of the relevant inferences. But are IA, EA and SSA really fictional assumptions required merely

---

[8] It was left there as an open question which of these two avenues (background or antecedent fictional assumptions) to take in the reconstruction of modellers' inferences. While this is an interesting issue for disputes regarding scientific realism, it is irrelevant to this paper's purposes.

for expedient inference? Or are they rather idealisations such that evidence can be provided for their suitable modification at least in some cases? Can we 'peek into' the interior of stars directly, that is, without presupposing that these fictional assumptions apply instrumentally in all our models? It turns out that we can, and such evidence is available.

7.4   Asteroseismology: The Observational Basis of Stellar Astrophysics Revisited

The second option is to find other observational means that allow us to probe deep into the interior of stars directly, without making any fictional assumptions regarding isolation, equilibrium, uniform composition, or spherical symmetry. The nowadays booming field of asteroseismology, which was merely nascent 20 years ago, provides such means to suitably modify the assumptions for a multitude of actual stars. The discovery and thorough analysis of observed seismic oscillations in stars has thus provided us with detailed knowledge of their interior. This is particularly the case for our local star, the Sun, and the claim that we know its interior rather well is nowadays supported by the new science of helioseismology.[9]

There is at any rate now a vast amount of data regarding stellar oscillations, some of it still awaiting full analysis. Most of it was collected in the CoRoT mission of the European Space Agency (ESA), which run between 2006 and 2013; and in NASA's successive Kepler missions, following on the initial launch of the Kepler space telescope in 2009 all the way to its retirement in 2018.[10] Such observational data has undoubtedly revolutionised our understanding of stellar structure models, imposing strong constraints on the idealising assumptions involved.[11] The study

---

[9] Helioseismology is the application of asteroseismological methods to the study of the oscillations of the Sun. It has been extremely successful mainly on account of the prevalence of pressure modes in the Sun, an average star in the main sequence, and the relative transparency of its radiation, which can moreover be resolved adequately for all nodes from Earth. Thus, our knowledge of the interior of the Sun is really quite astonishing, down to the detail of the radiative-convective border and the relative speed of the different layers. See Parker (2000), Christensen-Dalsgaard (2002) and Pijpers (2007) for excellent introductions. Nothing of the sort is available for distant stars, never mind the opaque interior of our own Earth which, obviously, does not radiate.

[10] This is ongoing observational work, anyway, and valuable astereoseismological data will be enhanced dramatically when ESA launches the new PLATO (PLAnetary Transits and Oscillations of Stars) space telescope in 2026. The PLATO mission is primarily devoted to the search for exoplanets, which require a careful study of the oscillations in brightness of relatively nearby stars. The same oscillations in brightness form the backbone of asteroseismological data, so the data collected will secondarily serve to advance research on seismic oscillations in a multitude of stars. See Aerts (2015); Rauer et al. (2016).

[11] The astounding asteoseismological data extracted from the Kepler missions is reported and discussed extensively in Chaplin et al. (2014); while the lessons from CoRoT, particularly as regards red giants, are succinctly discussed in De Ridder et al. (2009).

of asteroseismology thus provides us with insight into how astrophysical data is collected, analysed statistically, and represented in characteristically expanded HR diagrams. It turns out that the inferences from observed data to the data models – and those from these data models onto the parameter space in theoretical models – are predictably rich with modelling assumptions of their own. These assumptions in turn play critical roles in determining the quality of the data, and how precisely it weighs for and against the different idealisations employed in stellar structure models.

The basic phenomenon underlying all asteroseismology research is the regular pulsation in a star's observed brightness due to internal gravitational or acoustic oscillations caused by rotational or convection forces within it (Aerts 2015, 38–39). These investigations enable modellers to estimate both the opacity, or absorption coefficient through the star, and its hydrostatic equation state, while offering opportunities to measure diffusion through slow mixing; overshoot from convective cores of stars that have them (typically large stars, certainly larger in mass than the Sun); progressive mass loss, and near-surface convection in very old red giants (Christensen-Dalsgaard 1999, 1). Asteroseismologists study the oscillations in stars due to both pressure (in the convective layers of the star) and gravity (throughout the star). These are known as the g- and p- modes of oscillation; there are also mixed g-p modes that combine gravity and pressure waves and are most informative regarding the deep structure of the star (Aerts 2014, 155; Aerts et al. 2010, Ch. 7). For each of these modes, there are a range of oscillation nodes going from purely radial (i.e., arising at the core and expanding regularly outwards, as in a regularly pulsating sphere that expands and contracts repeatedly), to very non-radial (for instance when the star is literally deformed at two opposite ends of a quadrant at its surface, alternating north and south of the equator). In fact, stars show an extremely rich oscillatory pattern of nodes for both pressure and gravity modes, and to properly detect all this oscillatory behaviour requires very extensive longitudinal studies over many years. This is the reason why asteroseismology did not really take off until the launch of the CoRoT and Kepler missions. Only when such extensive longitudinal data for a single star is conveniently aggregated by Fourier transform methods, can we obtain an overall oscillatory profile for a star, such as that depicted in Fig. 7.3 for KIC 4726268, which aggregates the Kepler mission data for that star.

The existence of rotational and convection forces inside the star is contrary to at least two of the assumptions that run through most stellar structure modelling, namely the equilibrium (EA) and spherical symmetry (SSA) assumptions. If there is convection inside the star, that entails the energy transfer is not entirely radiative, but in some layers of the star at least energy gets transferred by means of convectional plasma movement (in essence: huge flows of parts of the stellar gas from some regions into other, presumably cooler, regions in the star). If this is so, the star cannot in fact be entirely in a state of thermal equilibrium, and the (EA) is not a convenient fiction but a false idealisation in at least some of the regimes within the star.

On the other hand, if there are rotational forces inside the star, it means that different regions of the star rotate at different speeds – and this ought to generate deformations of the layers of the star in different regions. It is nowadays known –

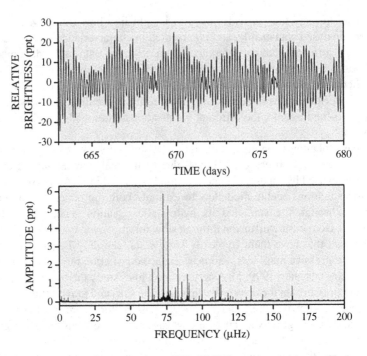

Fig. 7.3 The asteroseismical portrait of star KIC 4726268, as observed by the Kepler satellite (Reproduced from Aerts 2015, p. 39 with kind permission from Conny Aerts, as well as AIP publishing)

precisely out of helioseismology data regarding its pulsating oscillations – that the Sun experiences higher rotational speed of its radiating photosphere in the equator than the poles. The rotation period at the equator is about 25 days while that at the poles is about 35 days (Parker 2000, 27), which deforms the Sun into an oblong at the equatorial axis (i.e., the radius of the Sun is slightly longer to the equator than the pole). Since we now have good evidence in asteroseismology that stars experience similar differences in rotational speed due to divergent rotational forces, we know that the assumption of spherical symmetry (SSA) is not a convenient fiction but a false idealisation for most, if not all, stars.[12]

We have also by now gained – by similar asteroseismical methods – a lot of knowledge regarding the energy transfer mechanisms inside a multitude of stars and star types. The consensus nowadays is that young stars which have not yet burnt their hydrogen exhibit a certain pattern. The very small ones (with masses less than or equal to the solar mass) possess a purely radiative core in thermal

---

[12] The assumption has been known for a long time to be an idealisation in the case of binary stars – even Eddington's classic text (1926, 310–312) reports it, and Tayler (1994 [1970], Ch. 8) deals extensively with it. But these stars were supposed to be unique in experiencing very heavy reciprocal tidal forces.

equilibrium, but a large convective layer that experiences considerable rotational and tidal forces. This is represented accurately in Fig. 7.2 for our Sun, and already shows that the EA assumption is a very rough approximation to the outer layer of stars like ours. Those stars that are a little more massive (with masses between one and two solar masses), develop a convective inner core, and thus present a three-layered structure, with a small convective core, an extensive radiative layer in equilibrium, and a shallower outer convective layer. Finally, those stars that are very massive (above two solar masses) possess a small convective core and an often very large radiative outer layer only. Older stars that have burnt most of the fuel (i.e., red giants) possess some sort of convective envelope but its extent is not well known – they also often exhibit changing and irregular patches of convective and radiative energy transfer throughout (Aerts 2015, 37; Bedding 2011).[13] We see then that asteroseismology shows that the equilibrium, spherical symmetry, and uniform composition assumptions no longer operate as fictions, but that they turn out to be false idealisation in most cases. We even now have some good estimates, supported by evidence, of how they differ from the truth in many stars.

7.5 From Experimenter's Regress to Modeller's Nest

The vast amount of asteroseismology data that we now possess is deeply affecting our understanding of stellar interiors. The data both impose stringent constraints on some of the parameter values in stellar structure models, and they force us to modify some of their central assumptions. These data demonstrate that some of the assumptions that have characterised stellar astrophysics from its historical origin are demonstrably very far from the truth in many cases. At least for those stars for which we have recorded enough longitudinal data, over a long enough period, their oscillations are hardly compatible with the spherical symmetry and equilibrium assumptions. Therefore, asteroseismology rather dramatically expands our understanding of the evolutionary phases of stars too. This is true to the extent that the HR diagrams that we are most likely to see nowadays include shady areas representing the oscillations in luminosity and effective temperature for most stars (see Fig. 7.4).

Nevertheless, some legitimate worries concern circularity of inference: To what extent are we assuming the very models of stellar structure, including some of their formidable assumptions, in the study and statistical analysis of the asteroseismology data that we use to adjust the parameters and the assumptions in those models? A preliminary version of this worry is familiar to anybody who has been exposed

---

[13] Only the frequency oscillations of some white dwarfs can be modelled consistently with the hypothesis of a fully radiative and uniform medium, involving no rotation (Smart 2018) – and even here the lessons from asteroseismology are considerable, this time regarding the chemical composition of the star, which contains more oxygen and less carbon than previously expected.

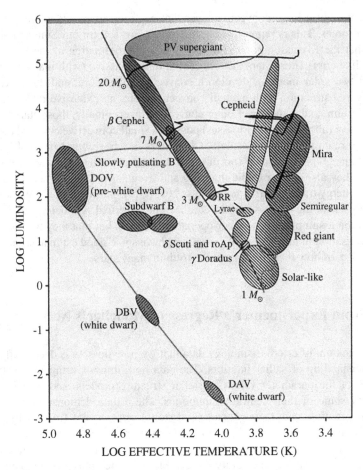

Fig. 7.4 A portion of the HR diagram accounting for characteristic oscillations revealed in asteroseismology (Reproduced from Aerts 2015, p. 37 with kind permission from Conny Aerts, as well as AIP publishing)

to the heated debates surrounding Harry Collins' (1992) "experimenters' regress". However, there is no threat of such a regress in the CoRoT and Kepler missions. The recording of the receiving star radiation by the space telescopes does not assume any specifics about the interiors of the objects that produce them. The oscillations in intensity recorded at the telescopes are insensitive to any of the features of the stars as modelled. It would make no difference, for instance whether we model the stars as having a convective or radiative core. It's rather the other way round – the recorded data set out constraints on how we can possibly model those stellar sources.

Furthermore, one advantage of a purely observational science like astrophysics is precisely that there is no material interaction with the object that is causing the recorded data, so there can be no experimental 'infection' of the source.

Similarly, though, for the recording instruments. Unlike, say, Collins' story with gravitational wave detection, the instruments themselves are not subject to the laws of the models under test. Rather, the assumptions and models that are under test in asteroseismology research into stellar interiors are themselves playing no role in the production or recording of the data. They are not cosmological assumptions, and they do not apply to any of the interstellar medium the radiation must travel through to reach the telescope nor to the space (space-time) that these instruments operate in.[14]

However, there are obvious modelling circularities at play. The data must be statistically analysed and modelled appropriately to generate informative diagrams such as those in Figs. 7.3 and 7.4. Most importantly, asteroseismology requires what is known as a *forward seismic model*: "a model that takes the physical properties of a star as input parameters and predicts the star's oscillations" (Aerts 2015, 38). That is, one must start with some description of the internal workings of the star and work one's way up to the photosphere luminosity and spectral class. This generates a sort of nested modelling practice, in which we first model the star, then deduce the oscillatory frequencies, then compare such oscillations to the actual observations, then correct the parameters and assumptions backwards as needed. While this is certainly worthy of further study, there is no prima facie offence to either logic or evidence in such a method, which is typical of a large range of our observational sciences.

Here the comparison with models of the interior of our planet in Earth science and geophysics is certainly instructive. Miyake (2015) describes how Preliminary Reference Earth Models (PREMS) are used to turn geological data into evidence for or against different values of parameters and assumptions in more sophisticated Earth models. There is some sense of circularity here: "Suppose I create an initial model, and then I study the deviations from this model. These deviations are then taken to be evidence for, say, casual factors that must be taken into account in the model. I then add these causal factors and improve the model" (Miyake 2015, 826). Similarly, in stellar astrophysics, we take the original HR diagram (Fig. 7.1), and the sort of preliminary stellar structure models that support it, as discussed in Sect. 7.3, and then study deviations from these models' predictions in asteroseismological data. Thus, it turns out, for example, that some stars' layers rotate at speeds different to and incompatible with some of the assumptions in the preliminary models. This leads us to reject such assumptions, to measurable or quantifiable degrees, and to correspondingly modify the HR diagram. The more complex version of the HR diagram (Fig. 7.4) is then fed back into more sophisticated models of stellar structure and interiors that answer to such asteroseismological data. While this 'bootstrapping' procedure may be surprising from a naïve hypothetico-deductive

---

[14] The CoRoT and Kepler telescopes must of course, be calibrated for the appropriate reception of starlight and radiation – but there is nothing in the procedures of calibration that cannot be accounted for as part of the *commissioning phase* of the experiment, as described by Boyd (2021).

point of view, it does not impugn the empirical status of the resulting scientific knowledge in any way.[15]

Nevertheless, as a methodological issue, the 'nested' nature of stellar structure models is remarkable. These models derive from earlier, simpler models, which they often subsume as limiting cases. And each is nowadays underpinned and severely constrained by asteroseismological data. The data itself has been modelled to render it intelligible; and to yield significant observational constraints on the parameters in the models. Thus, we find that stellar structure models are supported by further models, both in their historical development, and in their relation to the data that they account for. And, again, while there is no epistemic circularity involved here, the consequences for our understanding of the nature of those models and their relationship to their target phenomena are startling and deserve attention.[16] Let me here outline briefly three of these consequences that await further study.

First, Sibylle Anderl (2018) has defended the use of simple but highly idealised models in astrophysics, and this is a lesson that stellar structure models also bear out. In Sect. 7.2 I argued that the entire field of stellar astrophysics modelling emerged from some very straightforward interpretations of correlation data in HR diagrams as dynamical laws, together with highly simplifying fictional assumptions regarding stellar structure and evolution (the IA, UCA, EA, and SSA assumptions). Together these jointly yield a template for powerful models of stellar interiors accounting for such data. It is hard to imagine how stellar astrophysics could possibly have developed historically as a discipline without such straightforward, relatively simple, yet inferentially robust early models. Anderl (2018, 828) connects the intuitions lying behind such models with the pursuit of understanding – and this seems fitting here too. While the early models may have been convenient fictions, they provided richly layered understandings of the physics of stellar interiors. Astroseismology does not fundamentally alter that picture – we still very much understand the physical processes operating in stellar interiors in the terms laid out by the four equations described in Sect. 7.3 in this chapter. We are just adding significant detail deriving from the specific asteroseismological profile for each star.

Second, Daniela Bailer-Jones – a pioneer in the philosophy of astrophysics over 20 years ago – used to emphasise[17] that most models in astrophysics are composed of more specific sub-models aiming to capture different parts of the overall causal mechanism putatively responsible for their complex target phenomena. While Bailer-Jones' favourite cases of this sort of composition were cosmological models of galaxy formation, similar lessons apply to models of stellar structure too. For instance, asteroseismological models must consider both acoustic and gravity

---

[15] Roughly, where hypothetico-deductivism takes confirmation to flow with the entailment of data by theory, bootstrap accounts of confirmation take it rather to flow with the entailment of new models by data and background knowledge (Glymour 1980).

[16] There may, however, be an implied rejection of foundational epistemologies for scientific knowledge, in favour of a Neurathian 'sailing boat' picture of science in the making. But the Neurathian picture of scientific knowledge is nowadays prevalent and uncontroversial.

[17] In her (Bailer-Jones 2000) and in many conversations with the author over those years.

pressure modes in a typical star, as well as the range of mixed modes they generate. Each mode is responsive to a different mechanism, and the 'art of stellar modelling' requires a judicious choice of nodes in each mode for each star. In other words, it calls for judicious combinations of the underlying mechanisms.

Finally, and most tentatively, there is an issue concerning nested modalities within nested models in astrophysics, which Elena Castellani and Giulia Schettino explore in recent work (Castellani and Schettino, 2023). Inasmuch as a model describes a possible mechanism partly or fully responsible for a phenomenon, it lays down a possibility space. We may then wonder what sort of possibility is explored in nested models, where the possibility operator in the overall or ulterior model arguably ranges over the more primary possibility spaces of the simpler or antecedent models. One conjecture is that such possibility spaces obey a simple multiplication rule, and that degrees of possibility thus behave as classical probabilities. But how and on what grounds do we impose a data-sensitive probability measure on, say, the arrays of distinct models of convective flows in the convective envelope inside a star? Undoubtedly, these are issues that deserve further study.

7.6 Conclusions

In this chapter, I have aimed to (i) provide a historical introduction to the exciting new field of asteroseismology, and how the observational data coming from asteroseismological research significantly constrains stellar structure modelling; (ii) update my 15-years old account of fictions in structural modelling in the light of such new data, and (iii) outline some of the new avenues of philosophical research that open up in the wake of the 'nested' nature of the models involved. While stellar structure modelling throws out many interesting puzzles and distinctions, I have argued that there is no 'vicious circularity' arising from the nested nature of the models involved that could impugn the empirical knowledge that asteroseismology affords us. We do indeed know the interior of our Sun and the stars to a greater extent than we know many other systems in the observational or even experimental sciences.

Acknowledgements I thank the audiences at the *Symposium on The Philosophy of Astrophysical Practice*, at the Society for the Philosophy of Science in Practice annual conference in Ghent, in July 2018, as well as at the workshop on *The Concept of Evidence in Astrophysics* that I organised in Madrid in June 2019 for comments and suggestions on a distant ancestor of this paper. Bruno Merin, the head of the Science Data Centre at ESAC, directed me towards the most recent literature on asteroseismology and deserves special thanks. I also thank the editors of this volume, particularly Nora Boyd, and two anonymous referees for helpful and constructive comments. I declare financial support from the Spanish Ministry of Science and Innovation projects PGC2018-099423-B-100 and PID2021-126416NB-I00 and no further competing interests.

Figures 7.1 and 7.2 are in ESA/NASA public domains. Figures 7.3 and 7.4 are reproduced from Aerts, C. (2015), Asteroseismology, Physics Today 68, 5, 36 (2015) DOI: 10.1063/PT.3.2783 with kind permission from Conny Aerts, as well as AIP publishing.

References

Aerts, Conny. 2014. Massive Star Asteroseismology in Action. In *Proceedings of the International Astronomical Union Symposium*, ed. Georges Meynet, et al., Vol. 9, No. S307, 154–163.

———. 2015. Asteroseismology. *Physics Today* 68 (5): 36–42. https://doi.org/10.1063/PT.3.2783.

Aerts, Conny, Jorgen Christensen-Dalsgaard, and Donald W. Kurz. 2010. *Asteroseisemology*. Dordrecht: Springer.

Anderl, Sibylle. 2018. Simplicity and Simplification in Astrophysics Modelling. *Philosophy of Science* 85: 819–831.

Bailer-Jones, Daniela. 2000. Modelling Extended Extragalactic Radio Sources. *Studies in History and Philosophy of Modern Physics* 31 (1): 49–74.

Bedding, Tim, et al. 2011. Gravity Modes as a Way to Distinguish Between Hydrogen – and Helium – Burning Red Giant Stars. *Nature* 471 (608): 608–611.

Belot, Gordon. 2015. Down to Earth Underdetermination. *Philosophy and Phenomenological Research* XCI (2): 456–464.

Boyd, Nora. 2021. *Epistemology of Experimental Physics*. Cambridge Elements: Cambridge University Press.

Castellani, Elena, and Giulia Schettino. 2023. Nested Modalities and Astrophysical Modelling. *European Journal for Philosophy of Science* 13 (11). https//doi.org/10-1007/s13194-023-00511-2

Chaplin, William, et al. 2014. Asteroseismic Fundamental Properties of Solar-Type Stars Observed by the NASA *Kepler* Mission. *The Astrophysical Journal Supplement Series* 210: 1–22.

Christensen-Dalsgaard, Jorgen. 1999. Asteroseismology. *Astrophysics and Space Science* 261: 1–12.

———. 2002. Helioseismology. *Review of Modern Physics* 74: 1073–1129.

Collins, Harry. 1992. *Changing Order: Replication and Induction in Scientific Practice*. Cambridge: Cambridge University Press.

De Ridder, Joris, et al. 2009. Non-radial Oscillation Modes with Long Lifetimes in Giant Stars. *Nature Letters* 459: 398–400.

Eddington, Arthur S. 1920. The Internal Constitution of the Stars. *The Scientific Monthly* 11 (4): 297–303.

———. 1926. *The Internal Constitution of the Stars*. Cambridge: Cambridge University Press.

Gingerich, Owen. 2013. The Critical Importance of Russell's Diagram. In *Origins of the Expanding Universe: 1912–1932*, ASP Conference Series, Vol. 471, ed. Michael J. Way and Deidre Hunter. Astronomical Society of the Pacific.

Glymour, Clark. 1980. *Theory and Evidence*. New Haven: Princeton University Press.

Hacking, Ian. 1983. *Representing and Intervening*. Cambridge: Cambridge University Press.

———. 1989. Extragalactic Reality: The Case of Gravitational Lensing. *Philosophy of Science* 56 (4): 555–581.

Jacquart, Melissa. 2020. Observations, Simulations, and Reasoning in Astrophysics. *Philosophy of Science* 87: 1209–1220.

Maury, Antonia. 1897. Spectra of Bright Stars: Photographed with the 11-Inch Draper Telescope. *Annals of the Astronomical Observatory of Harvard College* XXVIII: Part I.

Miyake, Teru. 2015. Reference Models: Using Models to Turn Data into Evidence. *Philosophy of Science* 82: 822–832.

Parker, Eugene. 2000. The Physics of the Sun and the Gateway to the Stars. *Physics Today* 53 (6): 26–31.

Pijpers, Frank P. 2007. *Methods in Helio- and Asteroseismology*. Singapore: World Scientific Publishing Company.

Prialnik, Dina. 2000. *An Introduction to the Theory of Stellar Structure and Evolution*. Cambridge: Cambridge University Press.

Rauer, Heike, et al. 2016. The Plato Mission. *Astronomische Nachrichsten (AN)* 337 (8/9): 961–963. https://doi.org/10.1002/asna.201612408.

Smart, Ashley G. 2018. Asteroseismology Reveals the Structure of a White Dwarf. *Physics Today* 71 (30): 16–18.

Suárez, Mauricio. 2009. Scientific Fictions as Rules of Inference. In *Fictions in Science: Philosophical Essays on Modelling and Idealisation*, ed. Mauricio Suárez, 158–198. Routledge.

———. 2013. Fictions, Conditionals, and Stellar Astrophysics. *International Studies in the Philosophy of Science* 27 (3): 235–252.

———. 2024. *Inference and Representation: A Study in Modeling Science*, Chicago: University of Chicago Press.

Tayler, Roger J. 1994 [1970]. *The Stars: Their Structure and Evolution*. Cambridge: Cambridge University Press, 2nd ed., 1994.

Vaihinger, Hans. 1924. *The Philosophy of As-If: A System of the Theoretical, Practical and Religious Fictions of Mankind*. Trans. Charles Ogden (Reprinted 2021). London: Routledge.

Chapter 8
Idealizations in Astrophysical Computer Simulations

Melissa Jacquart and Regy-Null R. Arcadia

Abstract This chapter examines some of the philosophical literature on idealizations in science and the epistemic challenges idealizations potentially pose for astrophysical methodology, particularly its use of computer simulations. We begin by surveying philosophical literature on idealization connected to (1) kinds of idealizations deployed in science, (2) the aims of idealization in science, and (3) various strategies for de-idealization. Using collisional ring galaxy simulations as a case study, we examine how these three themes play out in the context of astrophysical computer simulations. Ultimately, we argue that deploying de-idealization strategies is central to bolstering epistemic confidence in simulations in astrophysics. We conclude with some remarks on the role of idealization in the context of astrophysical computer simulations more generally.

8.1 Introduction

Scientific models and computer simulations are indispensable to scientific practice.[1] Through their use, scientists can effectively learn about how the world works, and to discover new information. However, there is a challenge in understanding how scientists can generate knowledge from their use, stemming from the fact that models and computer simulations are necessarily incomplete representations and partial descriptions of their target systems (the real-world systems they aim

---

[1] For the purposes of this chapter and philosophical issues that are examined we will consider computer simulations as a specification of a kind of model, that is, computational model. In this sense, a computer simulation is the process of running certain model(s) (typically mathematical in nature) on a computer program over some amount of time in order to study and/or visualize the behavior or performance of some system. As such, we use the terms models and computer simulations interchangeably in this paper.

M. Jacquart (✉) · R.-N. R. Arcadia
Department of Philosophy, University of Cincinnati, Cincinnati, OH, USA
e-mail: melissa.jacquart@uc.edu

© The Author(s) 2023
N. Mills Boyd et al. (eds.), *Philosophy of Astrophysics*, Synthese Library 472,
https://doi.org/10.1007/978-3-031-26618-8_8

to represent). In order to construct a model or simulation, scientists must make idealizations, approximations, and abstractions. But what is the nature of these kinds of idealizations? How are these idealizations justified by scientists? Why are scientists epistemically justified in drawing conclusions about the nature of the real world from models and simulations when they contain idealizations, and are incomplete (and in some cases false) representations of real-world target systems?

This chapter examines the role of idealization in the context of astrophysical computer simulations. In the context of astrophysics, the use of models and computer simulations to study systems is pervasive. They are used to obtain a better understanding of small-scale astronomical objects (such as the evolution of stars or individual black holes), to explore astronomical interactions (such as the interactions of galaxy or galaxy cluster collisions), as well as to model and attempt to better understand the large-scale structure of the entire universe. Due to the complexity of these systems, and other epistemic challenges connected to astrophysics more generally, astrophysics provides an excellent opportunity to study the precise ways that idealization and representational trade-offs enter into the construction of simulations, and how they may determine values for simulation parameters.

Our goals in this chapter are three-fold. First, we aim to provide a survey of some of the existing philosophical literature connected to idealization. This, in part, will provide those who are interested in exploring the role of idealizations in the context of astrophysics a sense for what literature and philosophical problems might be relevant to their work. This also will allow us to, secondly, conduct philosophical analysis on a case study from astrophysics in which computer simulations play a central epistemic role, and examine the role of idealizations in this context. Ultimately, we use this work to argue in favor of the importance of using a variety of de-idealization strategies in addressing epistemic challenges connected to the use of computer simulations in the context of astrophysics.

8.2   Epistemic Challenges in Astrophysical Methodology

It is important to briefly discuss some of the background epistemic challenges astrophysics faces more generally before examining the role of idealizations in astrophysical computer simulations more specifically. Doing so will help highlight why philosophical analysis of idealizations specifically can aid in developing a better understanding of how idealizations aid or hinder knowledge development in the field of astrophysics, especially in the presence of computer simulations. First, one of the key limits to astrophysical methodology is its capacity to conduct direct experimentation on its object of study (Jacquart 2020; Weisberg et al. 2018). When comparing experimental access in astrophysics to the kind of access other sciences (such as biology or chemistry) have to their objects of study, these other sciences more frequently have the capacity to experiment on their object of study. Astrophysics, on the other hand, is generally not capable of experimenting on its objects of study (such as stars, galaxies, etc.) in such a direct or material matter (Jacquart 2020). Second, astrophysics also has a spatial-temporal limited vantage point; a significant amount of the phenomena of interest in astrophysics take place over a vast timespan and are only observable from one vantage point (such as a telescope in space near Earth). While some cosmic events like the death of stars or black hole mergers happen over shorter timespans, observations of these too are frequently confined to a series of snapshots of cosmic phenomena. This limited spatial-temporal vantage point leads to a sparseness-of-data issue (Jacquart 2020).

In light of these challenges, one of the central strategies used in astrophysics is deploying computer simulations in order to better understand the systems of inquiry. Computer simulations allow scientists to explore how various systems might evolve over time (in a way akin to long-time scale observations), or allow for manipulation of a system (in a way akin to experimentation). In the case where there is little (to no) direct access to a system itself (i.e., direct access to the object of study), incorporation of information or data one does have direct access to is critical. In the context of astrophysics, most simulations are developed based on the observational data astrophysicists do have access to, as well as various background theory. In the research areas in astrophysics where computer simulations are frequently used, astrophysical methodology faces epistemic challenges connected to computer simulation construction and evaluation. This includes broader issues related to verification and validation, the relationship between simulation and theory, and capacity for simulations to offer explanations (see, for example Kadowaki forthcoming; Winsberg 2010). It also includes issues connected to developing a scientific representation as a computer simulation, as well as the role of idealizations and approximations.[2] This latter set of challenges is where this paper will focus.

In the context of astrophysics, scientists are often trying to model systems ranging from individual stars, single galaxies, galaxy interactions, all the way up to the structure of the entire universe. Obviously, these systems rarely can be simulated in their entirety, for reasons connected to their sheer complexity as well as computational tractability. As such, idealizations (and approximations) are made about these systems in order to develop computer simulations representing these systems. Idealizations are intentional distortions or mis-representations of the target systems, often representing the system in some way in which it is not. Idealizations are "assumptions made without regard for whether they are true, generally with the full knowledge that they are false" (Potochnik 2017, 2). A model or computer simulation, then, is an idealized representation with respect to its target "when it fails to represent some important aspects of the target" (Weisberg 2013, 98). This raises

---

[2] Some discussions, such as Shech (2018), draw an important distinction between "broadly construed" and "narrowly construed" idealizations. Along these lines Norton (2012), for example, offers a narrow construal between an "approximation" and "idealization", and discussed implications for careful separation of the two concepts. For the purposes of this paper, we will take a broadly construed conception of idealization, in which it refers to anything that can reasonably and intuitively be called an idealization because it fails to meet some veridicality or accuracy condition (Shech 2018; Jacquart et al. forthcoming).

questions related to how simulations, in light of their deployment of idealizations, can obtain meaningful epistemic status to offer predictions or explanations about the real-world systems they proport to represent.

Given this web of epistemic challenges, the role of idealization in astrophysical simulations is in need of attention. There is a need to not only consider what kinds of idealizations occur in astrophysical simulations and the role they play in representing their real-world target systems, but also what idealizations are warranted, as well as how they are handled and mediated. In order to examine these concerns in detail, in the next section we provide a basic case study: collisional ring galaxy simulations. After providing this context, Sect. 8.4 will introduce some key ideas and themes connected to idealization, and their instantiation in this case study. We then use this discussion as a backdrop for examining the role of idealizations in astrophysical computer simulations and connection to epistemic claims.

8.3 Case Study: Collisional Ring Galaxies and Their Computer Simulations

Collisional ring galaxies are formed when a smaller galaxy passes, or collides, with the center of larger disk galaxy at relatively high speeds. Through this gravitational disruption, the smaller galaxy essentially collapses, with its gas and dust generating star formation (young blue stars) at the outer edge of the larger galaxy. This interaction then also affects the orbit of the larger galaxy, producing the ring-like structure (Appleton and Struck-Marcell 1996). The central means by which astrophysicists investigated this system and learned about its galactic formation was through the use of computer simulations.[3]

For these early simulations, the goal was simply to provide a general *how possibly account* for how these galaxies got their ring shape. With gravitational interaction is a primary driver in galaxy collisions, simulators decided that the masses of the two galaxies would be the critical features of the target systems, as well as the impact velocity and angle of the collision. The masses of the two galaxies, as well as the angle of collision, were varied as a means of exploring how the two galaxies might interact and to determine what conditions are necessary for these ring galaxies to obtain their ring shape. Simulations of these interactions also simplified the system to point particles, with the masses, or number of point particles, of the two galaxies varied to explore galaxy mass ratios (for instance, one galaxy having 600 particles with the other 150 particles) that would result in the ring galaxy phenomenon. Through this process they determined the ring shape occurs only in cases where a smaller compact companion galaxy and a larger disk system

---

[3] For additional philosophical discussion on collisional ring galaxies see Weisberg et al. (2018) and Jacquart (2020).

undergo a near head-on collision, with more pronounced rings occurring at higher impact speeds (Lynds and Toomre 1976; Appleton and Struck-Marcell 1996).

As computational capacities progressed, collisional ring galaxy simulations have been able to increase in complexity as well. Some contemporary collisional ring galaxy simulations for instance utilize GADGET—a code for cosmological N-body/Smoothed-particle hydrodynamics simulations, as well as GIZMO (building on GADGET) as a massively-parallel, multi-physics simulation code. Both of these allow for simulators to move beyond simple point particle simulations and include more refined physics and features such as hydrodynamics, magnetic fields, fluid dynamics, cosmological integrations, to name a few.[4] Research groups focused on galaxy simulations have taken these codes and expanded on them for their own purposes as well. For example, the FIRE (Feedback In Realistic Environments) project builds on GIZMO, and aims to improve the predictive power of individual galaxy formation simulations through including interstellar medium and star formation processes as critical drivers of single galaxy evolution. In the case of ring galaxy simulations, GIZMO+FIRE has been deployed as a means to explore the role star formation might play in the evolution of the galaxy collisions (Jacquart 2020). In future work, simulators working on collisional ring galaxies consider it necessary to model individual interacting galaxies such that it includes, at some level of approximation, stellar and gas dynamics of the multi-component galaxies with self-gravity, pressure and heating/cooling effects and will eventually require that the simulations include non-isothermal gas disks in both primary and companion galaxies (Appleton and Struck-Marcell 1996). While past simulations justified omitting these attributes and features due to computational tractability, when considering smaller-scale simulations of individual galaxies these attributes and features could have a significant impact on galaxy structure and evolution. As such they are now flagged by the community as relevant features that may turn out to be causally important.

Though we discuss collisional ring galaxy simulations specifically, we believe this case study has notable features shared across different kinds of simulations that occur in astrophysics. First, this case showcases a progression of simulation computational capacities. The first simulations were developed in the 1970s, when astrophysical computer simulations were primarily simple, small number point particle-based simulations governed almost exclusively by gravity. As computational power advanced, so too did the simulations to more complex N-body and hydrodynamical simulations. These later simulations also offer more refined gravity treatments of increasing complexity (particle-mesh, to tree particle-mesh, to fast multipole), and similarly with their hydrodynamics treatments (moving from adaptive-mesh refinement to smoothed particle hydrodynamics).[5] This progression

---

[4] See Springel et al. (2001), Hopkins et al. (2014) and Hopkins (2015) for additional details on simulation codes.

[5] See Vogelsberger et al. (2020) for further discussion of cosmological simulations of galaxy formation over time.

is seen not only in galaxy simulations on small scales (i.e. individual galaxies) but also in the large scale simulations (such as those used in large scale structure formation simulations like e.g. Millennium-II).

The collisional ring galaxy simulations also showcases variability in target system representation, that is, what features of the real-world target system the simulator chooses to include in the developed simulation. When modelling any galaxy formation there are several astrophysical processes that could be included: gas cooling, interstellar medium, star formation, stellar feedback, supermassive black holes, active galactic nuclei, magnetic fields, radiation fields, cosmic rays, etc.[6] These kinds of features (as will be discussed in the following section) are all also potential contributors to ring galaxy structure evolution and development. Representing all of these in one simulation is (at present) not possible, and so various idealizations (and approximations) are introduced. All of these present challenges for modeling ordinary baryonic matter. Additional challenges are also posed to modelling of dark matter in galaxy simulations due to the lack of knowledge regarding dark matter's precise nature (for example, if dark matter is weakly interacting massive particles (WIMPs), self-interacting (SIDM), or something else entirely).

8.4 Idealizations, De-idealizations, and Representation in Astrophysical Computer Simulations

We now turn to examine the role of idealization more closely in our case study. In Sect. 8.4.1 we provide an overview of kinds of idealizations that occur in developing scientific representations, and examples of what each kind of idealization looks like in the context of astrophysical computer simulations. Such taxonomies can be extremely useful for thinking through the use of idealizations in science, as specifying the kind of idealizations present not only can help reveal nuances to scientists' conceptualizations of their representational system, but they can also offer insight for the epistemic challenges and justifications for introducing them. In Sect. 8.4.2 we examine the aims of idealizations in scientific practice and introduce a framework for conceptualizing the aims of idealizations in the context of astrophysical simulations specifically. In Sect. 8.4.3 we connect this with strategies of de-idealization so that in Sect. 8.4.4 we can discuss connections between ideal-izations, de-idealization, and a common aim in models and simulations: developing more accurate representations of target systems in order to increase confidence in epistemic claims. Ultimately, we highlight how deploying de-idealization strategies is central to bolstering epistemic confidence in simulations.

---

[6] Again, see Vogelsberger et al. (2020) for extended discussion how these astrophysical processes contribute to galaxy formation and simulations.

8.4.1 Kinds of Idealizations in Astrophysical Computer Simulations

The importance of examining idealizations and their role in developing scientific representations has an extensive history within the philosophy of science and scientific modeling literature (see for example Nowak (1972), Cartwright (1983), McMullin (1985), Wimsatt (1987), and Giere (1988)). More recent analysis of this literature (such as Weisberg 2007, 2013; Elliott-Graves and Weisberg 2014; also discussed in Shech Forthcoming) suggests that there are three kinds of idealizations common in scientific modeling and simulations—Galilean idealization, minimalist idealization, and multiple-models idealization. Studying idealization requires examination of what activity is characteristic of that form of idealization (that is, what the representational goals are) and how that activity is justified (Weisberg 2013, 98).

Galilean idealization is the simplified representation of a target system for the sake of mathematical or computational tractability, and as such is justified pragmatically. Characterized most fully by McMullin (1985), the practice includes selecting a target system of interest, and then introducing distortions and simplifications (idealizations) that allow the scientist to simplify the system, and represent it in such a way to make progress on their problem of inquiry. These idealizations are meant to be temporary with the expectation of future de-idealization.

Considering our case study, we see nice examples of this project deploying Galilean idealization in that it's introducing distortions with the goal of simplifying to make the models and simulations computationally tractable. Very common to early astrophysical computer simulations (and even those developed today) is the need to simulate highly complex systems, such as a galaxy (and even large-scale structure of the universe). In these contexts, with past and current computational capacities, it is impossible to simulate the trajectory or interactions of every star, planet, gas. Instead, simplified point particle-based simulations are developed, letting a large number of particles stand in for the system as a whole. For instance, the 1976 simulations were pared down to a few hundred particles so that the simulations could run. Even the more contemporary simulations such as those utilizing GIZMO+FIRE have a limit in terms of how many particles can be included due to computational capacities. Galilean idealizations such as these (especially in domains of science that rely on simulations) are not only present, but prevalent. Over time, advances in computational power have allowed scientists to de-idealize, removing distortions and adding back in previously omitted details. As McMullin points out, the capacity and interest in doing so in fact "then serves as the basis for a continuing research program" (1985, 261). We will return to the topic of de-idealization in Sect. 8.4.3.

Let us turn next to another kind of idealization: minimalist idealizations aim to understand the core causal relations that give rise to a phenomenon (Weisberg 2013; Elliott-Graves and Weisberg 2014). Rather than trying to include all the details and complexities of a target system, minimal models include only those factors that are understood to be the core causal factors, or "difference makers" to the phenomenon investigated. This strategy introduces idealizations to eliminate all but the most significant causal influences which give rise to a phenomenon. With minimalist idealizations, justification is related to scientific explanation, and aiming to isolate the explanatorily causal factors either directly (Cartwright 1989 and Strevens 2011), asymptotically (Batterman 2002), or via counterfactual reasoning (Hartmann 1998) (see Weisberg 2013, 103 for extended discussion).

In connection to our case study, we also see minimal idealizations deployed, with the 1976 ring galaxy simulations demonstrating this the clearest. These first simulators were interested in understanding core causal relations that would allow a galaxy collision to produce the ring structure—they were interested in providing explanation for how the rings may have gotten their particular shape. In this context, the simulators included only the factors that make a difference to the occurrence and character of the phenomenon in question: mass ratios and angle of collision. In later simulations, such as those deploying GIZMO+FIRE, we also see simulation development through idealizations aimed at exploring if there are any other additional causal influences which could give rise to a phenomenon—that is in what way features like gas or stellar feedback might provide explanations for other structures or features in the rings.

We consider it worth noting at this point that simulations may not deploy one singular kind of idealization. There is a sense in which a simulation might deploy both a Galilean idealization in that it is simplifying and distorting a system to make it more tractable, while also aiming to isolate causal factors (and thus also motivated by aims akin to minimalist idealization practices). But what does seem clear is that there is a clear connection between the kinds of idealizations we deploy, and their purposes or aims for which the idealization is introduced. Idealizations are thus closely tied to, and require reflection on, the wide range of purposes or aims a model or simulation may be intended to serve.

Finally let us turn to a third kind of idealization, multiple-models idealization (MMI). MMI deploys several related but incompatible models together to shed light on a phenomenon. Each model "makes distinct claims about the nature and causal structure giving rise to a phenomenon", but with no expectation that a single best model will be generated, nor that de-idealization will occur (Weisberg 2013, 106). Central to the justification of MMI is necessary tradeoffs between varying representational goals and desiderata such as accuracy, precision, generality and/or simplicity. Multiple models are needed because no single model can achieve all representational goals while at the same time providing the highest achievement of all possible desiderata. Within the philosophical literature, there has been some discussion regarding how to interpret Weisberg's understanding of MMI (see for example Potochnik 2017 but also Rohwer and Rice 2013), either narrowly, in which multiple models might be employed within a single research program (akin to robustness analysis), or more broadly, in which multiple models are employed across the scientific enterprise as a whole and often focus on different aspects of phenomena, i.e. causal patterns (Potochnik 2017, 45–6).

In the context of astrophysical computer simulations, one might be tempted to think of a simulation's ability to run with various different parameter settings as an instance of MMI. As mentioned in connection to the case study, in the process of exploring possibility space in order to determine the conditions in which the ring phenomenon occurs, various parameters in the simulations are changed. One could consider each of these parameter specifications to be its own model, and thus the collection of these an instance of MMI. However, under both a narrow and broad reading of MMI, we do not consider this to be the sense in which "multiple models" is intended to apply as the overall idealizations that are made are unchanged. That is, there are no new idealizations or tradeoff of representational goals.

One might also consider MMI to occur when comparing the 1976 simulations to the more contemporary GIZMO+FIRE-based simulations.[7] In these instances, several simulations are employed together to shed light on a phenomenon, in this case, ring galaxies. This includes point-particle simulations to the more-complex-but-still-idealized simulations that include feedback and fluid dynamics. The simulations are testing the hypotheses of the rings obtaining their shape through these collisions, and if the cause is competent to produce it. Some simulations have more complexity, some have less. It is through different idealizing assumptions about the basic physical processes involved in ring galaxy formation that we determine under what conditions ring galaxies form as well as some of the more subtle features. There is a sense in which, when taken together, the simulations are not offering distinct claims about the nature and causal structure giving rise to a phenomenon. However, under both a narrow and broad reading of MMI the use of the multiple models bolster confidence in a more unified claim about the phenomena and its structures.

In astrophysical computer simulations, instances of MMI practices may be more likely to occur when considering issues of scale. The idealizations that are made in the case of simulating a single galaxy will almost certainly be in tension with idealizations made for large scale structure. Simulating single galaxies can help us understand what is occurring at the smaller scale, but it will be necessary to make different idealizations when examining how the interactions of single galaxies impact the larger scale structures.

8.4.2 Idealizations and the Aims of Astrophysical Computer Simulations

We have discussed three kinds of idealizations that can occur in developing scientific representations like computer simulations, the connected scientific goals

---

[7] The 1976 simulations and the GIZMO+FIRE-based simulations are related by way of their target system but have developed very different codes. In this case, the code is what houses claims about the nature and causal structure giving rise to a phenomenon. As such, we take them to be "incompatible" in the sense most relevant to MMI (i.e., they're "incompatible" in virtue of their codes).

and justification for introducing those idealizations, and provided some examples of instances of these kinds of idealizations in the context of astrophysical computer simulations. We turn next to discuss the aims of idealizations in scientific practice more broadly. Our intentions here are to, first, introduce a framework that may be of use for conceptualizing the aims of idealizations in the context of astrophysical simulations generally and, secondly, discuss how this applies in the context of our Sect. 8.2 case study specifically. For this discussion, we draw largely on Angela Potochnik's book, *Idealizations and the Aims of Science* (2017), in which she explicitly examines the role of idealizations in scientific endeavors.

According to Potochnik, science is a human enterprise best characterized as the search for causal patterns in nature's complexity. By causal patterns, she means dependencies between factors, revealed under manipulation, and which causal pattern emerges depends on our representational choices. The complexity of nature is what, in part, motivates science to make abstractions and idealizations. She describes abstractions as omissions "without consequence for the representation" (2017, 55). Idealizations on the other hand are *not* characterized as omissions or negative representational features, rather idealizations play a positive representational role. She defines idealizations as, "assumptions made without regard for whether they are true and often with full knowledge they are false" (ibid., 2, 42). For Potochnik, idealizations play an active role in scientific representations (such as models and computer simulations) of the world. By virtue of science being a human enterprise, causal patterns are identified in scientific representations as opposed to taken directly from the highly complex world. Scientists must then make choices in their representations of the world. These choices may be driven by the research projects, tractability, or simply by virtue of the scientists' know-how. In whatever way the representational choices are made, they have a direct impact on what causal patterns are derived from the representation. This point, taken in tandem with Potochnik's commitment to idealizations as assumptions, makes it salient that idealizations will play some active role in whatever causal pattern is derived in any given representation. Idealizations are actively selected for in a similar fashion that other representational choices are made. Much of this discussion mimics similar points we have detailed already in this paper, but it is worth noting the emphasis Potochnik places on connecting the deployment of idealizations to positively contribute to the identification of causal patterns.

Yet despite the vitalness of idealizations to science, Potochnik considers idealizations to be "rampant and unchecked" (ibid., 57). By *rampant* she means to draw attention to their pervasive nature within science—scientists employ idealizations all the time. By *unchecked* she means there is (1) little focus on eliminating idealizations (namely, conducting de-idealizations), or even (2) on controlling their influence. Potochnik is careful to note that unchecked does not necessarily mean unprincipled. Rather, it is that idealizations reflect the scientists' interests. And since idealizations play a positive representational role, the nature of the role must be appropriate for the focal causal pattern, causal details of phenomena, and aims and methods of the research (ibid., 60). What is less clear is the extent to which these features are reflected upon in practice. What we wish to do in this subsection is reflect on Potochnik's two components to "unchecked" idealizations in the context of astrophysical computer simulations.

With respect to (2), some philosophers (e.g., Batterman 2002; Strevens 2011; Weisberg 2007, 2013) see justification for these idealizations occurring only for insignificant features of a system, non-difference-makers, or details that, if wrong, are safely ignored; especially in instances when an idealization is permanent. Potochnik, on the other hand, "[permits idealizations] even of central causal influences, on a permanent basis, and without taking any steps to hold in check the resulting misrepresentation" (2017, 59). For Potochnik however, even mis-representation (representation as-if) positively contributes to the representation of actual systems. Her strong view of idealization allows for "the permanent use of idealizations in many roles, including a central role in representing actual phenomena, even when they stand in for significant causes and without measures taken to control their influence" (ibid.).

The initial idealizations in the 1976 simulations identified the causal patterns, and over time, these causal patterns were better and better understood by a process of developing more and more detailed simulations of the target. In considering the target system, the structure of even a single galaxy is highly complex. It consists of stars, stellar remnants, interstellar gas, dust, and dark matter. But even in this very simple simulation (i.e., from 1976) where we have idealized it to just mass and point particles, astronomers had identified the causal pattern of ring galaxies. Even with radical idealizations, astronomers had captured the relevant causal dependencies. Thus far, we think the role of idealizations in this context is very similar to the analysis Potochnik provides.

With respect to (1), for those who consider science aimed for truth, idealized representations must be de-idealized to achieve this aim. Potochnik (ibid., 92) points to Odenbaugh and Alexandrova (2011), who argue that without the removal of all idealizations (complete de-idealization) we have "no ground, beyond that of our background knowledge that informed the model, for claiming that the model specifies a causal relation" (765). Others like Wimsatt (2007) argue that idealized "false" models can be used to produce "truer" theories without recourse to de-idealization. Nevertheless, Potochnik points out that "when an idealization is present merely for temporary reasons, there may be a scientific benefit to de-idealization when those reasons no longer obtain. But this is uncommon" (2017, 60).

Two interesting lines of inquiry lie here. The first relates to whether one ought to consider the epistemic aim of science to be truth (Potochnik ultimately argues science isn't after truth, but rather understanding as its epistemic aim). For those who may consider science aimed at truth, idealizations (and their deliberate falsehoods) are likely to be seen as problematic, and as such they may place higher value on de-idealization. We are not going to consider this larger issue related to the scientific pursuit of truth in this chapter. What we wish to explore is a second line of inquiry connected to the role de-idealization might play more generally in the development of astrophysical computer simulations. While de-idealization is often brought up as a path to "truer" representations, we wish to explore what other possible roles de-idealization might play in scientific practices. To do so, we now introduce the reader to some further discussion of de-idealization.

8.4.3   De-idealizations & Astrophysical Computer Simulations

Tarja Knuuttila and Mary Morgan (2019) point out that the implicit view in the idealizations literature is that idealizations are, or potentially are, some kind of reversible process. That is to say, constructing a model or simulation is done through a *process*, which includes making simplifying assumptions, introducing abstractions, and idealizations. In fact, in the case of Galilean and minimalist idealizations, their conceptualizations crucially depends on the possibility and desirability of de-idealization (Knuuttila and Morgan 2019, 643–645). As discussed above, the capacity for de-idealization is seen by some as a desirable feature. Others see the ability for a model or simulation to be de-idealized as central way to distinguish between different kinds of idealizations. Yet despite the importance of de-idealization, there is little existing literature discussing this reversal, nor its desirability.

Knuuttila and Morgan argue that, when analyzed, it is clear de-idealization is not just a simple reversal process, rather that there are four categories of de-idealization processes: (i) recomposing, (ii) reformulating, (iii) concretizing, and (iv) situating. They consider these four to provide a framework for more effectively analyzing de-idealization that occurs the in scientific practice of model construction. Through discussion of these four distinct processes (and relevant examples) they illustrate that in fact de-idealization processes may often involve multiple of these strategies, and show that models are not simply decomposable and that philosophers of science must play closer attention to modeling heuristics. Thus, there is no easy "adding back in" or reversals of idealizations, and idealization as a simple, reversable process in science may be in itself, an idealization (ibid., 657). Let us look at each of these strategies a bit closer.

The first strategy is de-idealization via recomposing—reconfiguration of the parts of the model with respect to the causal structure of the world. Recomposing might be most akin to the idea of "adding back in" features into a model that were at one point idealized, previously ignored, or controlled for. That is, often the de-idealization process is considered in terms of the reversals of the various *ceteris paribus* conditions. But Knuuttila and Morgan (following Boumans 1999) consider there to be three processes of de-idealization, (*ceteris absentibus* factors, *ceteris neglectis* factors, and true *ceteris paribus* factors), which upon reflection, are more complex than a simple "adding" of a factor, and thus require more extensive recomposing of the model in order to de-idealize.

We want to attend to the details of these three further since we suspect that de-idealization via recomposing is how de-idealization is commonly conceptualized. The first is the de-idealization processes as adding back in factors that are normally assumed absent yet do have an influence (*ceteris absentibus*). These are likely to be causal factors, which may be quite significant, and adding such causal factors will significantly alter the existing model. Here a model can only be recomposited by knowledge of the rest of the elements (ibid., 647). Instances of this in our case study occur most notably through the inclusion of stellar feedback in modeling single galaxy structure and evolution. It is a factor that was absent in early simulations, but included later (i.e., FIRE-based simulations). Second is the de-idealization process of adding back in factors normally assumed of so little weight that they can be neglected (*ceteris neglectis*). Here Knuuttila and Morgan are concerned that even if individually these factors can safely be dismissed, jointly they could make a significant difference to the model. In our case study, this might be modeling both dark matter and baryonic matter as point particles—for some research goals, as long as the overall mass is accurate and proportional, idealizing these both as point particles may not matter. However, this is also something contemporary simulations aim to de-idealize. Finally, the de-idealization process of adding back variability in those factors that are present but whose effect in the model is neutral as they are assumed to be held constant (actual *ceteris paribus* factors). Knuuttila and Morgan explain why *ceteris neglectis* conditions are so central to modeling: they "smooth out variety to create stability and so enforce homogeneity" (ibid., 648). However, it is unclear how to reconstitute these variable factors back into models that have previously held them constant. This is in part because there might not be evidence of a real (de-idealized) value, "either because of absence of knowledge or because there are no possible equivalent deidealized values" (ibid.). While there may be challenges to finding values that de-idealize *ceteris neglectis* conditions, Knuuttila and Morgan point out that such de-idealization might be relatively easy, as in "replacing average values by probability distributions" (ibid.). In our case study, this might be current neglection of dark energy (and something that has yet to be "de-idealized").

Moving on to two other categories, de-idealization as reformulating and concretizing each deal with issues of model representation, focusing on two different sides of the abstractness of models: their symbolic and conceptual formation. Knuuttila and Morgan acknowledge that there are many different modes of representation scientists can choose for their model or simulation in order to convey their content. Each representational choice can provide advantages but can limit what can be represented too. De-idealization as reformulating addresses the mathematical formalism used in models. An example of this difference between the mathematical representation as either algebraic or geometric (ibid.). What starts to hint at de-idealization not being possible by a simple reversal in this context is that once choices related to mathematical modeling are made, they are not readily visible as other modeling choices. Given the integral nature of the mathematical formulation, de-idealization would then require a reformulating of the model. Since the mathematical construction bears on how the relevant set of elements is integrated, such reformulation, in an attempt to de-idealize, runs the risk of the model falling apart. In our case study this may be akin to simulations choosing to idealize gravity in non-relativistic ways. For the case of individual galaxies, Newtonian dynamics are generally permitted, even though it is considered to not accurately represent the actual causal structure of the world.[8] To de-idealize this component would require revising the very mathematical formulation of the simulations.

De-idealization as concretizing is related to the representational choices made by scientists that embed theoretical or conceptual commitments about either the system or elements of that system (ibid., 651). The de-idealization of these conceptual abstractions partly means making them operational, it also means including assumptions about the definition of those abstractions. How a system or elements of a system are concretized will depend on "specific purposes in theorizing or in application" (ibid., 651). It is key to note that though concretization is posed by Knuuttila and Morgan as a sort of de-idealization they also point out that concretization does not necessarily mean making a given model or its elements more realistic for even truer to observations about the target system. Rather concretized versions of conceptual abstractions will still be "wedded to their conceptual framing" (ibid., 652). This de-idealization may be more prevalent in the economic cases that are of concern to Knuuttila and Morgan, where decisions must be made about how to represent a utility maximizer. In the context of our case study, this may be seen in choices about how to model dark matter (most choosing non-interacting, yet this embeds some kind of theoretical commitment).

The final category is de-idealization as situating, which addresses the applicability of models to particular situations, and is concerned not just with how a model can be de-idealized to represent some determinable target situations, but how such a process enhances their use in theorizing (ibid., 646). In situating, scientists might use a model in many different but similar specific instances, using either statistical work or experimental work in lab or field. There is not any 'general' de-idealization, that takes place, but rather a different de-idealization for every different situation (such as time, place, or topic) (ibid., 656). In the context of our case study, this seems to be what takes place when specific parameter values (masses, velocities, angle of impact, other observational-based data from actual target systems) are entered into simulations. It is instances of de-idealization tied to specifics, rather than a kind of de-idealization occurring to the model or simulation as a whole.

---

[8] One might be tempted to think of this example as a kind of *ceteris absentibus,* i.e., ignoring relativistic effects on galactic scales and just assuming the Newtonian limit of GR. However, to undergo the de-idealization process it would not simply involve adding something ignored back into the system (the same way someone might, for example, add friction back in for an inclined plane). To de-idealize and properly account for GR in the context of these simulations would require fundamental mathematical reformulation of the simulations and code. We thank our reviewer for pressing us to clarify.

8.4.4 Idealizations, De-idealizations, and Epistemic Status of Simulations

Having detailed Knuuttila and Morgan's conceptions of de-idealizations, and provided some examples of each strategy we turn back to our larger goal of examining the aims and roles of de-idealization in astrophysics. If we take seriously Knuuttila and Morgan's conceptualization of de-idealization in this more complex matter (i.e., not as simple reversal) we see that use of simulation and code that is flexible enough to de-idealize representations plays a specific role in reasoning about results in the context of astrophysics. It's in these de-idealizations where a lot of the simulation's epistemic power lies in using simulations to connect a vast array of independent astronomical observations/phenomena to cosmologists' more global arguments.

More specifically, part of what is being done in the case study by deploying GIZMO+FIRE simulations is adding back in features of the target system to the simulations that had originally been idealized away, and which might actually be difference-making. GIZMO+FIRE simulations allow for exploration about these structures through de-idealizing, namely via including stellar feedback. On the scale of individual galaxies, this is a kind of difference-maker that matters for specific kinds of questions and complexity of questions that can be posed by scientists. The limits of scientific questions prior to GIZMO+FIRE were restricted to those of general structures, or general causal features. But as we have discussed, stellar feedback is critical to how individual galaxies develop and evolve over time; stellar feedback is a difference maker. In this process of de-idealizing these minimal causal models with more details, including stellar feedback, refinement in structures occurred. For instance, simulations allow scientists to now see what kind of stars are present (i.e., young hot blue stars vs older cooler red stars). These features emerge in the simulations only once you have the complexity of stellar feedback. Such features also allow scientists to gain more refined temporal information about the age of the ring galaxy because stellar structure contains this information.

Knuuttila and Morgan conceptualize the "menu" of de-idealization processes consisting of recompositing, reformulating, concretizing, and situating. We think embedded in these there is a useful set of processes-based dimension to de-idealization worth highlighting more explicitly than Knuuttila and Morgan have done. The first is de-idealizing within one context—the kind of de-idealization that occurs fitted to a specific case, data, or target system. The idea here is that some simulations, such as the GIZMO+FIRE simulation, allow for a basic setup, say, two galaxies of specific masses colliding, represented as point particles. Once a simulator has successfully set up this simulation, they can implement a variety of de-idealization strategies (adding in stellar feedback (FIRE), specifying a subset of those particles as stars, gas, etc.). The second is de-idealizations that occur across multiple projects in the way commonly demonstrated via robustness analysis, cross-comparison, and a plurality of "tests" most directly targeted towards identifying difference-makers. This can occur within one simulation instance (say, the set of

1976 point particle simulations), or across the history of simulation progress investigating a specific target system (such as investigations of ring galaxies comparing 1976 to contemporary GIZMO+FIRE simulations and knowledge regarding causal processes and relevant difference makers). The third is a de-idealization process that occurs over time, via progress on tractability. This requires taking models or simulations not individually but as a set, as an ongoing de-idealization through rebuilding simulations in their entirety. These allow for much more expansive "de-idealization" than others because it allows for the simulators to revise or return to representational choices and idealizations introduced into the system.

Knuuttila and Morgan emphasize that idealized models embed a scientist's theoretical or conceptual commitment about either the system, or elements of the system. Part of what one does in the process of de-idealization is think about how conceptual elements can be de-idealized in different ways, for different sites, and for different purposes. This is just the kind of story at play in simulation codes in astrophysics: namely a group of simulators will develop a code, and different research groups will put it to different purposes. In this process they de-idealize it for their context and goals. With different research groups doing this, it offers a plurality of tests of that simulation code. If it works out well for most groups that adds to the power of the simulation code, connecting a vast array of independent astronomical observations/phenomena to cosmologists' more global arguments made or embedded in the code. But if it fails to work through this process of de-idealizing, it highlights instances in which some critical representation has perhaps been overlooked, or is perhaps overtly flexible. Being too flexible is a worry in the context of astrophysics, partly because astrophysics and cosmology in particular is one of these cases in which scientists do not have a full understanding of the real world target systems of investigation, and thus what might even need to be in their model or simulation.

This brings us back to the central point raised by Potochnik regarding the connection between idealization, de-idealization, and representational choices. When considering the aims of idealization or even de-idealization, which causal pattern emerges depends on representational choices. Some features emerge because of scientists' representational choices, that is, their choice to include more features in the simulations than we included originally. But there also seems to be a process-based evolution to simulations: They often start with an idealized minimal model, that over time has undergone de-idealization, and an "adding back in" of features that may or may not be relevant to procuring the phenomenon. But this "adding back in" is not simple reversal, as highlighted by Knuuttila and Morgan, it is some kind of recompositing, reformulating, concretizing, and situating, which are ultimately informed by simulators' interests (goals for the simulation) connected to representational choices. Often, astrophysicists are aiming towards representations closer to the actual target system. Consequently, greater explanatory strength is added to their models. The process highlights a need to capture only that which might be causal.[9] Complexity achieved through de-idealization provides some of the simulation's inferential power; and attending to the way in which the de-idealization strategies are utilized and justified provides that epistemic support. Yet, a background concern is that of the computational tractability stage: there is give and take between what is included. This highlights a central tradeoff at the core of de-idealization between computational tractability, inclusion of aspects of the target system that make a difference for the goals of the scientists for the simulation.

8.5 Conclusion

In this chapter we have provided a survey of philosophical literature connected to idealization as it connects to (1) kinds of idealizations that occur in science, (2) the aims of idealization in science, and (3) various strategies for de-idealization in science. All of these topics and taxonomies can be deployed to obtain a better understanding of the relationship between model and simulator representational choices in developing their simulations, challenges of these representations necessarily being incomplete and partial descriptions of target systems, and what those simulators might then be merited in terms of epistemic claims. Throughout this discussion we have drawn on a simple case study of collisional ring galaxy simulations to help illustrate how these topics might connect to and apply in the context of astrophysical computer simulations. To this extent, our analysis has only skimmed the surface. We hope this chapter might inspire others to take a deeper dive.

Finally, let us consider the central themes discussed in Sect. 8.4, and the role of idealization in the context of astrophysical computer simulations more generally. First and foremost, it seems that the connection between the kinds of idealizations that are deployed in the development of models and computer simulations relies on a non-trivial awareness of the aims or purposes for which the model or simulation are being constructed. That is, the justification, or kind of idealization deployed in turn captures aspects of what the scientist views as the goal of the model or simulation more generally. As Potochnik points out, introduction of idealizations can go unchecked, but "unchecked" does not necessarily mean unjustified. Rather, introduction of idealizations does not always come with explicit justification by the scientist. But should this this justification be reflected on, there is a connection to aims of advancing human understanding and uncovering the causal patterns. In the context of our ring galaxy simulation case study, we see that at least these astrophysical simulations aim to more accurately represent target systems (e.g.,

---

[9] There are also connections to be drawn here between identifying causal features to providing causal explanations. However, engagement with this set of philosophical questions is beyond the scope of this paper. For preliminary discussion on the relationship between idealization and representation to explanation in the context of astrophysics see Kennedy (2012) and Jebeile and Kennedy (2015).

collisional ring galaxies), with hopes of having a resource or tool (the simulations) to aid better understanding of the system. Second, when an aim of the scientist is development of a further understanding of the system, it may serve an impetus to de-idealize. A central point to appreciate from Knuuttila and Morgan is these de-idealizations cannot be done as a simple reversal, they must happen via a variety of strategies. In turn, these strategies also reflect various aims and understanding goals. Four of these strategies are delineated by Knuuttila and Morgan, and we have highlighted the different process-dimensions also at play. These process-dimensions work to unpack more explicitly some of these aims and goals. Third, by attending to the aims and goals of introducing idealizations or attempts to de-idealize, we do not see one-to-one correspondence of kinds of idealizations originally made to a specific de-idealization strategy. Finally, though we made our case by way of the ring galaxy case study, we suspect generalizing our argument, at least partially, is possible to the use in other astrophysical contexts deploying idealizations.

References

Appleton, P.N., and C. Struck-Marcell. 1996. Collisional Ring Galaxies. *Fundamentals of Cosmic Physics* 16 (111–220): 18.

Batterman, R.W. 2002. Asymptotics and the Role of Minimal Models. *The British Journal for the Philosophy of Science* 53 (1): 21–38.

Boumans, Marcel. 1999. "Built-In Justification." In *Models as Mediators: Perspectives on Natural and Social Science*, ed Mary S. Morgan and Margaret Morrison, 66–96. Cambridge: Cambridge University Press.

Cartwright, N. 1983. *How the Laws of Physics Lie*. Oxford: Oxford University Press.

———. 1989. *Nature's Capacities and Their Measurement*. Oxford: Oxford University Press.

Elliott-Graves, A., and M. Weisberg. 2014. Idealization. *Philosophy Compass* 9 (3): 176–185.

Giere, R.N. 1988. *Explaining Science: A Cognitive Approach*. Chicago: University of Chicago Press.

Hartmann, S. 1998. Idealization in Quantum Field Theory. In *Idealization in Contemporary Physics*, ed. N. Shanks, 99–122. Amsterdam: Rodopi.

Hopkins, P.F. 2015. A New Class of Accurate, Mesh-Free Hydrodynamic Simulation Methods. *Monthly Notices of the Royal Astronomical Society* 450 (1): 53–110.

Hopkins, P.F., D. Kereš, J. Oñorbe, C.A. Faucher-Giguère, E. Quataert, N. Murray, and J.S. Bullock. 2014. Galaxies on FIRE (Feedback In Realistic Environments): Stellar Feedback Explains Cosmologically Inefficient Star Formation. *Monthly Notices of the Royal Astronomical Society* 445 (1): 581–603.

Jacquart, M. 2020. Observations, Simulations, and Reasoning in Astrophysics. *Philosophy of Science* 87 (5): 1209–1220.

Jacquart, M., E. Shech, and M. Zach. Forthcoming. Idealization, Representation, and Explanation Across the Sciences. *Studies in History and Philosophy of Science*.

Jebeile, J., and A.G. Kennedy. 2015. Explaining with Models: The Role of Idealizations. *International Studies in the Philosophy of Science* 29 (4): 383–392.

Kadowaki, K. Forthcoming. Verification, Validation, etc.

Kennedy, A.G. 2012. A Non Representationalist View of Model Explanation. *Studies in History and Philosophy of Science Part A* 43 (2): 326–332.

Knuuttila, T., and M.S. Morgan. 2019. Deidealization: No Easy Reversals. *Philosophy of Science* 86 (4): 641–661.

Lynds, R., and A.L.A.R. Toomre. 1976. On the Interpretation of Ring Galaxies: The Binary Ring System II Hz 4. *The Astrophysical Journal* 209: 382–388.

McMullin, E. 1985. Galilean Idealization. *Studies in History and Philosophy of Science Part A* 16 (3): 247–273.

Norton, J.D. 2012. Approximation and Idealization: Why the Difference Matters. *Philosophy of Science* 79 (2): 207–232.

Nowak, L. 1972. Theories, Idealization and Measurement. *Philosophy of Science* 39 (4): 533–547.

Odenbaugh, J., and A. Alexandrova. 2011. Buyer Beware: Robustness Analyses in Economics and Biology. *Biology and Philosophy* 26 (5): 757–771.

Potochnik, A. 2017. Idealization and the Aims of Science. In *Idealization and the Aims of Science*. Chicago: University of Chicago Press.

Rohwer, Y., and C. Rice. 2013. Hypothetical Pattern Idealization and Explanatory Models. *Philosophy of Science* 80 (3): 334–355.

Shech, E. 2018. Infinite Idealizations in Physics. *Philosophy Compass* 13 (9): e12514.

———. Forthcoming. Idealizations in Physics. In *Philosophy of Physics*, Cambridge Elements Series.

Springel, V., N. Yoshida, and S.D. White. 2001. GADGET: A Code for Collisionless and Gasdynamical Cosmological Simulations. *New Astronomy* 6 (2): 79–117.

Strevens, M. 2011. *Depth: An Account of Scientific Explanation*. Cambridge: Harvard University Press.

Vogelsberger, M., et al. 2020, January. Cosmological Simulations of Galaxy Formation. *Nature Reviews Physics* 2 (1): 42–66.

Weisberg, M. 2007. Three Kinds of Idealization. *Journal of Philosophy* 104: 639–659.

———. 2013. *Simulation and Similarity: Using Models to Understand the World*. New York: Oxford University Press.

Weisberg, M., M. Jacquart, B. Madore, and M. Seidel. 2018. The Dark Galaxy Hypothesis. *Philosophy of Science* 85 (5): 1204–1215.

Wimsatt, W.C. 1987. False Models as Means to Truer Theories. In *Neutral Models in Biology*, ed. M. Nitecki and A. Hoffman, 23–55. London: Oxford University Press.

———. 2007. *Re-engineering Philosophy for Limited Beings*. Cambridge: Harvard University Press.

Winsberg, E. 2010. Science in the Age of Computer Simulation. In *Science in the Age of Computer Simulation*. Chicago: University of Chicago Press.

Chapter 9
Simulation Verification in Practice

Kevin Kadowaki

Abstract With the increased use of simulations as investigative tools in various scientific fields, the question naturally arises as to how these simulations are epistemically justified. One natural approach is to insist that the numerical aspects of simulation justification be performed separately from the physical aspects, but Winsberg (2010) has argued that this is impossible for highly complex simulations. Based on a survey and close examination of a range of astrophysical MHD codes and their attendant literature, I argue that insisting on a strict separation of these aspects of simulation justification is neither epistemically necessary nor advisable.

9.1 Introduction

Given constraints on scientists' abilities to observe and experiment, simulations have become a crucial tool for investigating certain kinds of large-scale phenomena. These tools, however, do not come without costs, and naturally philosophers of science have raised a host of epistemic questions as to when simulations can be relied on and how this reliance can be justified. These questions are especially pressing in the case of highly complex simulations, where the efficaciousness of the various methods for sanctioning simulations—code comparisons, convergence tests, benchmarking—is often in question, due to nonlinearities and the sheer size of the simulation. In particular, the rise of simulation highlights the importance of understanding and guarding against the kinds of numerical error introduced by computational methods.

A common and *prima facie* intuitive approach to this problem is to insist that a proper epistemology of simulation will require a separation of the numerical or purely computational aspect of simulation justification from the process of comparing the simulation to real-world target system. The Verification and Validation

K. Kadowaki (✉)
Washington University in St. Louis, MO, USA, Irvine, CA, USA
e-mail: kevink@wustl.edu

(V&V) framework captures this intuition, conceptualizing a split between the purely numerical task of ensuring that the computer simulation adequately represents the theoretical model (verification), and the task of comparing the output of the computer simulation to the real-world target phenomenon (validation). Per the V&V account, these separate treatments are required to avert the epistemic risk that errors in one domain may "cancel" errors in the other, leading to false confidence in the adequacy of our scientific theories.

Eric Winsberg has argued that this prescription for strict separation between V&V is not followed—and indeed *cannot* be followed—as a matter of actual practice in cases of highly complex simulations (Winsberg 2010, 2018). In this paper, I will present further evidence showing that the prescription goes largely unheeded in the context of astrophysical magnetohydrodynamics (MHD) simulations. But even if Winsberg has successfully shown that simulationists *cannot* strictly separate these activities, we still must contend with the possibility that this has fatal epistemic consequences for simulation methods—after all, this strict separation is generally prescribed as a bulwark against an allegedly severe and systematic epistemic risk. In other words, it remains to be shown that methods that simulationists *do* use can mitigate this risk, despite the fact that they do not follow the strict V&V prescription. In what follows, I will argue that a careful examination of the development of simulation codes and verification tests allows us to develop just such an alternative account.

In Sect. 9.2, I present the survey of a range of representative MHD simulation codes and the various tests that were proffered in the literature to support and characterize them. In Sect. 9.3, I lay out the specifics of the V&V account and show that the survey results are incompatible with this account. To diagnose the problem, I examine a particular class of tests associated with the phenomenon of fluid-mixing instabilities, the circumstances under which this phenomenon became a concerning source of error, and the simulationists' response to these developments; on the basis of these and other considerations, I argue that this approach to complex simulation verification is more exploratory and piecemeal than philosophers have supposed. In Sect. 9.4, I examine some of the details of the purpose and implementation of these tests, and I argue that the mathematical and physical aspects of complex simulation evaluation cannot be neatly disentangled—and, in some cases, *should* not be disentangled.

9.2   A Survey of Galaxy MHD Simulation Codes

The survey here concerns *verification tests*, i.e. tests that involve running a simulation with specifically chosen initial conditions and comparing the output to a known analytic solution or some other non-empirical-data metric. Significant discrepancies are then generally taken to indicate some failure of the discretized simulation equations to mimic the original, non-discretized equations—e.g., if a set of hydrodynamic equations naturally conserve energy, but a test of the discretized simulation of these equations shows that energy is not conserved, one can conclude that the numerical methods implemented are the source of the error.

The primary codes examined for the present survey were FLASH (Fryxell et al. 2000), RAMSES (Teyssier 2002), GADGET-2 (Springel 2005), ATHENA (Stone et al. 2008), AREPO (Springel 2010), and GIZMO (Hopkins 2015). These simulations were chosen to span a range of years and MHD code types, focusing on simulations which were particularly influential and which had a substantive literature. ATHENA, for instance, uses a static grid-based Eulerian method; FLASH and RAMSES are also stationary grid-based methods, but use Adaptive Mesh Refinement (AMR) to refine the grid in places. GADGET-2 is a particular implementation of Smooth Particle Hydrodynamics (SPH), a Lagrangian method. AREPO combines elements of the AMR and SPH methods to create a "moving-mesh" code which allows for tessellation without stationary grid boundaries. GIZMO is similar to AREPO in that it combines advantages of the SPH and AMR methods, but it is roughly described as "meshless", as it involves a kind of tessellation akin to AREPO, but allows for a smoothing and blurring of the boundaries according to a kernel function.[1]

While some of the official public release versions of these codes included routines for tests not reported in the literature, the survey generally only looked to tests that were reported in published papers. This was for three reasons. First, I am primarily interested in tests that were considered important enough to be on display and described in some detail in the method papers presenting the code. Second, I am also interested in the analysis of the code's performance on particular tests; simply including a routine in the code suite does not indicate the significance of the test vis-à-vis particular kinds of error or whether the result of the routine measured up to some standard. Third, particular routines may have been included in either the initial or subsequent versions of the code; the papers, being timestamped, provide a better gauge of when tests were performed (or at least considered important enough to publish).

The two exceptions to this are FLASH and ATHENA. FLASH includes a bare minimum of tests in its initial release paper but provides many more tests and has an extensive amount of useful documentation in the User Guide (Flash User Guide). This user guide is also available in various editions corresponding to different release versions of FLASH, spanning version 1.0 from October 1999 to the most recent version 4.6.2 in October 2019; this allows us to track when the various test problems were introduced. A brief overview of this sequence will be discussed below as well. ATHENA includes a few additional fluid-mixing instability tests on a (now partially-defunct) webpage, and given my focus on these tests in Sect. 9.3, I have chosen to include them as well. Given that at least one fluid-mixing test was included in the methods paper (the Rayleigh-Taylor instability test), and given the

---

[1] Technically, GIZMO is able to facilitate a number of sub-methods, including "traditional" SPH. The new methods of interest here are the Meshless Finite-Volume and Meshless Finite-Mass described in Hopkins (2015).

timeline to be described in the next section, it is likely that the other fluid-mixing tests were performed around that time.

An overview of the various tests found in the initial documentation papers can be found in Table 9.1 (hydrodynamic tests), Table 9.2 (magnetohydrodynamics tests), and Table 9.3 (self-gravity tests) (FLASH is omitted from Table 9.2; for an overview of those MHD tests that were eventually included, see Table 9.4). Table 9.4 tracks the inclusion of tests over time in selected editions of the FLASH user guide. Based on the data laid out in the various tables, we can make a number of preliminary observations, some of which I will expand on in later sections.

Among those tests that are common to multiple codes, it is clear that there is a general accumulation of hydrodynamics tests as time progresses, with later-developed codes including far more tests than earlier codes. In many cases, the later codes will cite examples of the test as implemented in earlier codes, both among those surveyed here and elsewhere. While the tests are not all consistent, where possible I have cited to both the original paper that described or designed the test and indicated where authors used variants. As I will discuss in the next section, in some cases the appearance of a new test is a clear response to reported concerns about a particular source of error, especially where that source of error was a problem in prior codes and not particularly well-tracked by previously cited tests. In other circumstances, the overarching purpose for adding a new test is unclear— i.e., it may or may not be redundant with respect to the rest of the collection. This accumulation is also apparent in the history of the FLASH simulation, where many of the tests added in the two decades since its initial release overlap with the other surveyed codes and several even track with the times that they were introduced.

Where tests are not common among codes, they can roughly be divided into two categories. Some tests are unique to a particular code because they are generally inapplicable to other code types, which is to say they are tailored to test for numerical errors to which other code types are not susceptible. For example, FLASH and RAMSES both include unique tests of circumstances where the adaptive mesh refinement algorithm is forced to make sharp jumps in spatial resolution—these tests are obviously not applicable in the absence of AMR.

Other tests are not tailored in this manner, although this does not mean that they all serve disparate purposes—in some cases, different tests are probing the same kinds of phenomena, even while the setups and initial conditions are different. This is particularly unsurprising in the case of the myriad unique tests with full self-gravity, as there are few examples of problems with self-gravity where analytic solutions exist. Here, the broad aim is to simulate scenarios that are more "realistic" than the other highly simplified tests (albeit still fairly simple!), and consequently in these cases there is less emphasis placed on measuring the code's performance against straightforward rigorous quantitative standards such as analytic solutions. Further examination of multi-group code-comparison projects also shows that these projects are not always a straightforward exercise, often requiring a great deal of technical elaboration before comparisons can be drawn—and moreover, the various desiderata for these kinds of cross-code comparisons are often in tension with one another (Gueguen Forthcoming). The fact that these tests are not straightforward

Table 9.1 Hydrodynamics tests. Unless otherwise indicated, the test results as run by a particular code is recorded in the paper indicated at the top of each respective column column. The * citation indicates that a different test setup was cited

	FLASH Fryxell et al. (2000)	RAMSES Teyssier (2002)	GADGET-2 Springel (2005)	ATHENA Stone et al. (2008)	AREPO Springel (2010)	GIZMO Hopkins (2015)
One-dimensional wave[a]				✓	✓	✓
Sod shocktube[b]	✓	✓	✓	✓	✓	✓
Interacting blast waves[c]	✓			✓	✓	✓
Sedov-Taylor point explosion[d]	✓	✓			✓	✓
Noh problem[e]				✓	✓	✓
Gresho vortex[f]					✓	✓
Driven turbulence					✓[n]	✓
Keplerian disks					✓[o]	✓
Kelvin-Helmholtz			✓[p]		✓[q]	✓[r]
Rayleigh-Taylor[g]				✓	✓	✓*[s]
"Blob" test[h]						✓
"Square" test[i]						✓
Implosion[g]				✓		
Shu and Osher shocktube[j]				✓		
Forced AMR jump	✓	✓[t]				
Advection problem	✓					
Wind tunnel with step[k]	✓					
Strong shock[l]			✓			
Double Mach reflection[c]				✓		
Einfeldt strong rarefaction[m]				✓		
Moving boundary					✓	

[a] Stone et al. (2008)
[b] Sod (1978)
[c] Woodward and Colella (1984)
[d] Sedov (1959)
[e] Noh (1987)
[f] Gresho and Chan (1990)
[g] Liska and Wendroff (2003)
[h] Agertz et al. (2007)
[i] Heß and Springel (2010)
[j] Shu and Osher (1989)
[k] Emery (1968)
[l] Klein et al. (1994)
[m] Einfeldt et al. (1991)
[n] Bauer and Springel (2012)
[o] Pakmor et al. (2016)
[p] Stone (2019)
[q] Robertson et al. (2010)
[r] McNally et al. (2012)
[s] Abel (2011)
[t] Khokhlov (1998)

Table 9.2 Magnetohydrodynamics tests. As in Table 9.1, unless otherwise specified, the test results as run by a particular code is recorded in the paper indicated at the top of each respective column column. Each test is based on the setup given in the paper cited in the first column, with the exception of the MHD shocktube category: for those marked with *, the cited test was performed *instead*; for those marked with [†], the cited test was performed *in addition*

	RAMSES	GADGET-2	ATHENA	AREPO	GIZMO
	Fromang et al. (2006)	Dolag and Stasyszyn (2009)	Stone et al. (2008)	Pakmor et al. (2011)	Hopkins and Raives (2016)
MHD waves[a]			✓		✓
MHD shocktube[b]	✓*[i]	✓	✓	✓*[j]	✓[†k]
Orszag-Tang vortex[c]	✓	✓	✓	✓	✓
MHD rotor[d]		✓	✓		
Current sheet[e, f]	✓		✓[l]		✓
Loop advection[e]	✓		✓	✓	✓
Blast wave[d, g]		✓	✓	✓	✓
Magneto-rotational instabilities	✓				✓[m]
Kelvin-Helmholtz instability			✓[n]		✓
Rayleigh-Taylor instability			✓		✓
Circularly polarized Alfven waves[h]			✓		

[a] Stone et al. (2008)
[b] Brio and Wu (1988)
[c] Orszag and Tang (1979)
[d] Balsara and Spicer (1999)
[e] Gardiner and Stone (2005)
[f] Hawley and Stone (1995)
[g] Londrillo and Del Zanna (2000)
[h] Tóth (2000)
[i] Torrilhon (2003)
[j] Keppens (2004)
[k] Tóth (2000)
[l] Beckwith and Stone (2011)
[m] Guan and Gammie (2008)
[n] Stone (2019)

side-by-side comparisons, likely accounts for the fact that they do not display the same pattern of accumulation evident among the simpler hydrodynamics tests.

There are also some tests that are *prima facie* relevant to other codes, at least on the basis of the description provided—e.g., both ATHENA and GIZMO deploy a selective application of two Riemann solvers, including one (the Roe solver) that can give unphysical results if applied incorrectly, but only ATHENA presents the Einfeldt strong rarefaction test to establish that this will not cause a problem. This may simply be an indication that the problem is no longer of particular concern, or

Table 9.3 Self-gravity tests

	FLASH	RAMSES	GADGET-2	ATHENA	AREPO	GIZMO
	Fryxell et al. (2000)	Teyssier (2002)	Springel (2005)	Stone et al. (2008)	Springel (2010)	Hopkins (2015)
Zeldovich pancake[a]		✓			✓	✓
Santa Barbara cluster[b]			✓		✓	✓
Evrard collapse[c]			✓		✓	✓
Simple acceleration	✓					
ΛCDM acceleration	✓					
Spherical infall[d]	✓					
Isothermal collapse[e]			✓			
DM Clustering[f]			✓			
Galaxy collision					✓	
Galaxy disks						✓

[a] Zel'Dovich (1970)
[b] Frenk et al. (1999)
[c] Evrard (1988)
[d] Bertschinger (1985)
[e] Burkert and Bodenheimer (1993)
[f] Heitmann et al. (2005)

that the Roe solver was tested in GIZMO but the test was not considered important enough to include in the methods paper.

Additionally, some tests that are common among the various codes are nonetheless used for purposes that do not entirely overlap between codes. The most clear example of this is the distinct use of some common tests by stationary grid codes to test for artificial symmetry breaking along grid axes—e.g., the various shocktubes and blast waves are used in SPH and non-stationary grid codes to test their abilities to handle shocks and contact discontinuities, but in stationary grid codes they can be run both aligned and inclined to the static grid to test for artificial symmetry breaking along grid lines.

The magnetohydrodynamics tests do not display as clear a pattern of accumulation; unlike the hydrodynamics tests, there seems to be a common core of tests that have been more-or-less consistent over the span of years, with the notable exception of debut of the MHD Kelvin-Helmholtz and Rayleigh-Taylor instability tests. I speculate that the consistency apparent in magnetohydrodynamics tests is a function in part of the influence of J. Stone, who (with coauthors) proposed a systematic suite of test MHD test problems as far back as 1992 (Stone et al. 1992) and, together with T. Gardiner, wrote the 2005 paper (Gardiner and Stone 2005) that is either directly or indirectly (through his 2008 ATHENA method paper (Stone et al. 2008)) cited by all the MHD method papers in question.

Stone et al. (1992) is notable for being a standalone suite of MHD test problems without being connected to a particular code—in particular, this suite is not intended

Table 9.4 Tests included in various editions of the FLASH user guide

	1.0 (1999)	2.0 (2002)	2.5 (2005)	3.3 (2010)	4.6.2 (2019)
Sod shocktube[a]	✓	✓	✓	✓	✓
Shu and Osher shocktube[b]		✓	✓		✓
Interacting blast waves[c]	✓	✓	✓	✓	✓
Point explosion[d]	✓	✓	✓	✓	✓
Advection problem	✓	✓	✓		
Isentropic vortex[e]		✓	✓	✓	✓
Noh problem					✓
Wind Tunnel with step	✓	✓	✓	✓	✓
Driven turbulence				✓	✓
Relativistic Sod shocktube				✓	✓
Implosion test				✓	✓
Kelvin-Helmholtz					✓
Brio and Wu shocktube[f]		✓	✓	✓	✓
Orszag-Tang vortex[g]			✓	✓	✓
MHD rotor[h]				✓	✓
Current sheet[i]				✓	✓
Field loop advection[i]				✓	✓
Jeans instability[j]		✓	✓	✓	✓
Homologous dust collapse[k]		✓	✓	✓	✓
Huang-Greengard Poisson test[l]		✓	✓	✓	✓
Maclaurin test[m]				✓	✓
Zeldovich pancake			✓	✓	✓

[a] Sod (1978)
[b] Shu and Osher (1989)
[c] Woodward and Colella (1984)
[d] Sedov (1959)
[e] Yee et al. (2000)
[f] Brio and Wu (1988)
[g] Orszag and Tang (1979)
[h] Balsara and Spicer (1999)
[i] Gardiner and Stone (2005)
[j] Jeans (1902)
[k] Colgate and White (1966)
[l] Huang and Greengard (1999)
[m] MacLaurin (1801)

as a comprehensive collection of all known test problems, but rather as a minimal subset of essential tests, each corresponding to a different MHD phenomenon. As the field has progressed significantly since this suite was published, there is reason to believe that the specifics of this paper are out of date with respect to the surveyed code examples and the phenomena of interest. However, insofar as it lays out rationale, not only for each specific test, but also for the choice of the collection of tests as a whole, the paper provides a framework for thinking about how these tests might be understood to collectively underwrite simulations. In particular, while we may not be able to think of this framework as providing absolute sufficiency conditions for the adequacy of a given suite of test problems, this approach may still point us towards a more pragmatic notion of sufficiency, especially with respect to the current state of knowledge in the field. Admittedly, I have been unable to find similarly systematic proposals for test suites of hydrodynamic or self-gravity test problems; however, in anticipation of the argument that I will be making in Sect. 9.4, I will note that this emphasis on MHD *phenomena* as the guiding principle for test selection suggests an approach to these tests that goes beyond merely numerical considerations.

9.3 Fluid-Mixing Instabilities and Test Development

In the philosophical literature, the concept of simulation verification has been heavily influenced by the Verification & Validation (V&V) framework, which itself originated in a number of subfields within the sciences (Oberkampf and Roy 2010)—including computational fluid dynamics, which has some obvious theoretical overlap with the field of astrophysical magnetohydrodynamics. Despite this, with one exception (Calder et al. 2002), the V&V framework is not generally invoked in the field of astrophysical MHD simulations. Nonetheless, I will briefly outline the V&V framework to motivate a philosophical perspective on the proper approach to simulation verification, which I will then contrast with an examination of the tests as they are found in the above survey.

Within the V&V framework, a simulation is said to be *verified* when we are confident that the numerical methods employed in the simulation faithfully approximate the analytical equations that we intend to model; the simulation is said to be *validated* when the output of the simulation adequately corresponds to the phenomena in the world.[2] Together, these two components form a bridge between the phenomenon in the world and the analytical equations that constitute our attempts to theoretically capture that phenomenon, via the intermediary of the simulation code. Crucially, this means that verification and validation refer to correspondences over a range of simulation runs—see, e.g., various definitions of "validation" surveyed in (Beisbart 2019), where notions such as "domain of applicability" implicitly make clear that these concepts are not simply correspondences with respect to an individual system. Within this framework, the function of verification tests is to

---

[2] As Beisbart (2019) has shown, there is some ambiguity regarding the use of the term "validation" in the literature. Here, I will be using the term to refer to what he distinguishes as *computational model validation*, and for our purposes other distinctions are not relevant.

determine whether the numerically-implemented code is faithful to the analytical equations of the original model.

The epistemic challenge associated with this task stems from the two-part structure of V&V; in particular, the concern is that numerical errors could "cancel out" errors caused by an inaccurate model, leading to a simulation built on incorrect theory that nonetheless produces an output that corresponds to the phenomenon in question. This concern is compounded in highly complex simulations such as the ones at issue here, as the nonlinear regimes at issue make it difficult to assess whether an effect is numerical or physical. Ultimately, this epistemic concern has led some philosophers to stress the importance of a sequential ordering for these activities: first verification, then validation. If the simulationist ensures that the simulation code is free of numerical errors *independently* of any comparisons to the phenomena, then this should preempt any risk that we might accidentally fall prey to the cancellation of errors (Morrison 2015, 265); I will refer to this conception of simulation verification as the "strict V&V account."

With this framework in mind, one might then believe that the survey in Sect. 9.2 raises some serious concerns. As noted in the previous section, there has been a tendency for later-developed codes to include more tests than earlier-developed codes—this, in turn, would imply either that the new tests are superfluous, or that the old simulations were not adequately verified against certain kinds of numerical errors. The former possibility is unlikely, especially where newer tests show that new codes display marked improvement over the performance of prior codes. Thus, it would seem that earlier codes were not sufficiently verified. Moreover, absent some assurances that newer codes have remedied this issue, we have no particular reason to believe that the suite of tests is *now* comprehensive, and that future codes will not employ more tests that reveal shortcomings in our current standard codes. To be epistemically satisfied, it seems as if we should want something like a general account of how the various tests fit together into an overall framework, specifically in a way that provides good evidence that all relevant sources of error are accounted for once-and-for-all.

In the next section, I will argue that such a fully comprehensive, once-and-for-all approach to verification is unnecessary, and that the philosophical intuitions motivating the strict V&V account are misleading. To lay the groundwork for this argument, I will begin by discussing a particular class of tests—those concerning fluid-mixing instabilities—in more detail. Then, on the basis of these and other examples, I will argue that these tests as used here do not fit the above philosophical intuitions about simulation verification, and that we should (at least in some cases) think about simulation verification as a more piecemeal, exploratory process.

Fluid-mixing instabilities refer to a class of phenomena arising, naturally, in hydrodynamic contexts at the boundary between fluids of different densities and relative velocities. *Kelvin-Helmholtz* (KH) instabilities arise from a shear velocity between fluids, resulting in a characteristic spiral-wave pattern; *Rayleigh-Taylor* (RT) instabilities occur when a lighter fluid presses against a denser fluid with a relative velocity perpendicular to the interface, resulting in structures described variously as "blobs" or "fingers".[3] In the course of galaxy formation, these instabilities are also subject to magnetic fields, which can suppress the growth of small-scale modes and produce novel behavior if the strength of the magnetic field is in the right regime. The importance of these phenomena have been understood for some time—in particular, the presence of KH instabilities is thought to have a significant impact on the stripping of gas from galaxies via ram pressure, which may account for variations in the properties of galaxies (Close et al. 2013). Chandrasekhar's standard theoretical treatment of these instabilities, both in the presence and absence of magnetic fields, was first published in 1961 (Chandrasekhar 1961), and numerical studies of the same have been conducted at least since the mid-1990s (Frank et al. 1995; Jun et al. 1995).

Given the importance of these instabilities in galaxy formation processes, one might suppose that the ability of simulations to implement them properly would be an essential concern, and that the verification tests performed would reflect this. However, as noted in Tables 9.1 and 9.2, none of the codes prior to ATHENA (2008) included explicit tests of the KH or RT instabilities in their method papers, and only FLASH comments on the incidental appearance of KH instabilities in one of its tests. In addition to the surveyed codes, explicit KH and RT tests are also absent from the pre-2008 method papers for GASOLINE (TREE-SPH) (Wadsley et al. 2004), HYDRA (AP^{3}M-SPH) (Couchman et al. 1994), and ZEUS (lattice finite-difference) (Stone and Norman 1992). On the other hand, a brief perusal of post-2008 method papers such as RPSPH (Abel 2011), ENZO (AMR) (Bryan et al. 2014), GASOLINE2 ("Modern" SPH) (Wadsley et al. 2017), and PHANTOM ("Modern" SPH) (Price et al. 2018), shows that they all *do* cite to tests of these instabilities in various capacities.[4]

This disparity between pre- and post-2008 method papers with respect to their treatment of KH and RT tests can be traced (at least in significant part) to a code comparison project published in late 2007 (uploaded to arXiv in late 2006) by Agertz and other collaborators, including most of the authors of the various simulation codes already discussed (Agertz et al. 2007). In this hydrodynamic test, colloquially referred to as the "blob" test, a dense uniform spherical cloud of gas is placed in a supersonic wind tunnel with periodic boundaries and permitted to evolve, with the expectation that a bow shock will form, followed by dispersion via KH and RT instabilities. The dispersion patterns were compared to analytical approximations for the expected growth rate of perturbations, and the study concluded that, while Eulerian grid-based techniques were generally able to resolve these instabilities, "traditional" SPH Lagrangian methods tend to suppress them and artificially prevent the mixing and dispersion of the initial gas cloud.

These observations led to a number of discussions and disagreements in the literature regarding the precise nature and sources of these problems. Beyond

---

[3] Useful illustrations of both KH and RT instabilities, including time-series snapshots, are available in Heß and Springel (2010) and Hopkins (2015).

[4] Technically, ENZO only cites to Agertz et al. (2007), where it was used as one of the sample codes, but nonetheless the test is discussed in the method paper.

the normal issues with numerical convergence, the culprits were identified as insufficient mixing of particles at sub-grid scales (Wadsley et al. 2008) and artificial surface tension effects at the boundary of regions of different density caused by the specifics of SPH implementation (Price 2008). Eventually, these considerations led to other fluid-mixing tests aimed at addressing cited shortcomings with the "blob" test (Robertson et al. 2010; McNally et al. 2012).

Concurrent to and following the development of these tests, a number of new SPH formalisms and codes (so-called "Modern" SPH, in contrast to traditional SPH) have been developed to address these problems and subjected to these tests. The proposals themselves are quite varied, from introducing artificial thermal conductivity terms (Price 2008), to increasing the number of neighbor particles per computation (Read et al. 2010), to calculating pressure directly instead of deriving it from a discontinuous density (Hopkins 2013). But the common thread is that now, with the phenomenon established and its causes analyzed, the tests that were developed in response to these have (at least for the time being) become new standards for the field.

What observations can we draw from this narrative? First, it should be apparent that the process described here is incompatible with a strict V&V account of simulation verification. This is not to suggest that simulationists simply had no awareness that this area of their simulations might need more development—while the literature post-2008 certainly set the agenda and was the source for most of the key insights leading to the development of these tests, the problems with SPH were not entirely unknown before then. Indeed, while the specifics of the KH and RT instabilities were rarely referenced explicitly, SPH methods were known to have issues related to mixing and other instabilities at least as early as the 1990s (Morris 1996; Dilts 1999), and at least one variant of SPH was designed to address mixing issues as early as 2001 (Ritchie and Thomas 2001). Despite this, the tests did not generally make appearances in method papers until codes were already reasonably capable of handling them, at least in some regimes. This, in turn, raises a concern that an analogous situation holds in the case of our current codes, with respect to as-of-yet ill-defined or underreported sources of error.

Second, in response to this concern, we should note that these verification tests do not present themselves as obvious or canonical; rather, they are a product of experimentation. Obviously, any insistence that simulationists should have tested for these errors before the tests were developed is practically confused, but there is a deeper theoretical point to be raised against the more abstract epistemic objection: the tests themselves are not simply tests of a simulation's numerical fidelity, but are also tailored to probe at and attain clarity regarding the nature of particular vulnerabilities in specific code types. Hence, the tests for KH and RT instabilities are not just looking to reproduce the expected physics, but are also made specifically to expose the unphysical numerics associated with SPH tests as well. By itself, this may not satisfy a proponent of the strict V&V perspective, but it does suggest that these tests serve a purpose much broader than mere "verification" that numerical error is within tolerance levels for a given simulation—they are also giving simulationists tools to explore the space of simulation code types. I will discuss this in greater detail in the next section, but for now it is enough to note that this means that verification tests are doing far more than "verification" as strictly defined—and, indeed, the development of these tests is just as crucial to the progress of the field as the development of the simulation codes themselves.

9.4 Leveraging Both Physics and Numerics

Of course, while it may be suggestive, the narrative from the previous section does not show that this piecemeal and exploratory approach to simulation verification is epistemically sound. Certainly there is no sense in which these tests provide a patchwork cover of all possible situations wherein numerical error might arise, and thus they would fail to satisfy philosophers who stress the importance of complete verification upfront, per the strict V&V account. One might suppose that the above approach is simply the best that can be done, given the constraints of complexity and the current state of knowledge in the field, but even this would imply that the simulationists in question should be doing more to give more thorough accounts of how their tests fit together into the best-available suite given these constraints. In any case, I do not believe such an account would be particularly satisfactory in isolation. In this section, I want to argue that the approach taken by the surveyed astrophysical MHD codes is not just epistemically benign (at least in principle), but that limiting simulationists to the strict V&V approach would be an error of outsized caution. Specifically, I will argue that the risks incurred by simulationists are not radically different from those found in ordinary (i.e., non-simulation based) methods of scientific inquiry.

From the strict V&V perspective, the risk of physical and numerical errors "cancelling" each other out leads to the prescription that the verification and validation of simulations should be distinct and sequential—that is to say, that verification should be (strictly speaking) a purely numerical/mathematical affair, and that any evaluations in terms of physics should be confined to the validation phase. Of course, even in this case it would be permissible for a simulationist to incidentally cast verification tests in physical terms, e.g., in terms of specific physical initial conditions, but this would just be a convenience. But as I suggested above, verification tests are not simply convenient numerical exercises designed to check for generic numerical error. Rather, the tests serve as windows into the physics of the simulation, breaking down the distinction between physics and numerics and providing simulationists with a number of epistemic leverage points that would be obscured if we were to force them to regard verification tests as merely numerical in nature.[5]

In general, the tests provide the simulationist with a sense of the physical phenomena represented because simulationists can interpret and understand mathematical equations in terms of the physical phenomena they represent. In other words, simulationists are not simply checking to see if a given equation produces numerical error by means of comparison to an analytical solution, though that is a useful benchmark if it exists. Rather, terms in the simulation equations have physical significance, *including* terms that are artifacts of the discretization of the original continuous equations. In the case of fluid-mixing instabilities, e.g., the shortcomings of the traditional SPH methods were not simply referred to as "numerical errors"—the error term was specifically characterized as an "artificial surface tension" that became non-negligible in the presence of a steep density gradient (Price 2008). Where "fictions" such as artificial viscosity or artificial thermal conductivity terms are introduced, their justification is not cached out in numerical terms, but as appropriate physical phenomena whose inclusion will negate the influence of some other (spurious) error term, *because that error term behaves like a counteracting physical phenomenon*. Thus, on the one hand, the simulationist's preexisting physical intuitions about the appropriate behavior for the simulated system can serve to detect deviations that, upon investigation, may be determined to be numerical aberrations; on the other hand, the verification tests themselves enable the simulationist to develop this insight into the ways in which the simulation is functionally different from the corresponding real system.

Moreover, this insight into the physical significance of these numerical terms allows the simulationist to partition the space of possible simulation scenarios in a manner that is far more salient for the purposes of extracting scientifically useful confidence estimates. If, e.g., a simulationist wanted to know whether a particular simulation code is likely to give reliable results when they simulate a galaxy with a particular range of properties, estimates of performance in terms of the generic categories of "numerical error"—round-off error, truncation error, etc.—are not going to be particularly useful. But an understanding of the kinds of *physical* phenomena for which this code is particularly error-prone lends itself more naturally to judgements of this form. These judgements can even take a more granular form, where different aspects of a simulation could be gauged more or less reliable based on the strengths of the simulation code—e.g., a simulationist would presumably be somewhat hesitant to draw strong conclusions about aspects of galaxy formation that rely on KH or RT instability mixing on the basis of a traditional SPH code.

---

[5] This criticism should be distinguished from another prominent critique of V&V, by Oreskes et al. (1994). Oreskes and collaborators argue that verification is (strictly speaking) impossible given that real-world systems are not closed systems, and advocate instead for a model of confirmation by degrees. I am not unsympathetic to the spirit of this position. However, my argument does not commit to their abstract hypothetico-deductivist picture of confirmation, and moreover aims to give a concrete picture of how confidence-by-degrees is achieved in practice—and address the particular concerns about underdetermination that can be raised in by proponents of V&V.

But most importantly, this physical intuition allows for a kind of feedback loop, akin to the normal process of scientific discovery: we do our best to model complex systems by means of approximations, which in turn helps us understand how other, more subtle factors play an important role in the system; learning how to characterize and integrate these more subtle factors gives us a better, more robust model; and the process repeats. In this case, however, the object under investigation is not just the target system—we are also investigating the space of simulation code types, and experimenting with different ways to flesh out its properties by experimenting with various kinds of verification tests.

Of course, this approach is not foolproof. There will always exist the possibility that the simulationist is radically wrong about the adequacy of their simulation, that they have failed to account for some important phenomena. But this risk, while real, need not warrant wholesale skepticism of simulationist methods or embrace of the strict V&V account. In fact, this risk is analogous to the underdetermination risks incurred in the process of ordinary scientific inquiry—namely, that our theory might be incorrect or woefully incomplete, and that it only seems correct because some unaccounted-for causal factor is "cancelling out" the inadequacy of our theory. If we are going to regard this risk as defeasible in the context of the familiar methods of scientific inquiry, we should at least grant the possibility that the simulationist's risk is similarly benign.

Here, the proponent of the strict V&V approach may level an objection: namely, that the risks associated with simulation numerics "cancelling" other errors are potentially systematic in a way that the ordinary scientific risks of theory underdetermination by evidence are not. In the case of ordinary scientific theorizing, we regard this risk as defeasible because we have no reason to believe that the phenomena are conspiring to subvert our theorizing; even if we make mistakes given a limited set of data, we are confident that with enough rigorous testing we will eventually find a part of the domain where the inadequacies of the theory are apparent. In the case of simulation, however, one might worry that the risk may stem from a *systematic* collision between the numerical and physical errors, obfuscated by the complexities of the simulation—and if this is the case, further investigation will not allow us to self-correct, as continued exploration of the domain will not generally break this systematic confluence.

This objection makes some sense if we understand verification tests merely as straightforward tests of numerical fidelity. However, as I have tried to show, many verification tests are *not* of this simple character—by developing new kinds of tests to better understand the way simulation *codes* work, simulationists are simultaneously exploring the domain of possible real-world systems and probing the space of simulation code types. A particular verification test may be inadequate to the task of detecting or understanding certain kinds of errors—indeed, some argued in the literature that the original "blob" test proposed by Agertz et al. gave us a distorted picture of SPH's undermixing problem—but simulationists are not limited to a set of pre-defined tools. In the same way that we (defeasibly) expect that rigorous testing renders the risk of conspiracy tolerable in ordinary scientific contexts, the careful and targeted development of verification tests—in conjunction with the usual exploration of the domain of real systems—can mitigate the risk of conspiracy in the context of simulation.

With these considerations in mind, I would suggest that the best framework for thinking about these tests is as a collective network of tests roughly indexed to *phenomena*, specifically phenomena that, in the simulationist's estimation given the current state of knowledge in the field, are significant causal factors in the system under study. Under this picture, a simulation will be sufficiently (though defeasibly) verified just in case it produces tolerable results according to the full range of tests— which are themselves subject to scrutiny and modification as simulationists develop better understandings of how these tests probe their codes. This more pragmatic notion of sufficiency rejects the strict V&V insistence that simulations need to be verified against all sources of numerical error up front, but in exchange requires the simulationist to be sensitive to the various strengths and weaknesses of the code they are using—a sensitivity acquired in part by means of these tests, but also by general use of the code, and by familiarity with other codes and their strengths and weaknesses.

9.5 Conclusion

In this paper, I have presented a survey of the verification tests used in selected MHD codes, and drawn lessons about simulation justification on the basis of this real-world scientific practice. Notably, the pattern observed does not fit with the V&V framework's prescriptions, and a careful examination of the development and deployment of these tests shows that they serve epistemic functions beyond simply checking for numerical errors—they can be used to probe the differences between different code types and come to a deeper understanding of their strengths and weaknesses. By examining the case study of fluid-mixing instability tests, I traced this process in action and showed that the creation of these tests, the subsequent analysis, and the development of improved simulation codes is deeply entangled with our understanding of the underlying *physics*, not merely the numerics.

On the basis of this survey and case study, I argued that this process of improving our understanding of the target phenomena and the space of simulation code types can be understood to follow a pattern of incremental improvement similar to ordinary scientific theories in ordinary experimental contexts. I also addressed a skeptical objection that might be leveled by those convinced by the strict V&V approach—in particular, given this expanded understanding of how verification tests can inform our investigations, we can be reasonably confident that we are not exposing ourself to any severe underdetermination risks.

This wider understanding of the role of verification tests also has significant implications for how we characterize the role of the simulationist—in particular, the simulationist's knowledge of simulation methods and techniques is not merely *instrumental* for the goal of learning about the target phenomenon, because the simulationist's understanding of the target phenomenon is developed in tandem with their knowledge of simulation methods and techniques. This entanglement suggests that merely reproducing some target phenomenon by simulation is not sufficient for a full understanding of that phenomenon—the simulationist must also understand the principles by which the different specifics of the various code types yield this common result.

References

Abel, T. 2011. rpSPH: A novel smoothed particle hydrodynamics algorithm. *Monthly Notices of the Royal Astronomical Society* 413(1): 271–285.

Agertz, O., B. Moore, J. Stadel, D. Potter, F. Miniati, J. Read, L. Mayer, A. Gawryszczak, A. Kravtsov, Å. Nordlund, et al. 2007. Fundamental differences between sph and grid methods. *Monthly Notices of the Royal Astronomical Society* 380(3): 963–978.

Balsara, D.S., and D.S. Spicer. 1999. A staggered mesh algorithm using high order Godunov fluxes to ensure solenoidal magnetic fields in magnetohydrodynamic simulations. *Journal of Computational Physics* 149(2): 270–292.

Bauer, A., and V. Springel. 2012. Subsonic turbulence in smoothed particle hydrodynamics and moving-mesh simulations. *Monthly Notices of the Royal Astronomical Society* 423(3): 2558–2578.

Beckwith, K., and J.M. Stone. 2011. A second-order Godunov method for multi-dimensional relativistic magnetohydrodynamics. *The Astrophysical Journal Supplement Series* 193(1): 6.

Beisbart, C. 2019. What is validation of computer simulations? Toward a clarification of the concept of validation and of related notions. In *Computer simulation validation*, ed. Beisbart, C., and Saam, N., pp. 35–67. Berlin: Springer.

Bertschinger, E. 1985. Self-similar secondary infall and accretion in an einstein-de sitter universe. *The Astrophysical Journal Supplement Series* 58: 39–65.

Brio, M., and C.C. Wu. 1988. An upwind differencing scheme for the equations of ideal magnetohydrodynamics. *Journal of Computational Physics* 75(2): 400–422.

Bryan, G.L., M.L. Norman, B.W. O'Shea, T. Abel, J.H. Wise, M.J. Turk, D.R. Reynolds, D.C. Collins, P. Wang, S.W. Skillman, et al. 2014. Enzo: An adaptive mesh refinement code for astrophysics. *The Astrophysical Journal Supplement Series* 211(2): 19.

Burkert, A., and P. Bodenheimer. 1993. Multiple fragmentation in collapsing protostars. *Monthly Notices of the Royal Astronomical Society* 264(4): 798–806.

Calder, A.C., B. Fryxell, T. Plewa, R. Rosner, L. Dursi, V. Weirs, T. Dupont, H. Robey, J. Kane, B. Remington, et al. 2002. On validating an astrophysical simulation code. *The Astrophysical Journal Supplement Series* 143(1): 201.

Chandrasekhar, S. 1961. *Hydrodynamic and hydromagnetic stability*. London and New York: Clarendon Press.

Close, J., J. Pittard, T., Hartquist, and S. Falle. 2013. Ram pressure stripping of the hot gaseous haloes of galaxies using the k–ε sub-grid turbulence model. *Monthly Notices of the Royal Astronomical Society* 436(4): 3021–3030.

Colgate, S.A., and R.H. White. 1966. The hydrodynamic behavior of supernovae explosions. *The Astrophysical Journal* 143: 626.

Couchman, H., P. Thomas, and F. Pearce. 1994. Hydra: An adaptive–mesh implementation of PPPM–SPH. astro-ph/9409058.

Dilts, G.A. 1999. Moving-least-squares-particle hydrodynamics—I. Consistency and stability. *International Journal for Numerical Methods in Engineering* 44(8): 1115–1155.

Dolag, K., and F. Stasyszyn. 2009. An MHD GADGET for cosmological simulations. *Monthly Notices of the Royal Astronomical Society* 398(4): 1678–1697.

Einfeldt, B., C.-D. Munz, P.L. Roe, and B. Sjögreen. 1991. On Godunov-type methods near low densities. *Journal of Computational Physics* 92(2): 273–295.

Emery, A.F. 1968. An evaluation of several differencing methods for inviscid fluid flow problems. *Journal of Computational Physics* 2(3): 306–331.

Evrard, A.E. 1988. Beyond n-body-3d cosmological gas dynamics. *Monthly Notices of the Royal Astronomical Society* 235: 911–934.

Flash User Guide. 2019. https://flash.rochester.edu/site/flashcode/user_support/. Accessed: 2020-11-14.

Frank, A., T.W. Jones, D. Ryu, and J.B. Gaalaas. 1995. The mhd kelvin-helmholtz instability: A two-dimensional numerical study. *The Astrophysical Journal* 460: 777.

Frenk, C., S. White, P. Bode, J. Bond, G. Bryan, R. Cen, H. Couchman, A.E. Evrard, N. Gnedin, A. Jenkins, et al. 1999. The santa barbara cluster comparison project: A comparison of cosmological hydrodynamics solutions. *The Astrophysical Journal* 525(2): 554.

Fromang, S., P. Hennebelle, and R. Teyssier. 2006. A high order Godunov scheme with constrained transport and adaptive mesh refinement for astrophysical magnetohydrodynamics. *Astronomy & Astrophysics* 457(2): 371–384.

Fryxell, B., K. Olson, P. Ricker, F. Timmes, M. Zingale, D. Lamb, P. MacNeice, R. Rosner, J. Truran, and H. Tufo. 2000. Flash: An adaptive mesh hydrodynamics code for modeling astrophysical thermonuclear flashes. *The Astrophysical Journal Supplement Series* 131(1): 273.

Gardiner, T.A., and J.M. Stone. 2005. An unsplit Godunov method for ideal mhd via constrained transport. *Journal of Computational Physics* 205(2): 509–539.

Gresho, P.M., and S.T. Chan. 1990. On the theory of semi-implicit projection methods for viscous incompressible flow and its implementation via a finite element method that also introduces a nearly consistent mass matrix. Part 2: Implementation. *International Journal for Numerical Methods in Fluids* 11(5): 621–659.

Guan, X., and C.F. Gammie. 2008. Axisymmetric shearing box models of magnetized disks. *The Astrophysical Journal Supplement Series* 174(1): 145.

Gueguen, M. Forthcoming. A tension within code comparisons. *British Journal for the Philosophy of Science*. http://philsci-archive.pitt.edu/id/eprint/19227

Hawley, J.F., and J.M. Stone. 1995. MOCCT: A numerical technique for astrophysical MHD. *Computer Physics Communications* 89(1–3): 127–148.

Heitmann, K., P.M. Ricker, M.S. Warren, and S. Habib. 2005. Robustness of cosmological simulations. I. Large-scale structure. *The Astrophysical Journal Supplement Series* 160(1): 28.

Heß, S., and V. Springel. 2010. Particle hydrodynamics with tessellation techniques. *Monthly Notices of the Royal Astronomical Society* 406(4): 2289–2311.

Hopkins, P.F. 2013. A general class of Lagrangian smoothed particle hydrodynamics methods and implications for fluid mixing problems. *Monthly Notices of the Royal Astronomical Society* 428(4): 2840–2856.

Hopkins, P.F. 2015. A new class of accurate, mesh-free hydrodynamic simulation methods. *Monthly Notices of the Royal Astronomical Society* 450(1): 53–110.

Hopkins, P.F., and M.J. Raives. 2016. Accurate, meshless methods for magnetohydrodynamics. *Monthly Notices of the Royal Astronomical Society* 455(1): 51–8.

Huang, J., and L. Greengard. 1999. A fast direct solver for elliptic partial differential equations on adaptively refined meshes. *SIAM Journal on Scientific Computing* 21(4): 1551–1566.

Jeans, J.H. 1902. I. The stability of a spherical nebula. *Philosophical Transactions of the Royal Society of London. Series A, Containing Papers of a Mathematical or Physical Character* 199(312–320): 1–53.

Jun, B.-I., M.L. Norman, and J.M. Stone. 1995. A numerical study of rayleigh-taylor instability in magnetic fluids. *The Astrophysical Journal* 453: 332.

Keppens, R. 2004. Nonlinear magnetohydrodynamics: Numerical concepts. *Fusion Science and Technology* 45(2T): 107–114.

Khokhlov, A.M. 1998. Fully threaded tree algorithms for adaptive refinement fluid dynamics simulations. *Journal of Computational Physics* 143(2): 519–543.

Klein, R.I., C.F. McKee, and P. Colella. 1994. On the hydrodynamic interaction of shock waves with interstellar clouds. 1: Nonradiative shocks in small clouds. *The Astrophysical Journal* 420: 213–236.

Liska, R., and B. Wendroff. 2003. Comparison of several difference schemes on 1d and 2d test problems for the euler equations. *SIAM Journal on Scientific Computing* 25(3): 995–1017.

Londrillo, P., and L. Del Zanna. 2000. High-order upwind schemes for multidimensional magnetohydrodynamics. *The Astrophysical Journal* 530(1): 508.

MacLaurin, C. 1801. *A treatise on fluxions: In two volumes,* vol. 1. London: W. Baynes and W. Davis.

McNally, C.P., W. Lyra, and J.-C. Passy. 2012. A well-posed Kelvin-Helmholtz instability test and comparison. *The Astrophysical Journal Supplement Series* 201(2): 18.

Morris, J.P. 1996. A study of the stability properties of smooth particle hydrodynamics. *Publications of the Astronomical Society of Australia* 13: 97–102.

Morrison, M. 2015. *Reconstructing reality: Models, mathematics, and simulations.* Oxford Studies in Philosophy. Oxford: Oxford University Press.

Noh, W.F. 1987. Errors for calculations of strong shocks using an artificial viscosity and an artificial heat flux. *Journal of Computational Physics* 72(1): 78–120.

Oberkampf, W.L., and C.J. Roy. 2010. *Verification and validation in scientific computing.* Cambridge: Cambridge University Press.

Oreskes, N., K. Shrader-Frechette, and K. Belitz. 1994. Verification, validation, and confirmation of numerical models in the earth sciences. *Science* 263(5147): 641–646.

Orszag, S.A., and C.-M. Tang. 1979. Small-scale structure of two-dimensional magnetohydrodynamic turbulence. *Journal of Fluid Mechanics* 90(1): 129–143.

Pakmor, R., A. Bauer., and V. Springel. 2011. Magnetohydrodynamics on an unstructured moving grid. *Monthly Notices of the Royal Astronomical Society* 418(2): 1392–1401.

Pakmor, R., V. Springel, A. Bauer, P. Mocz, D.J. Munoz, S.T. Ohlmann, K. Schaal, and C. Zhu. 2016. Improving the convergence properties of the moving-mesh code AREPO. *Monthly Notices of the Royal Astronomical Society* 455(1): 1134–1143.

Price, D.J. 2008. Modelling discontinuities and Kelvin–Helmholtz instabilities in SPH. *Journal of Computational Physics* 227(24): 10040–10057.

Price, D.J., J. Wurster, T.S. Tricco, C. Nixon, S. Toupin, A. Pettitt, C. Chan, D. Mentiplay, G. Laibe, S. Glover, et al. 2018. Phantom: A smoothed particle hydrodynamics and magnetohydrodynamics code for astrophysics. *Publications of the Astronomical Society of Australia* 35:e031.

Read, J., T. Hayfield, and O. Agertz. 2010. Resolving mixing in smoothed particle hydrodynamics. *Monthly Notices of the Royal Astronomical Society* 405(3): 1513–1530.

Ritchie, B.W., and P.A. Thomas. 2001. Multiphase smoothed-particle hydrodynamics. *Monthly Notices of the Royal Astronomical Society* 323(3): 743–756.

Robertson, B.E., A.V. Kravtsov, N.Y. Gnedin, T. Abel, and D.H. Rudd. 2010. Computational Eulerian hydrodynamics and Galilean invariance. *Monthly Notices of the Royal Astronomical Society* 401(4): 2463–2476.

Sedov, L. 1959. *Similarity and dimensional methods in mechanics.* New York: Academic. cahill me and taub ah, 1971. *Commun. Math. Phys* 21(1).

Shu, C.-W., and S. Osher. 1989. Efficient implementation of essentially non-oscillatory shock-capturing schemes, II. In *Upwind and high-resolution schemes,* pp. 328–374. Berlin: Springer.

Sod, G.A. 1978. A survey of several finite difference methods for systems of nonlinear hyperbolic conservation laws. *Journal of Computational Physics* 27(1): 1–31.

Springel, V. 2005. The cosmological simulation code gadget-2. *Monthly Notices of the Royal Astronomical Society* 364(4): 1105–1134.

Springel, V. 2010. *E pur si muove*: Galilean-invariant cosmological hydrodynamical simulations on a moving mesh. *Monthly Notices of the Royal Astronomical Society* 401(2): 791–851.

Stone, J.M.: The athena code test page. https://www.astro.princeton.edu/~jstone/Athena/tests/. Accessed: 2019-11-30.

Stone, J.M., T.A. Gardiner, P. Teuben, J.F. Hawley, and J.B. Simon. 2008. Athena: A new code for astrophysical MHD. *The Astrophysical Journal Supplement Series* 178(1): 137.

Stone, J.M., J.F. Hawley, C.R. Evans, and M.L. Norman. 1992. A test suite for magnetohydrody-namical simulations. *The Astrophysical Journal* 388: 415–437.

Stone, J.M., and M.L. Norman. 1992. ZEUS-2D: A radiation magnetohydrodynamics code for astrophysical flows in two space dimensions. I-the hydrodynamic algorithms and tests. *The Astrophysical Journal Supplement Series* 80: 753–790.

Teyssier, R. 2002. Cosmological hydrodynamics with adaptive mesh refinement-a new high resolution code called ramses. *Astronomy & Astrophysics* 385(1): 337–364.

Torrilhon, M. 2003. Uniqueness conditions for riemann problems of ideal magnetohydrodynamics. *Journal of Plasma Physics* 69(3): 253.

Tóth, G. 2000. The $\nabla \cdot b = 0$ constraint in shock-capturing magnetohydrodynamics codes. *Journal of Computational Physics* 161(2): 605–652.

Wadsley, J., G. Veeravalli, and H. Couchman. 2008. On the treatment of entropy mixing in numerical cosmology. *Monthly Notices of the Royal Astronomical Society* 387(1): 427–438.

Wadsley, J.W., B.W. Keller, and T.R. Quinn. 2017. Gasoline2: A modern smoothed particle hydrodynamics code. *Monthly Notices of the Royal Astronomical Society* 471(2): 2357–2369.

Wadsley, J.W., J. Stadel, and T. Quinn. 2004. Gasoline: A flexible, parallel implementation of treesph. *New Astronomy* 9(2): 137–158.

Winsberg, E. 2010. *Science in the age of computer simulation.* Chicago: University of Chicago Press.

Winsberg, E. 2018. *Philosophy and climate sciencde.* Cambridge: Cambridge University Press.

Woodward, P., and P. Colella. 1984. The numerical simulation of two-dimensional fluid flow with strong shocks. *Journal of Computational Physics* 54(1): 115–173.

Yee, H.C., M. Vinokur, and M.J. Djomehri. 2000. Entropy splitting and numerical dissipation. *Journal of Computational Physics* 162(1): 33–81.

Zel'Dovich, Y.B. 1970. Gravitational instability: An approximate theory for large density pertur-bations. *Astronomy and Astrophysics* 5: 84–89.

Chapter 10
(What) Do We Learn from Code Comparisons? A Case Study of Self-Interacting Dark Matter Implementations

Helen Meskhidze

Abstract There has been much interest in the recent philosophical literature on increasing the reliability and trustworthiness of computer simulations. One method used to investigate the reliability of computer simulations is code comparison. Gueguen, however, has offered a convincing critique of code comparisons, arguing that they face a critical tension between the diversity of codes required for an informative comparison and the similarity required for the codes to be comparable. Here, I reflect on her critique in light of a recent code comparison investigating self-interacting dark matter in two computer simulation codes. I argue that the informativeness of this particular code comparison was due to its targeted approach and narrow focus. Its targeted approach (i.e., only the dark matter modules) allowed for simulation outputs that were diverse enough for an informative comparison and yet still comparable. Understanding the comparison as an instance of eliminative reasoning narrowed the focus: we could investigate whether code-specific differences in implementation contributed significantly to the results of self-interacting dark matter simulations. Based on this case study, I argue that code comparisons can be conducted in such a way that they serve as a method for increasing our confidence in computer simulations being, as Parker defines, adequate-for-purpose.

10.1 Introduction

Following influential works questioning the epistemic standing of computer simulations (see, e.g., Winsberg 2010), the recent philosophical literature has turned to investigations of the reliability of computer simulations and methods of increasing this reliability (see, e.g., Mättig 2021; Boge forthcoming). Some philosophers have begun discussing the limitations of popular proposals for increasing the epistemic

H. Meskhidze (✉)
Department of Logic and Philosophy of Science, University of California, Irvine, CA, USA
e-mail: emeskhid@uci.edu

© The Author(s) 2023
N. Mills Boyd et al. (eds.), *Philosophy of Astrophysics*, Synthese Library 472,
https://doi.org/10.1007/978-3-031-26618-8_10

standing of computer simulations. One example of a proposal that has been found to be lacking is robustness analysis. Though robustness analysis is a central practice in modeling, Gueguen (2020) has outlined its limitations. She argues that robustness analysis, at least in the context of large-scale simulations, is insufficient to ground the reliability of such simulation results. This is because "robust but mutually exclusive predictions can obtain in N-body simulations" (Gueguen 2020, 1198). Indeed, numerical artifacts—or errors introduced by the numerical prescription adopted by the simulation code—may themselves be at the heart of a robust prediction. Thus, she argues, a prediction being robust is insufficient to warrant our trust in it.

While some like Gueguen investigate existing methods for increasing reliability, others put forward proposals for new methods. Smeenk and Gallagher (2020), for example, consider the possibility of using eliminative reasoning. They begin with the recognition that the convergence of simulation results is insufficient to ground trust in those results. This is because cosmologists do not have the required ensemble of models (the "ideal ensemble") over which they would require convergence (Smeenk and Gallagher 2020, 1229–30); the parameter space over which their actual ensembles span is quite narrow.[1] Furthermore, even if they did find a convergence of results, that convergence would not immediately indicate a reliable feature as the convergence could be due to numerical artifacts, as Gueguen argues. Instead, Smeenk and Gallagher argue that we should shift to an eliminative approach where we find and avoid sources of error.

Yet another means of increasing our confidence in computer simulations is through code comparisons. These are comparisons of the results of different computer simulation codes, which often feature different implementations of some processes of interest. Gueguen (forthcoming) provides the only discussion of code comparisons in astrophysics found in the philosophical literature. She, quite compellingly, argues that the diversity of parameters and implementations needed for an informative code comparison ultimately undermines the feasibility of the comparison: incorporating the necessary diversity makes the codes incomparable but making the codes comparable eliminates the necessary diversity.

In this chapter, I reflect on a recent code comparison project that I was a part of—one in which we investigated two different implementations of self-interactions amongst dark matter (DM) particles in two computer simulation codes (Meskhidze et al. 2022). I argue that the informativeness of our comparison was made possible due to its targeted approach and narrow focus. In particular (as elaborated in Sect. 10.3), this targeted approach allowed for simulation outputs that were diverse enough for an informative comparison and yet still comparable. Understanding

---

[1] Smeenk and Gallagher are considering the kind of robustness analysis Gueguen argues against—the kind presented by Weisberg. One might wonder to what extent their critiques would generalize to the style of robustness analysis proposed by Schupbach (2018). Though Schupbach's proposal seems much closer to eliminative reasoning, Smeenk and Gallagher argue that the demand for diversity with respect to some particular features may still be too strong, especially in light of the lack of modularity evident in cosmological simulations (2020, 1231).

the comparison as an instance of eliminative reasoning narrowed the focus. We could investigate whether code-specific differences in implementation contributed significantly to the results of self-interacting dark matter (SIDM) simulations. I take this code comparison project to be a proof-of-concept: code comparisons can be conducted in such a way that they serve as a method for increasing our confidence in computer simulations. Indeed, they may be used as part of a larger project of eliminative reasoning but may also be seen as ensuring that particular simulation codes are, as Parker (2020) defines, adequate-for-purpose.

I begin (Sect. 10.2) by discussing previous code comparisons conducted in astrophysics. These are the subject of Gueguen's (forthcoming) critique and helped inform the methodology adopted in the code comparison discussed in this paper. I then outline our methodology in the comparison and the results of the comparison (Sect. 10.3). I conclude by reflecting on what enabled the success of this latter comparison (Sect. 10.4).

10.2 Code Comparisons in Astrophysics

Code comparison projects in astrophysics can be traced back to Lecar's (1968) comparison of the treatment of a collapsing 25-body system by 11 codes. Such comparison projects began in earnest with the Santa Barbara Cluster comparison project at the turn of the century (Frenk et al. 1999). The Santa Barbara Cluster comparison project demonstrated the benefits of adopting the same initial conditions across a variety of codes and comparing the results. This project—and nearly all subsequent astrophysics code comparison projects—was especially interested in the differences that might be found between the two most common ways of implementing gravitational interactions: particle-based approaches and mesh-based approaches. Particle-based approaches track particles' movements and interactions in the simulation volume while mesh-based approaches divide the simulation volume into a mesh and track the flow of energy through the mesh. These methods of implementing gravitational interactions are referred to as a simulation's underlying "gravity solver." The strengths and weaknesses of the different gravity solvers have been the subject of much study (see, e.g., Agertz et al. 2007).

Despite their different scopes and purposes, contemporary code comparisons follow one of two methodologies. They either (1) adopt the same initial conditions, evolve the codes forward with their preferred/default configurations without any tuning or calibration, and compare the results, or (2) adopt the same initial conditions, calibrate the codes to be as similar as possible, and compare the results. The first methodology is exemplified by the Aquila code comparison project while the second is exemplified by the Assembling Galaxies Of Resolved Anatomy (AGORA) code comparison project.

The two different methodologies ground different types of conclusions. The methodology of the Aquila project allowed the authors to identify the "role and importance of various mechanisms" even when the codes did not agree (Scannapieco et al. 2012, 1742). Take, for instance, when they compared the simulated stellar masses.[2] They observed large code-to-code scatter with mesh-based codes predicting the largest stellar masses. This indicated that the feedback implementations[3] in these mesh-based codes were not efficient since, typically, feedback suppresses star formation and yields a smaller stellar mass overall (Scannapieco et al. 2012, 1733). The conclusion drawn from this analysis applied to all three of the mesh-based codes tested in the Aquila comparison. Further, the conclusion is about the overall implementation of a physical process (feedback) in the codes, not about any particular parameter choices.

In the case of the AGORA project, the authors learned about the internal workings of each individual code and even "discovered and fixed" some numerical errors in the participating codes (Kim et al. 2016, 28). An example of the style of analysis prominent in their project can be found in their discussion of supernova feedback implementations as well.[4] They noted that hot bubbles can be seen in the results of one simulation code (CHANGA) but not in another (GASOLINE), even though both codes are particle-based. The AGORA authors argued that the cause of the observed difference was that the two codes implemented smoothing in their hydrodynamics differently: CHANGA uses 64 neighbors in its calculation while GASOLINE uses 200 (Kim et al. 2016, 13, fn72). But what conclusion could be drawn from this discrepancy? Certainly, one could tune the parameters, bringing these two particular simulation results into better agreement but tuning the parameters for this particular result would likely make other predictions diverge. Further, one would not develop any further insight into the physical process being modelled merely by tuning the parameters. Indeed, the AGORA authors do not recommend that either code adopt the other's smoothing parameters. They even note that the resolution of their simulations "may not be enough for the particle-based codes to resolve [the area under consideration]" (Kim et al. 2016, 13). In sum, unlike what we saw with Aquila, the conclusions drawn by the AGORA authors relate to specific codes and are about particular parameter choices for the physical processes under investigation.

Let us now step back to assess the comparisons themselves. The stated goal of the Aquila comparison project was to determine whether the codes would "give similar results if they followed the formation of a galaxy in the same dark matter halo" (Scannapieco et al. 2012, 1728). How did it fare with respect to that goal? Not only did the simulated galaxies show a "large spread in properties," "none of them [had] properties fully consistent with theoretical expectations or observational constraints" (ibid., 1742). The substantial disagreements amongst the codes led the authors to claim: "There seems to be little predictive power at this point in state-of-

---

[2] I.e., the stellar/baryonic component of a galaxy, in contrast to the dark matter component.

[3] Here, "feedback" is a catch-all term for various mechanisms that inject energy into the system, heating it and preventing it from overcooling.

[4] Supernova feedback corresponds to the energy of a star exploding and, like feedback generally, injects energy into the system.

the-art simulations of galaxy formation" (ibid.). In sum, the results of the Aquila project were neither convergent, nor did they show consistency with theoretical expectations or observational constraints.

What about the AGORA project? Its stated goal was "to raise the realism and predictive power of galaxy simulations and the understanding of the feedback processes that regulate galaxy 'metabolism'" (Kim et al. 2014, 1). Understood very modestly, perhaps they did achieve this goal. After all, they found and eliminated numerical errors in the codes they were comparing. However, the parameter tuning required to bring the codes into agreement ought to make us skeptical of the comparison substantially increasing the realism or predictive power of the codes beyond the very narrow conditions tested.

One might be tempted to reassess the Aquila project in terms of the goals of the AGORA project. Though the simulations considered in the Aquila comparison did not yield similar results, the project did seem to be better positioned to "raise the realism and predictive power" of the simulations. This is because it did not tune the simulations to yield similar results but instead focused on better understanding the impact(s) of various physical mechanisms in the codes. If we were to assess the Aquila comparison with respect to these goals (as opposed to those stated by the Aquila authors themselves), the comparison seems much more fruitful.

Gueguen's assessment of the Aquila project is that it fails because it "fails to compare similar targets" (forthcoming, 22). Her assessment of the AGORA project is that it fails because the infrastructure required to conduct the comparison "itself becomes an unanalyzed source of artifacts" (ibid.)[5] and because the parameter tuning required to conduct an "apples-to-apples" comparison such as AGORA is unconstrained by theoretical considerations. This leads her to argue that there is a 'tension' between the diversity of implementations necessary to ground trust and the similarity of implementations necessary to carry out a code comparison itself. Given this fatal analysis of two significant code comparisons in the astrophysics literature, we must ask: are code comparisons futile? By using a case study comparing two implementations of self-interacting dark matter, I hope to demonstrate that such comparisons can be fruitful.

10.3 Comparing Self-Interacting Dark Matter Implementations

It is now accepted by astrophysicists and cosmologists that contemporary simulations of gravitational systems must incorporate some form of dark matter. The prominent cold dark matter paradigm (CDM) began to face challenges on smaller scales in the 2010s (Bullock and Boylan-Kolchin 2017). Initially, these issues

---

[5] This infrastructure included a common initial conditions generator and platform to complete the final analysis of the codes.

included the "core-cusp problem," the "missing satellites problem," and the "too-big-to-fail problem."[6] Many have since argued that incorporating baryonic effects in simulations can satisfactorily solve some of these small-scale issues (see, e.g., Sawala et al. 2016). Even so, some issues remain for CDM. Today it is argued that CDM cannot explain the range of observations of the inner rotation curves of spiral galaxies, an issue dubbed the "diversity problem." In particular, observational evidence shows a significant spread in the inner DM distribution of galaxies in and around the Milky Way but simulations of CDM with baryons do not capture this diversity (Kaplinghat et al. 2019).

In addition to exploring the effects of incorporating baryonic feedback in simulations, astrophysicists are exploring alternative models of DM. SIDM models were proposed in the literature as early as the 1990s (see, e.g., Carlson et al. 1992). However, Spergel and Steinhardt (2000) were the first to propose self-interactions in response to the small-scale challenges outlined above. In very broad terms, self-interactions allow energy transfer from the outermost hot regions of a halo inwards, flattening out the velocity dispersion profile of the halo and "cuspy" density profiles. This allows a DM halo to achieve thermal equilibrium and have a more spherical shape (Tulin and Yu 2018, 8). Though the ability of SIDM to create DM cores in simulations is still valued, more recently, SIDM has been investigated as a solution to the diversity problem (Robles et al. 2017; Fitts et al. 2019). Thus, the *purpose* of simulations with SIDM nowadays is determining whether SIDM can be used to alleviate the diversity problem and so, according to Parker's (2020) framework, simulations modeling SIDM must be adequate for that purpose.[7]

Depending on a simulation's underlying treatment of gravitational interactions, there are various methods for implementing self-interactions amongst the DM particles. There had not been a comparison of the results of these different implementations prior to the paper serving as the case study here. It was with this lack of any prior comparison of SIDM implementations as well as Gueguen's critiques of prior code comparisons that our team began. Working as an interdisciplinary team made up of astrophysicists and philosophers, we designed our comparison methodology in a way that we hoped would avoid the tension Gueguen describes and

---

[6] The "core-cusp problem" refers to the discrepancy between simulation results (which yield "cuspy" halos with density profiles that go as r^{-1}) and observations of dark matter halos (which are more "cored" with density profiles that go as r^0). The "missing satellites" problem refers to the large discrepancy between the low number of observed luminous satellite galaxies and the high number of dark matter substructures predicted by CDM. The "too-big-to-fail" problem refers to the further issue that those dark matter substructures predicted are too big to fail to form stars and become observable galaxies.

[7] Parker proposes the adequacy-for-purpose framework as an alternative to a view that holds that a model's quality is a function of how accurately and completely it represents a target. On her account, what is required of a model may depend on the target, the user, the methodology, the circumstances, and the purpose/goal itself (2020). This is, clearly, a broader class of considerations than just a model's representational capacity. As noted, I take the purpose of SIDM simulations to be investigating whether SIDM can alleviate the diversity problem and the users to be astrophysicists (not particle physicists).

would also be useful for deepening our understanding of SIDM. I turn to presenting our methodology and results below.

10.3.1   SIDM in Gizmo and Arepo

For our comparison, we required two distinct astrophysical simulation codes implementing SIDM. Our group had members who were well acquainted with one simulation code: Gizmo. Gizmo adopts a mesh-free finite mass, finite volume method for its underlying gravity solver. This means the fundamental elements are cells of finite mass and volume representing "particles" distributed throughout the simulation volume (Hopkins 2015). The other simulation code we chose to compare it to was Arepo. Unlike Gizmo, Arepo adopts a moving, unstructured mesh where the simulation volume is partitioned into non-regular cells, and the cells themselves are allowed to move/deform continuously (Springel 2010). These codes are both popular amongst simulators and thus worthwhile to compare: papers introducing the new iteration of the codes receive hundreds of citations and both have been the subject of former code comparisons. They were both, for example, included in the AGORA project.

Given the differences in their gravity solvers, we knew that the codes' implementations of SIDM would be sufficiently different to allow for a fruitful comparison. Beyond this, the SIDM treatments themselves also differ in their approach to the underlying "particles." In N-body simulation codes, the "particles" are not meant to represent individual dark matter particles but rather patches of phase space representing collections of such particles. Further differences in implementation arise from different means of handling this underlying fact in Gizmo vs. Arepo.

In Gizmo, one begins by setting the (distance) range over which DM particles can interact (the "smoothing length"). Then, one calculates the rate of interaction between the particles. This is a function of, amongst other variables, the interaction cross-section, the mass of the target particle, and their difference in velocity. The most important parameter is the interaction cross-section. There have been many projects investigating what cross-section is required for SIDM to recreate observations, (see Vogelsberger et al. 2012 or Rocha et al. 2013). Some have even proposed a velocity-dependent cross-section (Randall et al. 2008).

Once the rate of interaction is calculated, the simulation must determine whether an interaction actually takes place. To do so, a random number is drawn for each pair of particles that are sufficiently nearby such that their probability of interaction is non-zero. Finally, if an interaction does occur and the pair scatters, a Monte Carlo analysis is done to determine the new velocity directions. As Rocha et al. write, "If a pair does scatter, we do a Monte Carlo for the new velocity directions, populating these parts of the phase space and deleting the two particles at their initial phase-space locations" (2013, 84). Note here that these authors are taking the nature of the particles as phase space patches quite literally; these "particles" can simply be deleted and repopulated with new, updated properties.

Arepo's implementation of SIDM begins with a search of the nearest neighbors of a particle; one must specify the number of neighbors to search when running the code. The probability of interaction with those neighbors is then calculated as a function of the simulation particle's mass, the relative velocity between the particles being compared, the interaction cross-section, and the smoothing length. Like Gizmo, Arepo then determines if an interaction takes place by drawing a random number between zero and one. The particles are then assigned new velocities based on the center-of-mass velocity of the pair. As Vogelsberger writes: "once a pair is tagged for collision we assign to each particle a new velocity . . ." (2012, 3). Clearly, the procedure used by Arepo is distinct from that of Gizmo: whereas the Gizmo authors appealed to the phase-space locations of the macro-particles to interpret interactions, the Arepo authors seem to think about the particles in their simulation more directly as particles that collide and scatter off one another.[8]

In sum, the two codes do have some common SIDM parameters—the interaction cross-section and the mass of a DM particle—but they also have parameters that are specific to their particular SIDM implementation—the smoothing length in Gizmo vs. the number of neighbors searched in Arepo. Beyond particular parameters, the treatment of SIDM in the two codes is different. In a review article reflecting on these distinct SIDM implementations, Tulin and Yu write "It is unknown what differences, if any, may arise between these various methods" (2018, 27), suggesting that a comparison of the methods would indeed be valuable.

10.3.2 Methodology of Our Code Comparison

Having chosen our two simulation codes and verified that comparing their SIDM implementations would be fruitful, we next needed to decide on a comparison methodology. From the beginning, we decided to make our scope much narrower than that of Aquila and AGORA: our goal was to compare SIDM implementations, not entire astrophysical codes. More specifically, though Gizmo and Arepo can model baryonic effects and more complex systems, we only used their gravitational physics and SIDM modules and modeled an isolated dwarf halo. This narrow scope, we hoped, would allow us to avoid the issues with code-comparison projects that Gueguen (forthcoming) outlined.

Following both the Aquila and AGORA projects, we decided to adopt identical initial conditions. For the SIDM cross-section per unit mass, we adopted identical values. However, for parameters that were different between the two codes and arose due to the codes' different gravity solvers and treatment of SIDM, we chose to follow Aquila's comparison methodology and have each code adopt its preferred

---

[8] For more details on the SIDM implementation in each code, see Meskhidze et al. (2022). For the papers outlining each code's SIDM methodology, see Rocha et al. (2013) and Vogelsberger et al. (2012).

(default) parameters.[9] This was because we thought it would be more informative with regard to the physical processes being modeled to allow each simulation to adopt its preferred parameters. Finally, because we had access to both simulation codes ourselves,[10] we were able to run and analyze the results of the simulations using an identical framework: the same computing cluster and plotting scripts.

The comparison proceeded with a dynamic methodology. This was partly because we were unsure what types of comparisons would be most fruitful and partly due to the interdisciplinary nature of our team. The different (disciplinary) perspectives we brought often overlapped but rarely coincided. Frequently, one comparison would prompt some members to propose another comparison but other members to propose a different way of presenting the results of the first comparison.[11]

Let me offer some concrete examples of questions we grappled with that shaped the methodology. Though we knew we wanted to model an isolated DM halo, when we began, we were unsure what kind of density profile to adopt for our halo. Should we adopt a more conservative halo profile similar to ones found in the literature? Or, should we model something more extreme—perhaps a very concentrated halo—and investigate how each SIDM implementation would handle such a halo? We ended up choosing the former because we were more interested in understanding differences between the codes in the parameter space most relevant to simulators. We faced a similar question regarding the range of SIDM cross-sections to simulate. Should we simulate those of interest in the literature? Or try to push the codes to their extremes and adopt a much wider range of SIDM cross-sections? Again, we chose to model the narrower range as we expected it to be more informative to the investigation of the differences that arise between the codes when modeling contexts of interest.

In the end, our write-up contained the results of about 20 distinct simulations, though we likely ran three to four times this many simulations overall.[12] We presented the results of the initial conditions evolved through 10 billion years for

---

[9] See Table 1 of Meskhidze et al. (2022) for a detailed list of these parameters and the values we adopted.

[10] Gizmo is open-source and Arepo's public release incidentally coincided with our project timeline. Each code is available online; Gizmo is available at https://bitbucket.org/phopkins/gizmo-public/src/master while Arepo is available at https://gitlab.mpcdf.mpg.de/vrs/arepo. Note, however, that the public release of Arepo does not include the SIDM implementation which we requested separately.

[11] Some may object that the methodology outlined above seems ad hoc. While it is true that the questions we asked on our project evolved, the answers to those questions are nonetheless well-justified. Indeed, I would argue that some dynamicism in methodology is an inevitable feature of scientific investigations.

[12] The discrepancy between the number of simulations run and the number of simulations compared in the final paper was partly because of our dynamic methodology: some of the simulations we ran did not lead to interesting or informative comparisons. However, other simulations functioned as checks along the way. For instance, we ran some simulations to check our results against those reported in the literature and reran some simulations to ensure our results were replicable with the same parameters but a different random seed for the initial conditions.

10 different sets of parameters in each of the two simulation codes. The results spanned two resolutions: a baseline simulation suite with one million particles and a high-resolution suite with five million particles. We tested 3 different SIDM cross-sections (1, 5, and 50 cm^2 g^{-1}) in addition to the CDM simulations. We also compared the results of increasing/decreasing the probability of SIDM self-interactions (via the smoothing length and neighbors searched in Gizmo and Arepo respectively) from their default values.

10.3.3 Results of Our Code Comparison

Below, I very briefly outline the results of the comparison. Some may, of course, find the results scientifically interesting, but my goal here is to highlight the types of conclusions we were able to draw. Overall, we found good agreement between the codes: the codes exhibited better than 30% agreement for the halo density profiles investigated. In other words, at all radii, the density of a halo in one code was within 30% of the other code's prediction. This is considered quite remarkable agreement between the codes, especially considering that the error was often a sizable portion of the difference.[13]

Our comparison found a few other notable trends:

1. Increasing the SIDM cross-section in both codes flattened out the density and velocity dispersion profiles in the innermost region of the halo. The density profile become more "cored" as more energy was transferred from the outermost regions of the halo inwards.
2. Increasing the resolution (from one to five million particles) brought the results of the two simulation codes into better agreement.
3. Neither code exhibited core-collapse behavior across the cross-sections tested, despite our group initially anticipating that they would.[14]
4. The number of self-interactions in the codes scaled nearly linearly with the cross-section. For example, the simulations that adopted SIDM interaction cross-sections of 50 cm^2 g^{-1} exhibited 8 times as many interactions as those that adopted 5 cm^2 g^{-1}. Similarly, those with interaction cross-sections of 5 cm^2 g^{-1} exhibited 4 times as many self-interactions as those with 1 cm^2 g^{-1}.
5. Changing the code-specific SIDM parameters (i.e., the smoothing factor in Gizmo and the number of neighbors considered in Arepo) did change the inner

---

[13] The error we used was the Poisson error, calculated as the density at each histogram bin divided by the square root of the number of particles in the bin.

[14] Gravitothermal core collapse of SIDM halos has been investigated by Sameie et al. (2020) and Turner et al. (2021) among others. In this process, the DM core grows until it reaches the outer parts of the halo. This growth prevents the halo from reaching thermal equilibrium and leads to its collapse, which then creates a very steep cusp in the density profile. It is thought that SIDM cross-sections >10 cm^2 g^{-1} lead to gravothermal collapse.

halo profile somewhat (about a 10% difference at $r < 300$ pc) but there was no general trend evident with the increase/decrease of those parameters.

6. The degree of agreement between the codes (30%) is smaller than what can be observationally constrained. In other words, observations (and their systematic errors) are not precise enough to detect the degree of difference we find between the two codes.

7. Finally, and most significantly, the differences between the results of the two codes (understood in terms of their density and velocity dispersion profiles as well as the number of DM self-interactions) for any particular interaction cross-section were much smaller than the differences between the various SIDM cross-sections tested.

The results listed above (especially 6 and 7) led us to conclude that "SIDM core formation is robust across the two different schemes and conclude that [the two] codes can reliably differentiate between cross-sections of 1, 5, and 50 cm^2 g^{-1} but finer distinctions would require further investigation" (Meskhidze et al. 2022, 1). In other words, if the goal is to use these codes to constrain the SIDM cross-section by comparing the simulation results to observations, the agreement between the codes is strong enough to support differentiating between the results of adopting a cross-section of 1, 5, or 50 cm^2 g^{-1} but not, e.g., a cross-section of 1 vs. 1.5 cm^2 g^{-1}.

10.4 Discussion

10.4.1 Avoiding Tensions

Let us now consider the methodology and results of the code comparison more broadly. SIDM simulations generally should answer the question "Can we reliably predict the effects of self-interactions on DM halos?" However, I do not understand our comparison to have answered this question fully. Indeed, I would caution against using our results to argue that we can reliably model any SIDM halo with a cross-section of 1, 5, or 50 cm^2 g^{-1}. Our project may be one step towards such a conclusion, but a full response would require us to model many of the other relevant physical processes, establish their validity, and ensure that the modules are all interacting properly. Furthermore, answering such a question would require us to ensure that our results were not the consequence of some numerical artifact shared between the two codes. While such an artifact is unlikely since the only modules we used were written separately and there is no overlap between these parts of the codes, it is nonetheless possible.

Though we did not establish the general validity of the SIDM modules through our code comparison, our narrow focus did enable us to avoid the tension that Gueguen describes in code comparisons: that one cannot achieve the necessary diversity of codes required for an informative comparison while maintaining enough similarity for the codes to be comparable (forthcoming). (How) were we able to

avoid this tension and conduct an informative comparison? By only comparing the
SIDM modules of the two codes (i.e., not including baryonic feedback, cooling,
hydrodynamics, etc.), we could span the necessary diversity of implementations
with just two codes. This is because the diversity required was in the SIDM
implementation. Differences in SIDM implementation built on differences in the
underlying gravity solver of the simulations—whether they used a particle-based
vs. mesh-based solver. Thus, using two simulation codes, each of which was based
on a different gravity solver, we could span the necessary diversity in SIDM
implementation. Though we did not have an "ideal ensemble" over which to
consider convergence, we did have representative codes of each distinct approach.
In sum, the codes we used were still diverse enough for an interesting comparison
but, by only looking at their SIDM implementations, we were able to ensure that the
results were still comparable.

It is worth noting explicitly that to avoid Gueguen's critique, we had to radically
restrict our scope. Unlike the Aquila and AGORA comparisons whose far-reaching
goals included checking agreement amongst various codes and raising the realism
and predictive power of the simulations respectively, we only wanted to test whether
differences in SIDM implementations would impact the results of our simulations.
The costs of our limited scope are of course that the conclusions we can draw from
our code comparison are much narrower than those of the Aquila and AGORA
comparisons. There may be other ways of conducting code comparisons that allow
one to avoid the pitfalls Gueguen outlines, perhaps even without requiring such a
narrowing of scope. My goal here is only to present the methodology of a case study
that was able to avoid the dilemma and the conclusions one can draw with such a
methodology. To better understand what our conclusions were and why I argue that
this code comparison was successful, let us revisit Smeenk and Gallagher's proposal
for a methodology of eliminative reasoning.

10.4.2   The Eliminative Approach

As mentioned in Sect. 10.1, Smeenk and Gallagher discuss the limits of convergence
(2020). They acknowledge that convergence over an ideal ensemble of models is
often unrealistic and convergence over the set of models that we do have "does not
provide sufficient evidence to accept robust features" (2020, 1230). Their proposal
steps back to ask what the purpose of identifying robust features was to begin with.
They write:

> To establish the reliability of simulations—or any type of inquiry—we need to identify
> possible sources of error and then avoid them. It is obviously unwise to rely on a single
> simulation, given our limited understanding of how its success in a particular domain can
> be generalized. Robustness helps to counter such overreliance. But there are many other
> strategies that simulators have used to identify sources of error and rule them out (Parker
> 2008). First we must ask what are the different sources of error that could be relevant? And
> what is the best case one can make to rule out competing accounts? (Smeenk and Gallagher
> 2020, 1231)

In other words, we ought to consider the use of other methods if we do not have access to ideal ensembles over which to find convergent results. These other methods are taken to fall in the general approach of "eliminative reasoning." The goal of such projects is to identify possible sources of error and either rule them out or avoid them.

What do projects that are part of this approach look like? Their paper offers one example in which the simulator steps away from considering simulation scenarios that may be tuned to match observations and instead considers a simple setup that can be compared to an analytic solution. They warn, however, that the simulator must ensure that whatever is concluded about the simple case will extend to the complex setups required for research, that the "differences between the simple case and complex target systems do not undermine extending the external validation to the cases of real interest" (Smeenk and Gallagher 2020, 1232).

While the above example is obviously a type of eliminative reasoning, Smeenk and Gallagher's description affords a lot of flexibility in how to conduct such a project. Indeed, the method could be seen as satisfied by the benchmarking astrophysicists do with test problems. Such tests often involve highly idealized systems with analytic solutions. Alternatively, one might imagine the "crucial simulations" described by Gueguen (2019) as an example of eliminative reasoning. To conduct a crucial simulation, a researcher must identify all physical mechanisms and numerical artifacts capable of generating some scrutinized property to ensure the simulation result is indeed reliable (2019, 152). Both these projects—benchmarking and crucial simulations—seem to be concrete examples of the general method described above. What I hope to now show is that the code comparison project outlined in this paper is another concrete example of a project of eliminative reasoning.

10.4.3 Code Comparison as Eliminative Reasoning

We are now in a better position to articulate what the SIDM code comparison was able to show: it eliminated code-to-code SIDM implementation differences as a possible source of error. In particular, it showed that whether one implements SIDM as Gizmo does or as Arepo does, one can still reliably differentiate amongst the SIDM cross-sections explored in the code comparison. This result, in turn, means that no code-to-code variations in implementation will undercut the adequacy of the simulations for determining whether SIDM can be used to alleviate the diversity problem.

The conclusions drawn based on the simulations carried out as part of the code comparison are defeasible as there are further parameters to explore and eliminate as possible sources of error. Indeed, we considered SIDM halos in isolation so issues may arise when generalizing our results to systems incorporating many such halos and/or baryonic effects. As mentioned above (Sect. 10.4.1), one cannot generally claim that these simulations reliably model any SIDM halo with cross-sections of 1, 5, or 50 cm^2 g^{-1}. Given these limitations, the code comparison conducted above does not seem to satisfy the requirements proposed by Gueguen for a crucial simulation.[15] Nonetheless, one can claim that differences between implementations in the two codes will not contribute meaningfully to the results. Another way of putting this conclusion is that the minimal differences in the outputs of the codes indicate that either of the two simulation codes is adequate for the purpose of distinguishing the effects of CDM from SIDM as well as distinguishing the effects of various cross-sections of SIDM.[16] In conclusion, code comparisons provide a fruitful, concrete example of eliminative reasoning. Insofar as eliminative reasoning increases our trust in the results of computer simulation, code comparisons do as well.

10.5 Conclusion

Motivated by Gueguen's recent critique of code comparisons, we (an interdisciplinary group of philosophers and astrophysicists) designed a project to compare the self-interacting dark matter modules to two popular simulation codes. Here, I argued that this project reflects a fruitful methodology for code comparisons: narrowing one's focus allows for an informative code comparison between two codes whose results remain comparable. More broadly, I showed that code comparisons can be used as part of a broader methodology of eliminative reasoning in grounding our trust in simulation results.

References

Agertz, Oscar, Ben Moore, Joachim Stadel, and others. 2007. Fundamental Differences Between SPH and Grid Methods. *Monthly Notices of the Royal Astronomical Society* 380 (3): 963–978. https://doi.org/10.1111/j.1365-2966.2007.12183.x.

---

[15] This is not to say that code comparisons cannot in principle be used as crucial simulations. Recall that a crucial simulation tests "a numerical hypothesis about the origin of a prediction [. . .] against a physical explanation" (Gueguen 2020, 1207). I agree with Gueguen's suggestion to use code comparisons as "a new source of hypotheses to test in crucial simulations" (Gueguen 2019, 158) and also believe that, were one to conduct a crucial simulation, it would be advisable to perform the test with multiple simulation codes, some sharing the background assumption under investigation and some not. I thank an anonymous reviewer to pressing me to clarify this point.

[16] Of course, these simulations are only adequate-for-purpose in light of some background assumptions. These background assumptions include the idea that the two methods of implementing SIDM studied here are the two relevant ones, that the conclusions drawn here about isolated DM halos will generalize, and many more.

Boge, Florian Johannes. forthcoming. Why Trust a Simulation? Models, Parameters, and Robustness in Simulation-Infected Experiments. *The British Journal for the Philosophy of Science.* https://doi.org/10.1086/716542.

Bullock, James S., and Michael Boylan-Kolchin. 2017. Small-Scale Challenges to the ΛCDM Paradigm. *Annual Review of Astronomy and Astrophysics* 55: 343–387. https://doi.org/10.1146/annurev-astro-091916-055313.

Carlson, Eric D., Marie E. Machacek, and Lawrence J. Hall. 1992. Self-Interacting Dark Matter. *The Astrophysical Journal* 398 (October): 43. https://doi.org/10.1086/171833.

Fitts, Alex, Michael Boylan-Kolchin, Brandon Bozek, James S. Bullock, Andrew Graus, Victor Robles, Philip F. Hopkins, et al. 2019. Dwarf Galaxies in CDM, WDM, and SIDM: Disentangling Baryons and Dark Matter Physics. *Monthly Notices of the Royal Astronomical Society* 490 (1): 962–977. https://doi.org/10.1093/mnras/stz2613.

Frenk, C.S., S.D.M. White, P. Bode, and others. 1999. The Santa Barbara Cluster Comparison Project: A Comparison of Cosmological Hydrodynamics Solutions. *The Astrophysical Journal* 525 (2): 554–582. https://doi.org/10.1086/307908.

Gueguen, Marie. 2019. *On Separating the Wheat from the Chaff: Surplus Structure and Artifacts in Scientific Theories.* Electronic Thesis and Dissertation Repository, 6402.

———. 2020. On Robustness in Cosmological Simulations. *Philosophy of Science* 87 (5): 1197–1208. https://doi.org/10.1086/710839.

———. forthcoming. A Tension within Code Comparisons. *British Journal for the Philosophy of Science.*

Hopkins, Philip F. 2015. A New Class of Accurate, Mesh-Free Hydrodynamic Simulation Methods. *Monthly Notices of the Royal Astronomical Society* 450 (1): 53–110. https://doi.org/10.1093/mnras/stv195.

Kaplinghat, Manoj, Mauro Valli, and Yu. Hai-Bo. 2019. Too Big to Fail in Light of Gaia. *Monthly Notices of the Royal Astronomical Society* 490 (1): 231–242. https://doi.org/10.1093/mnras/stz2511.

Kim, Ji-hoon, Tom Abel, Oscar Agertz, Greg L. Bryan, Daniel Ceverino, Charlotte Christensen, Charlie Conroy, et al. 2014. The AGORA High-Resolution Galaxy Simulations Comparison Project. *Astrophysical Journal Supplement Series* 210 (1): 14. https://doi.org/10.1088/0067-0049/210/1/14.

Kim, Ji-hoon, Oscar Agertz, Romain Teyssier, Michael J. Butler, Daniel Ceverino, Jun-Hwan Choi, Robert Feldmann, et al. 2016. The AGORA High-Resolution Galaxy Simulations Comparison Project. II. Isolated Disk Test. *Astrophysical Journal* 833 (2): 202. https://doi.org/10.3847/1538-4357/833/2/202.

Lecar, Myron. 1968. A Comparison of Eleven Numerical Integrations of the Same Gravitational 25 Body Problem. *Bulletin Astronomique* 3 (91): 1221–1233. https://doi.org/10.1086/710627.

Mättig, Peter. 2021. Trustworthy Simulations and Their Epistemic Hierarchy. *Synthese* 199: 14427–14458.

Meskhidze, Helen, Francisco J. Mercado, Omid Sameie, Victor H. Robles, James S. Bullock, Manoj Kaplinghat, and James O. Weatherall. 2022. Comparing Implementations of Self-Interacting Dark Matter in the Gizmo and Arepo Codes. *Monthly Notices of the Royal Astronomical Society* 513 (2): 2600–2608. https://doi.org/10.1093/mnras/stac1056.

Parker, W.S. 2008. Franklin, Holmes, and the epistemology of computer simulation. *International Studies in the Philosophy of Science* 22 (2): 165–183.

Parker, Wendy S. 2020. Model Evaluation: An Adequacy-for-Purpose View. *Philosophy of Science* 87 (3): 457–477. https://doi.org/10.1086/708691.

Randall, Scott W., Maxim Markevitch, Douglas Clowe, Anthony H. Gonzalez, and Marusa Bradač. 2008. Constraints on the Self-Interaction Cross Section of Dark Matter from Numerical Simulations of the Merging Galaxy Cluster 1E 0657-56. *The Astrophysical Journal* 679 (2): 1173–1180. https://doi.org/10.1086/587859.

Robles, Victor H., James S. Bullock, Oliver D. Elbert, Alex Fitts, Alejandro González-Samaniego, Michael Boylan-Kolchin, Philip F. Hopkins, Claude-André Faucher-Giguère, Dušan Kereš, and Christopher C. Hayward. 2017. SIDM on FIRE: Hydrodynamical Self-Interacting Dark Matter

Simulations of Low-Mass Dwarf Galaxies. *Monthly Notices of the Royal Astronomical Society* 472 (3): 2945–2954. https://doi.org/10.1093/mnras/stx2253.

Rocha, Miguel, Annika H.G. Peter, James S. Bullock, Manoj Kaplinghat, Shea Garrison-Kimmel, Jose Oñorbe, and Leonidas A. Moustakas. 2013. Cosmological Simulations with Self-Interacting Dark Matter – I. Constant-Density Cores and Substructure. *Monthly Notices of the Royal Astronomical Society* 430 (1): 81–104. https://doi.org/10.1093/mnras/sts514.

Sameie, Omid, Hai-Bo Yu, Laura V. Sales, Mark Vogelsberger, and Jesús Zavala. 2020. Self-Interacting Dark Matter Subhalos in the Milky Way's Tides. *Physical Review Letters* 124 (14): 141102. https://doi.org/10.1103/PhysRevLett.124.141102.

Sawala, Till, Carlos S. Frenk, Azadeh Fattahi, and others. 2016. The APOSTLE Simulations: Solutions to the Local Group's Cosmic Puzzles. *Monthly Notices of the Royal Astronomical Society* 457 (2): 1931–1943. https://doi.org/10.1093/mnras/stw145.

Scannapieco, C., M. Wadepuhl, O.H. Parry, and others. 2012. The AQUILA Comparison Project: The Effects of Feedback and Numerical Methods on Simulations of Galaxy Formation. *Monthly Notices of the Royal Astronomical Society* 423 (2): 1726–1749. https://doi.org/10.1111/j.1365-2966.2012.20993.x.

Schupbach, Jonah N. 2018. Robustness Analysis as Explanatory Reasoning. *The British Journal for the Philosophy of Science* 69 (1): 275–300. https://doi.org/10.1093/bjps/axw008.

Smeenk, Chris, and Sarah C. Gallagher. 2020. Validating the Universe in a Box. *Philosophy of Science* 87 (5): 1221–1233. https://doi.org/10.1086/710627.

Spergel, David N., and Paul J. Steinhardt. 2000. Observational Evidence for Self-Interacting Cold Dark Matter. *Physical Review Letters* 84 (17): 3760–3763. https://doi.org/10.1103/PhysRevLett.84.3760.

Springel, Volker. 2010. E Pur Si Muove: Galilean-Invariant Cosmological Hydrodynamical Simulations on a Moving Mesh. *Monthly Notices of the Royal Astronomical Society* 401 (2): 791–851. https://doi.org/10.1111/j.1365-2966.2009.15715.x.

Tulin, Sean, and Yu. Hai-Bo. 2018. Dark Matter Self-Interactions and Small Scale Structure. *Physics Reports* 730: 1–57. https://doi.org/10.1016/j.physrep.2017.11.004.

Turner, Hannah C., Mark R. Lovell, Jesús Zavala, and Mark Vogelsberger. 2021. The Onset of Gravothermal Core Collapse in Velocity-Dependent Self-Interacting Dark Matter Subhaloes. *Monthly Notices of the Royal Astronomical Society* 505 (4): 5327–5339. https://doi.org/10.1093/mnras/stab1725.

Vogelsberger, Mark, Jesus Zavala, and Abraham Loeb. 2012. Subhaloes in Self-Interacting Galactic Dark Matter Haloes. *Monthly Notices of the Royal Astronomical Society* 423 (4): 3740–3752. https://doi.org/10.1111/j.1365-2966.2012.21182.x.

Winsberg, Eric. 2010. *Science in the Age of Computer Simulation*. Chicago: University of Chicago Press.

Chapter 11
Simulation and Experiment Revisited: Temporal Data in Astronomy and Astrophysics

Shannon Sylvie Abelson

Abstract The ongoing debate in philosophy of science over whether simulations are experiments has so far operated at too high a level of generality. I revisit this discussion in the context of simulation in astronomy and astrophysics, arguing that a specific subclass of simulations that include a significant amount of empirically obtained temporal data count as experiments. This subclass will be a small one, as the majority of simulations in astronomy and astrophysics will still suffer from a sparseness of data. But it remains the case that there exist examples of simulations that are experiments.

11.1 Introduction

The legitimacy of simulation as experiment has received much attention in the philosophy of science. On the one hand, there are those who object to the treatment of simulations as experiments, either because they are merely formal exercises or because they are representationally inferior to traditional experiment (Guala 2006; Morgan 2005). On the other hand, there are proponents of epistemic equality between experiments and simulations who argue that the traditional objections to simulations misunderstand their structure and their role in science (Parker 2009; Morrison 2009). While I have reservations about whether it is necessary to meet the experimental threshold to do justice to the epistemology involved in astronomical and astrophysical (hereafter, A&A) practice, it is undeniable that the status of simulations as experiments has captured philosophical interest. In the majority of the literature, the debate concerns the use of simulation as one particular type of experiment: what Allan Franklin has called a conceptual experiment, one that engages in theory testing and/or prediction (Franklin 2016; Franklin 1981).

S. S. Abelson (✉)
Indiana University Bloomington, Bloomington, IN, USA
e-mail: abelson@iu.edu

N. Mills Boyd et al. (eds.), *Philosophy of Astrophysics*, Synthese Library 472,
https://doi.org/10.1007/978-3-031-26618-8_11

I argue that this debate has thus far operated at too high a level of generality. As in many areas of complex scientific practice, a general answer to the epistemic significance of some activity is likely to miss important details in individual cases and contexts. Instead, I focus on a key ingredient in the representational validity of a certain subclass of simulations that has undergone comparatively little scrutiny. I argue that the inclusion of temporal data has the potential to inform simulations in A&A in a way that meets representational adequacy constraints and sidesteps concerns about materiality. This practice can permit empirically rich simulations of evolving systems to count as conceptual experiments.

In what follows I briefly review a selective history of the simulation-experiment debate, focusing on the disputes over materiality and representational adequacy. I then describe the role of temporal data as a substantial property in dynamical simulations, which when used to inform simulations of evolving systems serves to increase representational adequacy. I build upon recent work in the growing field of philosophy of A&A by Melissa Jacquart (2020), Sibylle Anderl (2016, 2018), Siska De Baerdemaeker (2022), Siska De Baerdemaeker and Nora Boyd (2020), Jamee Elder (forthcoming), Katia Wilson (2016, 2017), Chris Smeenk (2013), and Michelle Sandell (2010) and defend a novel view in which simulation of the temporal evolution of a system constitutes a necessary, and together with certain other contextually variable measures of empirical adequacy, jointly sufficient condition for conceptual experimentation on dynamical systems. This conclusion is both pessimistic and optimistic: on the one hand, a significant majority of simulations in A&A will not fulfill these conditions and thus are not experimental. However, there does exist a specific subclass of simulations that do meet these criteria and should therefore be seen as conceptual experiments.

The question of whether simulations can count as experiments can be contextualized as hinging on the question:

CQ: Do simulations connect to the world in the relevant way?

I suggest that the answers to this question might be less obvious than it might initially seem, and that the inclusion of specific kinds of data (namely, temporal data) may be illuminating. In Sect. 11.1, I address the challenges inherent in achieving representational adequacy in simulation. It is almost always acknowledged that if simulations are to approach the epistemic productivity of experiments, they must achieve a level of representational adequacy sufficient for empirical accuracy. In Sect. 11.2 I will discuss the importance of temporal data with respect to this question, arguing that instantiations of temporal evolution serve as a so-far philosophically neglected relationship between simulations and the target system. Before that, however, it is important to take a step back and inventory the requirements placed on experiments in the literature and what these conditions indicate about the epistemology of experiment.

11.2 Epistemology of Simulations and Experiments

Allan Franklin, perhaps the most authoritative voice on experiment in the philosophical literature, lays out an array of ways in which experiments can be epistemically positive. They can be exploratory endeavors operating largely independently of previously accepted theory, they can be exercises in further clarifying the consequences of an assumed theory or model, they can test predictions or model components, they can be collections of measurements, they can also be rhetorically useful fictional exercises that serve to propel further research. Experiments may also concern the testing and perfection of new methodologies, rather than direct investigation of some target system (Franklin 2016, 1–4). This list is non-exhaustive, and the goals contained in it are not mutually exclusive. We may for example simultaneously seek to probe a new phenomenon in a model-independent context, while also testing a new measurement technique. The epistemology of experiment captured by these various activities points to a pragmatic picture of experimental activity. The role experiments play is deeply connected to the aims with which scientific research is conducted. Experiments can serve a wide variety of roles connected to scientific inquiry, so long as those roles are governed by endeavors that, as Franklin identifies, "add to scientific knowledge or [are] helpful in acquiring that knowledge" (Franklin 2016, 300–301).

There is nothing in the above-sketched picture of the epistemology of experiment that to my mind *prima facie* excludes simulation. I would argue instead that this picture probably applies to much of the work done using simulations, but certainly it specifically applies to the subclass of simulations that incorporate a fully developed model of temporal evolution. I think it is likely that when philosophers of science dispute the legitimacy of simulation as experiment, they have in mind a narrower view of experimentation somewhere along the lines of what Franklin calls "conceptually important" and "technically good" experiments. The former are experiments designed to test theories and predictions, while the latter are experiments that attempt to improve accuracy and precision in measurement (Franklin 1981, 2016, 2). A conceptual experiment can be understood as one that falls into the category of traditional Baconian experiment, where a researcher attempts to isolate a specific aspect of a system in order to test assumptions (whether those be specific variables, parameters, or predicted effects). It is an experiment that allows the refinement of theoretical concepts. Thus, another way to characterize the project set out here is a defense of simulations in A&A as conceptual experiments: when temporal data is properly instantiated in a simulation to represent a complete picture of the event/process under scrutiny, then the conditions for conceptual experimentation have been met.

11.3 Materiality and Representation

Often the question is raised whether A&A can be classified as experimental sciences. Ian Hacking infamously argued that the observational nature of A&A precluded them from operating as experimental sciences, and thereby impoverishes their epistemic significance (Hacking 1989). He lamented both the limited interventive ability of A&A and their observational nature, stating that, "galactic experimentation is science fiction, while extra- galactic experimentation is a bad joke," and that A&A cannot facilitate realism about their postulated entities (Hacking 1989, 559). Since then, the situation has changed in some ways for A&A, and not in others. We still cannot directly intervene on galactic and extragalactic systems, but the use of simulations to fill this gap has become ubiquitous. Hacking's challenge now requires an answer that addresses both the objections to the use of simulation as epistemically sufficient for experimentation, and the question of whether a largely observational science can still exhibit experimentation.

Much of the debate over whether computer simulations are representationally meaty enough to count as experiments has turned on the role of what has been called, "materiality." The central idea, first expressed by Francesco Guala (2006) and subsequently critiqued by Parker (2009), is that the relationship between a traditional experiment (e.g., a swinging pendulum) and its target system of study is both a formal and material one. By this Guala means that there are certain formal similarities (e.g., physical laws) and material similarities (e.g., physical constituents) that obtain between the contents of the experiment and the content of the target system. An experimental measurement of gravitational acceleration using a simple pendulum as a harmonic oscillator has material content in common with the target system. Conversely, on Guala's view simulations retain only formal similarity with their target system. They lack the common material substratum found between traditional experiments and a system, and thus cannot bear as substantively on research questions about how the world is really composed and structured. A similar view is expressed by Mary Morgan (2005), who claims that traditional experiments that share "ontological" composition with the target system are more epistemically powerful (Morgan 2005, 326).

Wendy Parker (2009) has criticized this view of experiment as too narrow and as stipulating a mutually exclusive relationship between simulations and experiments where one need not obtain. Rather, she argues that simulations should be understood as "time-ordered sequences of states," with computer simulations specifically understood as such a sequence undertaken by a "digital computer, with that sequence representing the sequence of states that some real or imagined system" exhibits (Parker 2009, 487–488). She additionally defines a "computer simulation study" as that simulation plus all the attendant research activities that usually accompany the use of simulations in scientific practice, including development and analysis (ibid., 488) . Under this view, simulations very often qualify as experiments, and moreover can license generalization from conclusions about the simulation system to conclusions about a real-world material system, *if* that simulation properly represents the content of the world, including the causal relations between objects.[1] I find Parker's view largely convincing and a good starting point for the central role of temporal data on which I will focus. While it is surely right to regard simulations as "time-ordered sequences," one must be careful not to assume that the temporal dimension of simulations is merely formal. I will explain this idea in detail in the next section.

The central goal of Parker's account is to disentangle discussions of simulations and experimentation from a focus on shared materiality. She argues that what is important in this discussion is the relevant similarity between the simulation and the material system simulated, given the research question of the study. Unless that research question is specifically about how to reconstruct a physical system in a lab setting, then material similarity need not be understood exclusively as common material composition (Parker 2009, 493). She cites meteorological simulation cases as prime examples of settings in which trying to construct a simulation made of the same material as the target system would be fruitless and impossible (Parker 2009, 494). Parker's account does not prohibit the epistemic superiority of traditional experiment over simulation on ontological grounds *in some cases*; rather she rejects the generalization that such superiority should obtain across the board.

Margaret Morrison's (2009) view is an attempt to further elucidate the way in which all experiments are almost always highly dependent on modeling. Traditional experiments are just as highly dependent on models for their epistemic context (Morrison 2009, 53–54). She likens the construction and tuning of parameters in a computer simulation to the calibration of equipment in a traditional experiment (Morrison 2009, 55). On this view, models themselves are "tools" for experimental inquiry and thus play both a formal and material role. Simulations are first and foremost models, and simulation studies with their attendant computational equipment and pre-and post-hoc analysis are not substantively different from traditional experiments with much of the same modeling infrastructure (ibid., 55). It is precisely because the simulations are built from data models of the phenomena in question, which themselves include volumes of indirect observational data, that these simulations can be said to "attach" to the physical system they are used to explain.

This point has also been nicely made by Katia Wilson, who describes astrophysical simulations as being composed of many pieces of empirical data, including processual data, morphological data, parameters from best fit, and empirical data included in the attendant background theory informing the simulation (Wilson 2017). Wilson stops short of considering this composition of simulations as sufficient to ground a view of them as fully experimental, largely because in many simulations there is insufficient data to fully represent the system. This

---

[1] Parker eschews talk of "systems" and "target systems" in her discussion because she wants to avoid an account where what defines a target of experiment is dependent upon the intent of the researcher (Parker 2009, 487). I retain this terminology because it is more consistent with that used in scientific literature.

concern, that simulations might not be reliable guides to new information about the world, still holds considerable sway among philosophers (Gelfert 2009; Roush 2017). This concern might be well placed when it comes to many simulations of A&A phenomena, particularly when, as I will discuss later, the body of data is subject to what James Peebles (2020) and Melissa Jacquart (2020) have called the "snapshot problem"—where data is assembled from multiple entities in an attempt to reconstruct a picture of single type of entity.

11.3.1 Intervention and Observation

I now turn briefly to the role of intervention (also called "manipulation") in simulations. There may be some consensus that simulations are amenable to intervention. It is common practice to intervene on certain parameters or model components in search of a detectable change in effect. The role of intervention has been most influentially explicated in the philosophical literature by James Woodward (2003; 2008) in connection to causation. Briefly, Woodward's idea is that if one were to intervene on a cause (manipulate it in some way) a corresponding change in an effect should be observed (Woodward 2008). In the context of experiment, intervention is often viewed as one of the defining traits: an experiment is seen as a controlled attempt to intervene on different features of a structure in order to observe the corresponding effects. Intervention is what largely characterizes the purposeful nature of experiment: experiments do not occur naturally but are the product of direct intervention by an experimenter, usually with a specific research goal in mind. Morgan, Guala, Parker, and Morrison all endorse some version of intervention as a necessary condition for experiment. Allan Franklin counts intervention as a hallmark of "good" experimentation, one that increases confidence in both the predicted effects of the experimental intervention and the experimental apparatus itself (Franklin 2016). Intervention should be understood as supervening on representation, as any intervention that occurs in an experiment that is not appropriately related to the target system of interest is not epistemically productive (a point well-captured by Franklin (2016)'s other desiderata).

It is a common position that A&A do not lend themselves to the same kind of intervention as other sciences because of their observational nature. This point was most forcefully made by Hacking (1989) in his discussion of the importance of realism and the epistemic hurdles that largely observational sciences face in producing experiments that can explain phenomena of interest. It is easy to see why such an argument can plague A&A: we cannot intervene on a target system of the Universe when we have yet to send probes much further than our own Solar System.

Simulation has been the answering methodology to the challenge of intervention in A&A. We *can* intervene on simulations, even highly complex ones. Specifically, simulations in which in the parameters and variables, and the relationships between them, are well understood (i.e., many hydrodynamical simulations, simulations of celestial movement and proto-planet evolution, etc.) are those in which finely grained intervention can and does take place. Intervention is, in many ways, one of the main goals of much simulation. Thus, the real challenge in my view to the legitimacy of A&A simulation as experimentation is not the lack of intervention, but clearly defining those cases in which the inner workings and relationships of the simulation are understood well enough to license the appropriate inferences. It is in this way that I believe the question of interventive potential in simulations supervenes on the question of representational adequacy. A simulation that counts as a conceptual experiment must include a sufficient amount of empirical data, and I argue that those that include a substantial amount of empirical temporal data will meet this requirement. It is important to be upfront that this demand *will* rule out many simulations in A&A as insufficiently representative and non-interventive, and therefore non-experimental. Simulations with highly uncertain dynamics or assumptions about model relations will not lend themselves to experimentation. This would therefore exclude many astrophysical models of as-yet poorly understood processes, such as black hole seed formation models.[2]

11.4 A&A Simulation and Temporal Data

The importance of temporal data in their own right, rather than as mere modifiers or structures for other data, has only very recently begun to merit serious consideration from philosophers of science. David Danks and Sergey Pils have recently argued for the importance of considering measurement timescales in discussions of evidence amalgamation (Danks and Pils 2019). Julian Reiss has argued that time series data must be considered in accounts of causation, and moreover that it presents a challenge to the applicability of certain accounts (Reiss 2015). Dynamical sciences, those that concern the development and changes in a system over time, are deeply dependent on the acquisition of data that reports on the *rate* at which such processes occur, the *order* in which they occur, and the temporal *duration* of the system as a whole. Explanations of evolving systems cannot proceed without this kind of data, which is often acquired through varied and robust evidential sources.

Meanwhile, the fields of A&A have historically occupied a fringe position in philosophy of science. While discussions of discovery and theory change have sometimes engaged with cases in A&A (e.g., Kuhn 1957, 1970), the underlying conception of the discipline among philosophers appears to have largely adhered to Hacking's (1989) dismissal of astronomy as a purely observational, and thus philosophically sparse, endeavor. Recent attempts to rehabilitate the significance of A&A for philosophy of science have highlighted the ways in which contemporary

---

[2] Ricarte and Natarajan have shown that existing models of black hole seed formation designed to predict electromagnetic detection signatures are actually unable to distinguish seed signatures from effects of background assumptions regarding accretion mechanics. Models predicting gravitational wave signatures fare better because they do not require accretion assumptions, but they fall victim to highly uncertain dynamics (Ricarte and Natarajan 2018).

A&A go beyond the realm of pure description to offer predictions, explanations, and confirmations of A&A models. Most importantly for this discussion, much of data-driven A&A is concerned with the consideration of dynamical systems: observations of star mergers, galaxy rotations, cosmic inflation, planet composition, etc. all require the consideration of temporally evolving systems. Following the 2010 Decadal Survey, the cyclical report considered the roadmap for A&A science and the result of collaboration by a large panel of researchers, astronomers Graham et al. identify a key shift in A&A practice from stagnant "panoramic digital photography" to "panoramic digital cinematography," where the time domain becomes the necessary setting for studying a large swath of A&A phenomena (Graham et al. 2012, 374). The recently released 2021 Decadal Survey, identifies time-domain astronomy as "the highest priority sustaining activity" in space research (NASEM 2021, 1–17), and states that,

> Time-domain astronomy is now a mature field central to many astrophysical inquiries...The recent addition of the entirely new messengers—gravitational waves and high-energy neutrinos—to time-domain astrophysics provides the motivation for the survey's priority science theme within *New Messengers and New Physics*. (NASEM 2021, 1–6)

This understanding of A&A, and by extension A&A simulation, indispensably involves time series data, requiring the analysis of temporal development of a system over time. Temporal data feature prominently and crucially in contemporary A&A, for small, medium, and large A&A entities alike. Contemporary, data-driven A&A must synthesize vast quantities of evidence to analyze and draw inferences regarding the behavior of astronomical bodies. In so doing, these sciences are deeply engaged in not merely the taxonomy and composition of phenomena, but dynamical modeling and explanation of the evolution of those phenomena over time.

11.4.1 The Nature of Temporal Data

Temporal data, both the content found in time series data and the structures created with time steps and timescales, have always played a foundational role in the practice of A&A (e.g., in measurements of the length of day and constellations), but formal development of analysis techniques coincided with the influx of new technological infrastructure, particularly more advanced telescopes, interferometers, and arrays (Scargle 1997). Nowadays, advanced statistical analysis of time series data is conducted using Fourier analysis, autoregressive modeling, Bayesian periodicity, and other techniques. Recent advances have also introduced parametric autoregressive modeling techniques to analyze astronomical light curves in order to accommodate irregular time series data sets (Feigelson et al. 2018). There are interesting questions connected to the epistemic status of individual analysis techniques, but I will bracket those for the forthcoming discussion. The important thing to note is that the contemporary analysis of time series data is now a complex, formalized, and multiplatform endeavor that often synthesizes data collected from numerous detectors and research teams. Simulation suites invoking empirical time series data have already been developed for upcoming and recently launched detectors, such as the JexoSim exoplanet transit simulation program for the James Webb Space Telescope (Sarkar et al. 2020).

Time series data can be broadly divided into two types: periodic and aperiodic. Periodic time series data describe the regular behavior of phenomena, such as the orbit of planets and pulsars. The periodicity of the system is a specific measurable value, necessary to classify the system. Aperiodic data pertain to phenomena that have a beginning and an end, such as the death of stars. Aperiodic time series data can be further divided into stochastic and transient data. The former describes phenomena like the accretion of stars and galaxies, which involves highly irregular and non-deterministic processes. Transient systems are those that undergo a discrete transformation or set of transformations from an initial to a final state (i.e., thermonuclear death of stars and binary mergers). Most time series data collection concerns the measurement of light from a source, though the advent of multimessenger astronomy has broadened the category of sources to include neutrinos, gravitational waves, and cosmic rays. Time series data collection from a light source is almost always used to generate a light curve, which is a representation of the brightness of a source at or over specific times. As astronomer Simon Vaughan describes it,

> . . . the astronomer is usually interested in recovering the deterministic component and testing models or estimating parameters, e.g., burst luminosity and decay time, rotation period, etc. In other cases, the 'noise' itself may represent the fundamentally stochastic output of an interesting physical system, as in turbulent accretion flows around black holes. Here, the astronomer is interested in comparing the statistical properties of the observations with those of different physical models, or using the intrinsic luminosity variations to 'map out' spatial structure. These projects, and many others, are completely dependent on time-series data and analysis; our only access to the properties of physical interest is through their signature on the time variability of the light we receive. (Vaughan 2013, 3)

Vaughan's characterization of the way in which time series analysis is indispensable to astronomy underscores what I identify as the necessary condition for conceptual experimentation on evolving systems. It is entirely impossible to generate a representation that is faithful to the nature of an evolving system without the introduction of temporal data.[3] This data also requires a different collection and interpretation methodology than other aspects of empirical representation. It must be further synthesized into a meaningful variable of the system (e.g., a light curve, leading to a brightness or luminosity estimate), that is an (at least) two-dimensional value, rather than a one-dimensional one (e.g., solar mass).

---

[3] It is possible and sometimes desirable to generate fake or otherwise simplified time series data sets. This practice may be particularly useful (or necessary) when we are confronted with processes that cannot be observed in their entirety and are part of systems that we do not yet understand well enough to extrapolate from background theory. These data sets, when they are largely randomly generated rather than empirically derived, should not be understood to play the representational role I discuss here.

One of the principal goals in the collection of time series data is the generation of a *timescale*, which is a numerical representation of the rate at which a type of process occurs. For example, in order to have a comprehensive understanding of stellar evolution, it is not enough to simply observe the temporal evolution of individual stars; that is one step of many. A *theory* of stellar evolution will explain the rate at which entities that conform to certain constraints (i.e., types of stars) evolve; it is applicable to all tokens of a type. The generation of a timescale thus permits future classification of other systems.

Already one might have anticipated the way in which I seek to characterize the role of temporality in simulation. The philosophically common view of simulations (as largely formal exercises that do not attach to the target system in any way substantial enough to qualify as representations) risks conflating the differences between time series data and timescales. Timescales *are* formal, mathematical constraints on what counts as an instance of a type of temporally encoded process. Time series data *are not*: they are empirical data. This is because time series data are inherently observational, and thus representative of the target system. It is in this way that temporal data operates as a specific subclass of representation, one that is essential for an accurate representation of an evolving system. This kind of representation is often glossed over or otherwise neglected in philosophical treatments of representation, whereas it is actually a necessary condition for accurately representing a system that changes over time. Moreover, the ways in which temporal data is collected, synthesized, and instantiated in a simulation are complex and varied and cannot be subsumed under the umbrella of other means of instantiating empirical data in models.

11.4.2 Examples

Aperiodic timescales for transient phenomena range from the very long (Gyr or even longer, if we consider the cosmological scale) to the very short (days to weeks, in the case of supernovae). A prime example of aperiodic simulation study is that of supernovae, the explosive deaths of stars. A set of recent studies using 2-D and 3-D simulations of neutron-driven supernovae make substantial use of time series data in order to draw conclusions about the dynamical evolution of this type of stellar process (Scheck et al. 2008; Melson et al. 2015). In Scheck, et al.'s first study, both high and low energy (neutrino velocity) models of stellar accretion and subsequent core-collapse explosion are integrated into the simulations, which each feature different light curve variables (Scheck et al. 2008, 970). Light curves (again, the numerical representation of brightness variation over time) are statistically represented as time series in simulations. These light curves are determined from a computation of the magnitude of the target object, in this case a star of 9.6 $M_\odot$, plotted as a function of time (see Fig. 11.1). Scheck et al.'s results provide early indication that a "hydrodynamic kick mechanism" initiates ~1 second after the core bounce of the supernovae. Interestingly, they "unambiguously" state that this

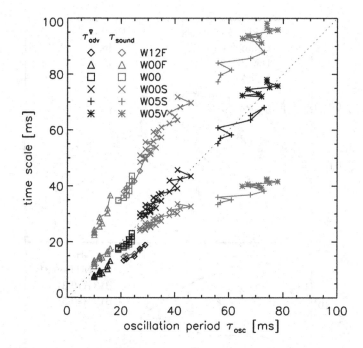

Fig. 11.1 Scheck et al.'s plot of advection time τ_{adv}^{∇} of shock-dispersed fluid and acoustic times (τ_{sound}) against oscillation period (τ_{osc}) for 6 of the models within the simulation. (Credit: Scheck et al., Astronomy and Astrophysics, vol 931, 477, 2008, reproduced with permission © ESO)

observation yields a specific testable follow-up prediction, that the neutron velocity is measured against the direction of the outflow of supernova ejecta (Scheck et al. 2008, 985).

Melson et al. build upon Scheck's 2-D simulation, importing the physical assumptions, to generate a more advanced 3-D simulation. Their study, using the same instantiation of light curve variation, shows that the post-shock turbulence of neutrino-driven supernovae produces measurable effects of "reduced mass accretion rate, lower infall velocities, and a smaller surface filling factor of convective downdrafts" (Melson et al. 2015, 1). These studies demonstrate the pivotal role of time series data in the construction of simulations, which is itself derived from empirical measurement. Moreover, in this case specific conclusions about the target systems are drawn. Thus, it seems clear that in the A&A context, simulations that use time series are being treated as a kind of experiment.

Yet another aperiodic case, this time at the even larger scale of galaxy evolution, illustrates the indispensability of time series data. Dubois et al. developed a large-scale, zoom-in hydrodynamical simulation, NEWHORIZON, of galaxy evolution using the adaptive mesh refinement RAMSES code. Time series data taken from observations and synthesized with theory of star formation rates (SFR) are used to inform empirical parameters for thermal pressure support (α_{vir}) and the instan-

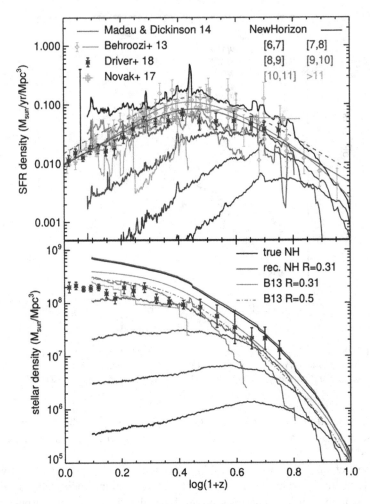

Fig. 11.2 Dubois et al.'s comparison of NEWHORIZON cosmic SFR density and stellar density evolution over time as a function of redshift with observational results (Credit: Dubois et al., Astronomy and Astrophysics, vol A109, 651, 2021, reproduced with permission © ESO)

taneous velocity dispersion (σ_g). They also employ a time-integrated value for cosmic SFR density, which they generated by taking the individual mass of all particles (stars) and summing over them to attain a cosmic SFR density value. They then compare this value to observational measurements of cosmic SFR density, demonstrating the use of time series data to constrain temporal representation in the simulation (see Fig. 11.2) (Dubois et al. 2021). Importantly, they do flag uncertainties in these rates, but point out that these are consistent with uncertainties in observations due to cosmic variance.

Dubois et al. unambiguously laud the ability of simulations to improve theoretical understanding of galactic dynamics:

> Therefore, cosmological simulations are now a key tool in this theoretical understanding by allowing us to track the anisotropic non-linear cosmic accretion... in a self-consistent fashion. (Dubois et al. 2021, 1)

> ... large-scale hydrodynamical cosmological simulations with box sizes of ~50–300 Mpc have made a significant step towards a more complete understanding of the various mechanisms (accretion, ejection, and mergers) involved in the formation and evolution of galaxies... (Dubois et al. 2021, 3)

The NEWHORIZON simulation constitutes another case of the indispensability of time series data, this time both to inform and generate content in the simulation and to check the validity of the simulation values afterward. The tasks described by Dubois et al. as being performed by the simulation fall squarely into the tasks described by Franklin as the province of conceptual experiments: the testing and refinement of theory.

11.4.3 Challenges

One might argue that this view only works if one grants that simulations count as measurements, i.e., if they attached to the target system rather than function as exercises in detached theorizing. Guala has suggested that this kind of representation obtains when the experimental constituents are made of the same stuff as the target system (Guala 2006). But Parker and Morrison have emphasized how this requirement is too strong. I suggest the situation for simulations with significant amounts of temporal data is more complex (Parker 2009; Morrison 2009). Simulations of the kind most commonly used in A&A represent and attach to the world by instantiating substantial amounts of observational temporal data. In short, simulations of this kind are *heavily* constrained by observational data. As Morrison and Parker point out, simulations are almost never isolated computer exercises, but rather instantiate well-supported data models and are embedded in larger studies that include the sum of available observation and analysis on the research question at hand (Parker 2009; Morrison 2009). They are observationally constrained, which means that they effectively represent the target system in such a way that the system is instantiated in the simulation. Even more decisive is the way in which a simulation which accurately represents the temporal evolution or periodicity of a system instantiates one of that system's most important properties: its temporal features. Usage of temporal data, as I have outlined above and as it is explained by Vaughan (2013), is a necessary condition of representing many astronomical systems. Insofar as such representation does not cut any corners that would have been covered in a laboratory setting, there is no real philosophical difference with respect to materiality between the representation found in simulations and those that could be produced in an ideal lab setting. In sum, simulations of dynamical systems *do* attach to the world, and a major part of how they do so is their instantiation of temporal data.

One might also object to the characterization of temporal data I have given by arguing that phenomena do not have any intrinsic timescale, but rather such constraints are placed on them by our analysis (Griesemer and Yamashita 2005). This argument is one that supposes that temporal constraints are the product of an observer, and thus are an artifact of scientific practice, not an inherent feature of the phenomena that can be used to ground claims about the representational relationship between models and target systems. It is straightforwardly true that phenomena do not have inherent timescales. A timescale, however, is the numerical rate of change of a given system. It is synthesized *from* time series data. A timescale is the formal constraint distilled from multiple sets of time series data. It can be understood as the formula that permits identification of future tokens as members of a type (e.g., if a repeated observations of specific events such as kilonovae show an emission period of roughly 2 weeks, then we can estimate the timescale for this kind of event to be 2 weeks and that window may be used to help classify future events). Thus, it is quite obviously a constraint on phenomena imposed by the observer. The worry expressed by Griesemer and Yamashita fails to adequately distinguish between time series data sets and the timescales created from them.

The more interesting question is whether temporal order is inherent in phenomena. This is a complicated question, one that requires addressing ongoing debates about the irreducibility of the arrow of time in evolving phenomena. But I do counter that linear temporal evolution of systems is not reducible in many sciences, so it's unobvious why that would present a problem for the representational adequacy of simulations of evolving systems. It is a well-known problem that phenomenal laws of thermodynamics cannot be stripped of their linear directionality.

11.4.4 Discussion

A significant challenge to the integration of aperiodic temporal data in A&A simulations comes from what Jacquart and Peebles have discussed as "the snapshot" view—where data is assembled from multiple entities in an attempt to reconstruct a picture of single type of entity (Jacquart 2020; Peebles 2020). Essentially, it is often the case that researchers cannot observe aperiodic temporal changes in the same (i.e., same token) object over time, but must rather compile "snapshots" of evolutionary processes from different objects in different stages of the process. This kind of piecemeal assembly is required for models and simulations of most transient events, planetary and stellar evolution, gas clouds, etc. The reason why this problem permeates aperiodic temporal modeling is straightforward: almost all A&A systems evolve over *very long* timescales. This means that the timescales at which these entities evolve are so large that we could never hope to observe even a substantial part of them throughout the whole of human history, much less observe them entirely during normal research programs. Stars and planets evolve over tens of thousands of years, and so we must resort to discontinuous means to assemble continuous observations of them (Jacquart 2020). Jacquart explains that if one assumes continuity of a target object of study over time is necessary for true experimentation, then the kind of "cosmic experiments" performed in A&A are problematic (ibid.).

This problem is partly mitigated by the wide range of A&A entities at our viewing disposal, such that we can and do assemble largely complete pictures of temporal evolution from disparate entities. Other aperiodic events have comparatively much shorter timescales (e.g., ~2 weeks for the final stage of a neutron star merger). These short-timescale events do not struggle with the snapshot problem because they can be observed continuously in their entirety. But there is still a large catalogue of objects for which our temporal pictures are incomplete. And moreover, as Jacquart has pointed out, there are serious epistemic concerns with the snapshot approach to observing temporal evolution. It is not a given that the assemblage of data from different tokens, at different stages of development, even if they are of the same type, can stand in for a single continuous observation (ibid.).

The first thing to note about this problem is that it is not entirely unique. Similar problems are found in climate science, geology, and paleontology. To the extent that a problem pervades many sciences, there is *prima faci* reason to reject the suggestion that such a problem creates doubt as to the experimental status of one particular science. And more directly to the challenge, I think that the way we overcome this kind of problem rests in the practice of using simulations. We simulate the temporal evolution as it would appear if we were able to look at the real system in a time lapse (recall Graham et al.'s description of time-domain astronomy as "digital cinematography" (Graham et al. 2012; 374). This process does not need to be perfect to count as experimentation. Evidence produced by a simulation need not be definitive or ineluctable to count as evidence from experiment. Very few researchers assume that simulation outputs are the final word on any research question. Rather they often hope that their simulation results might be further corroborated by empirical evidence in the future, whether by additional simulations constructed from different data sets (a robustness condition) or by different types of empirical evidence (a variety of evidence condition). Their reason for optimism is the same reason why we would desire robust corroboration from traditional experiments. But the challenge with these simulations is that often the needed independent data is either not available, or else is drawn from the same data set used to construct the simulation, which creates a circularity problem for their empirical accuracy. The potential for circularity can be ameliorated by "splitting" data sets, in which some portions of datasets are used for model construction and other portions are kept separate for subsequent testing (Lloyd 2012, 396).

But this does not mean that simulations are in principle qualitatively different from and inferior to traditional experiments in the way often argued. Rather it means that there are practical constraints limiting the use of simulations as ideal representations of the target system. These same practical constraints exist in the context of many, if not most, traditional experiments. Because this a practical constraint, the situation might be expected to improve with the launches of more powerful detectors (the recently launched James Webb Space Telescope, and the upcoming Vera C. Rubin Observatory (f/k/a Large Synoptic Survey Telescope), Nancy Grace Roman Telescope, LISA, etc.).

A further but related challenge to the understanding of simulations as conceptual experiments may target the epistemic output of simulations when they are used to test models or model aspects. Can the output of a simulation provide a definitive answer to a research question, especially when that question requires an empirical, not merely logical, answer? Wilson stops short of considering the empirical composition of simulations sufficient to ground a view of them as fully experimental, largely because in many simulations there is insufficient data to fully represent the system (Wilson 2017). Wilson notes, drawing on Winsberg (2010) that there are cases in which much of the output data collected from the simulation synthesizes information already possessed, where the epistemic contribution of the simulation lies in its ability to illuminate hidden relationships in the data. But this problem is not insurmountable. Simulations can admit of empirical answers to research questions when the simulations are constructed with a sufficient amount of empirical data.

This view of simulation construction and our confidence in simulation results coheres with what Elisabeth Lloyd has called the "complex empiricism" approach to simulations and models. Rather than a one-to-one, veridical testing relationship between a single assumption and some piece of "raw" data (what Lloyd calls the "direct empiricist" approach and identifies as a descendant of the Hypothetico-Deductive view of explanation), simulations are more appropriately evaluated as complex entities that are constructed using a body of theoretical assumptions, background empirical evidence, and informed decision-making:

> This updated view of model evaluation focuses on independent avenues of theoretical and observational support for various aspects of the simulation models, as well as the accumulation of a variety of evidence for them . . . Additionally, the provision of independent observational evidence for various aspects and assumptions of the models—such as measuring parameter values and relations between variables—increases the credibility of claims made on behalf of models. This support can go beyond or replace the provision of empirical support that might otherwise be provided by matching the predictions with observational datasets. There are, in other words, many more ways to empirically support a model than through predictive success of a single variable. (Lloyd 2012, 393–396)

Under this view, simulations attach to the world they are intended to represent through complex network of background evidence. Representation in simulations is not the unedited copying of complete pictures of the empirical world, but rather the complicated process of constructing an empirically informed patchwork from pieces. With this view of simulation in mind, the snapshot problem becomes an understood reality of modeling practice that can be addressed and worked around, rather than an insurmountable epistemic shortcoming.

One might still argue that the empirical data instantiated in these simulations is still piecemeal and therefore necessarily discontinuous. My answer here requires the recognition of the importance of regularity and law-governedness in A&A. We are often able to generalize beyond the empirical background data in a simulation to piece together discontinuous bits of new data because we are dealing with physical phenomena at (usually, and hopefully) well-understood scales and governed by background theory that has already earned confidence. This point is also underscored by Lloyd in the context of climate science, in which,

> ... the derivation of aspects of model structure from physical laws adds to the modelers' convictions that some of the basic structure, proportions, and relations instantiated in models are fundamentally correct, and are unlikely to be challenged or undermined by datasets that themselves embody potentially arbitrary assumptions. (Lloyd 2012, 396)

In these cases, we can feel confident in generalizing outwards because we have reasonable theoretical grounds to assume certain regularities. Most importantly, this practice can help us make the epistemic jump from shorter to longer timescales, provided that we have sufficient background information to assume a continuity of physical constraints between systems. If we are able to continuously observe a shorter timescale event, like a star merger, in its entirety then we may proceed with more confidence in constructing simulations of longer timescale events, such as the prior evolution of those stars and their journey toward merger. While the latter will require the piecemeal assembly of the snapshot view, the primary worry associated with this methodology, that we may choose the *wrong* snapshots, can be ameliorated. We can generalize from well-understood physics of short timescale events to long timescale events to get around the epistemic uncertainty of the snapshot problem. Simulations that can overcome the snapshot hurdle by instantiating enough temporal data can therefore serve as conceptual experiments, *if* we are confident enough in the means by which their temporal data was assembled to license generalization to other cases. This is the condition for conceptual experimentation Franklin sets out (that theory and models may be tested), and I submit that the role played by temporal data in A&A simulations is sometimes representationally robust enough to meet it.

Importantly though, this response does not automatically apply to less well understood scales, such as the very large (cosmological) and very small (quantum).[4] For those cases, I suggest that the possible remedy once again lies in the use and instantiation of temporal data in simulation. It is *not* the case that only deterministic or simplified models of temporal data are used for simulation. Simulations can and are conducted using stochastic models of temporal data (e.g., Parkes Timing Array data for pulsars as seen in Reardon et al. 2021). It is these cases that suggest a possible way forward for more accurate and well-understood representations of simulations in A&A.

11.5 Conclusion

In the preceding discussion I have examined a brief history of the simulation/experiment debate and argued that much of the objections to simulations

---

[4] There must also be added complexity for the representation of systems with stochastic or indeterminate properties.

as experiments do not apply to dynamical simulations of temporal systems in A&A. These simulations, because they instantiate a significant amount of empirical temporal data and achieve a higher level of representational adequacy, can serve as conceptual experiments in sense discussed by Franklin.

References

Anderl, Sibylle. 2016. Astronomy and Astrophysics. In *The Oxford Handbook of Philosophy of Science*, ed. Paul Humphreys. Oxford: Oxford University Press.

———. 2018. Simplicity and Simplification in Astrophysical Modeling. *Philosophy of Science* 85: 819–831.

De Baerdemaeker, Siska. 2022. Method-Driven Experiments and the Search for Dark Matter. *Philosophy of Science* 88 (1): 124–144.

Danks, David, and Sergey Pils. 2019. Amalgamating Evidence of Dynamics. *Philosophy of Science* 196 (8): 3213–3230.

De Baerdemaeker, Siska, and Nora Mills Boyd. 2020. Jump Ship, Shift Gears, or Just Keep on Chugging: Assessing the Responses to Tensions Between Theory and Evidence in Contemporary Cosmology. *Studies in History and Philosophy of Science Part B: Studies in History and Philosophy of Modern Physics* 72: 205–216.

Dubois, Yohan, et al. 2021. Introducing the NewHorizon Simulation: Galaxy Properties with Resolved Internal Dynamics Across Cosmic Time. *Astronomy and Astrophysics* 651: A109.

Elder, J. Forthcoming. Theory Testing in Gravitational-Wave Astrophysics. In *Philosophy of Astrophysics: Stars, Simulations, and the Struggle to Determine What is Out There*, ed. Nora Mills Boyd, Siska De Baerdemaeker, Vera Matarese, and Kevin Heng. Synthese Library.

Feigelson, Eric D., G. Jogesh Babu, and Gabriel A. Caceres. 2018. Autoregressive Times Series Methods for Time Domain Astronomy. *Frontiers in Physics* 6: 80. https://doi.org/10.3389/fphy.2018.00080.

Franklin, Allan. 1981. What Makes a 'Good' Experiment? *British Journal for the Philosophy of Science* 32 (4): 367–374.

———. 2016. *What Makes a Good Experiment?* Pittsburgh: University of Pittsburgh Press.

Gelfert, Axel 2009. Rigorous results, cross-model justification, and the transfer of empirical warrant: the case of many-body models in physics. *Synthese* 169 (3): 497–519.

Graham, M.J., S.G. Djorgovski, A. Mahabal, et al. 2012. Data Challenges of Time Domain Astronomy. *Distrib Parallel Databases* 30: 371–384.

Griesemer, James and Yamashita, G. 2005. "Zeitmanagement bei Modellsystemen: drei Beispiele aus der Evolutionsbiologie." In *Lebendige*, by H. Schmidgen. Berlin: Kulturverlag Kadmos.

Guala, Francesco. 2006. *The Methodology of Experimental Economics*. Cambridge: Cambridge University Press.

Hacking, Ian. 1989. Extragalactic Reality: The Case of Gravitational Lensing. *Philosophy of Science* 56 (4): 555–581.

Jacquart, Melissa. 2020. Observations, Simulations, and Reasoning in Astrophysics. *Philosophy of Science* 87 (5): 1209–1220.

Kuhn, Thomas. 1957. *The Copernican Revolution: Planetary Astronomy in the Development of Western Thought*. Cambridge: Harvard University Press.

———. 1962/1970. *The Structure of Scientific Revolutions*. Chicago: University of Chicago Press.

Lloyd, Elisabeth A. 2012. The Role of 'Complex' Empiricism in the Debates about Satellite Data and Climate Models. *Studies in History and Philosophy of Science* 43: 390–401.

Melson, Tobias, Hans-Thomas Janka, and Andreas Marek. 2015. Neutrino-Driven Supernova of a Low-Mass Iron-Core Progenitor Boosted by Three-Dimensional Turbulent Convection. *The Astrophysical Journal Letters* 801 (2): L24.

Morgan, Mary S. 2005. Experiments versus models: New phenomena, inference and surprise, *Journal of Economic Methodology*, 12 (2), 317–329.

Morrison, Margaret. 2009. Models, Measurement, and Computer Simulation: The Changing Face of Experimentation. *Philosophical Studies 143 (1)*: 33–57.

National Academies of Sciences, Engineering, and Medicine. 2021. *Pathways to Discovery in Astronomy and Astrophysics for the 2020s*. Washington, DC: The National Academies Press. https://doi.org/10.17226/26141.

Parker, Wendy. 2009. Does Matter Really Matter? Computer Simulations, Experiments, and Materiality. *Synthese 169 (3)*: 483–496.

Peebles, P.J.E. 2020. *Cosmology's Century*. Princeton University Press.

Reardon, D., et al. 2021. The Parkes Pulsar Timing Array Second Data Release: Timing Analysis. *Monthly Notices of the Royal Astronomical Society* stab1990.

Reiss, Julian. 2015. Causation, Evidence, and Inference. Routledge.

Ricarte, Angelo, and Priyamvada Natarajan. 2018. The Observational Signatures of Supermassive Black Hole Seeds. *MNRAS* 481 (3): 3278–3292.

Roush, Sherrilyn. 2017. The Epistemic Superiority of Experiment to Simulation. *Synthese 195*: 4883–4906.

Sandell, Michelle. 2010. Astronomy and Experimentation. *Techné* 14 (3).

Sarkar, Subhajit, Nikku Madhusudhan, and Andreas Papageorgiou. 2020. JexoSim: a Time-Domain Simulator of Exoplanet Transit Spectroscopy with JWST. *MNRAS* 491: 378–397.

Scargle, J. D. 1997. Astronomical Time Series Analysis. In *Astronomical Time Series*, by Dan, Sternberg, Amiel, Leibowitz, Elia M. Maoz, 1–12. Springer Netherlands.

Scheck, L., K. Kifonidis, H. Janka, and E. Müller. 2008. Multidimensional Supernova Simulations with Approximative Neutrino Transport. *Astronomy & Astrophysics* 457: 963–986.

Smeenk, Chris. 2013. Philosophy of Cosmology. In *The Oxford Handbook of Philosophy of Physics*, ed. Robert Batterman. Oxford: Oxford University Press.

Vaughan, Simon. 2013. Random Time Series in Astronomy. *Philosophical Transactions of the Royal Society A* 371: 20110549. https://doi.org/10.1098/rsta.2011.0549.

Wilson, Katia. 2016. Astrophysics in Simulacrum: The Epistemological Role of Computer Simulations in Dark Matter Studies. PhD thesis. University of Melbourne.

———. 2017. The Case of the Missing Satellites. *Synthese* 145.

Winsberg, E. 2010. *Science in the Age of Computer Simulation*. Chicago: University of Chicago Press.

Woodward, James F. 2003. Making things happen: a theory of causal explanation. New York: Oxford University Press.

———. 2008. Causation and manipulability. Stanford Encyclopedia of Philosophy.

Chapter 12
What's in a Survey? Simulation-Induced Selection Effects in Astronomy

Sarah C. Gallagher and Chris Smeenk

Abstract Observational astronomy is plagued with selection effects that must be taken into account when interpreting data from astronomical surveys. Because of the physical limitations of observing time and instrument sensitivity, datasets are rarely complete. However, determining specifically what is missing from any sample is not always straightforward. For example, there are always more faint objects (such as galaxies) than bright ones in any brightness-limited sample, but faint objects may not be of the same kind as bright ones. Assuming they are can lead to mischaracterizing the population of objects near the boundary of what can be detected. Similarly, starting with nearby objects that can be well observed and assuming that objects much farther away (and sampled from a younger universe) are of the same kind can lead us astray. Demographic models of galaxy populations can be used as inputs to observing system simulations to create "mock" catalogues that can be used to characterize and account for multiple, interacting selection effects. The use of simulations for this purpose is common practice in astronomy, and blurs the line between observations and simulations; the observational data cannot be interpreted independent of the simulations. We will describe this methodology and argue that astrophysicists have developed effective ways to establish the reliability of simulation-dependent observational programs. The reliability depends on how well the physical and demographic properties of the simulated population can be constrained through independent observations. We also identify a new challenge raised by the use of simulations, which we call the "problem of uncomputed alternatives." Sometimes the simulations themselves create unintended selection effects when the limits of what can be simulated lead astronomers to only consider a limited space of alternative proposals.

S. C. Gallagher (✉)
Department of Physics and Astronomy and Institute for Earth and Space Exploration, Western University, London, ON, USA
e-mail: sgalla4@uwo.ca

C. Smeenk
Rotman Institute of Philosophy and Department of Philosophy, Western University, London, ON, USA

N. Mills Boyd et al. (eds.), *Philosophy of Astrophysics*, Synthese Library 472,
https://doi.org/10.1007/978-3-031-26618-8_12

12.1   Introduction

Scientists have increasingly come to rely on computer simulations as an essential component of empirical research. Philosophers have studied the epistemological role simulations play in a handful of fields, including high-energy physics and climate science. They have discovered that recent research in both areas has blurred the boundaries between measurement, observation, experiment, and simulation. This prompts a general question: What are the risks associated with treating not just experience, but experience enhanced through simulations, as our primary epistemic authority and guide?

Philosophers have approached this general question by giving detailed assessments of the use of simulations in different domains. Parker (2020) considers the practice of data assimilation in climate science, in which empirical measurements are combined with simulations to generate a more complete characterization of the state of the atmosphere. Based in part on a liberal account of what constitutes a measurement, Parker defends treating a description of the atmospheric state constructed in this fashion as a "measurement" even though it incorporates simulation outputs. Several recent studies of the use of simulations in high-energy physics describe their essential role in designing and interpreting experiments. Discoveries such as that of the Higgs boson at the Large Hadron Collider rely on intricate simulations of both the events that occur in the beam pipe, and how the decay products produced by these events interact with detectors. These simulations are needed to characterize the background against which a novel signal can be detected and to select appropriate candidate events from the detectors. Philosophers have debated the precise contributions of simulations, such as the extent to which the Higgs discovery logically or causally depends on them (Morrison 2015; Massimi and Bhimji 2015; Boge 2021). But it is not controversial that these cases illustrate the thorough integration of simulation into experiment and observation.

We also take these studies to show that, at least in some cases, scientists have overcome simulation-dependence to achieve reliable results. But exactly how reliability can be established depends on what role the simulations play in research. In Parker's case study, for example, the simulation has to generate a description of the atmospheric state that is sufficiently close to the unknown true atmospheric state for the relevant purposes. The main challenge to establishing the reliability of using the simulated state is that of calibration. To play a role similar to that of measurements, the simulation needs to provide not just an estimate of the state but also of the associated uncertainties. By contrast with instrumental results, however, atmospheric scientists generally do not have well-motivated uncertainty estimates for the simulated states (Parker 2020, Sect. 7). Assessing how simulations contributed to the Higgs discovery, and their reliability, involves a quite different set of issues. To find out whether we can preserve reliability while integrating simulations into observations, we need to first clarify what role simulations actually play in a given field. Our aim below is to highlight and assess a distinctive role simulations play in astrophysics.

Astrophysicists have used simulations to treat selection effects, which often arise in scientific fields that rely primarily on passive observations. What inferences we can draw from an observed sample depend on whether it is a fair sample from the overall population, in the relevant respects. A selection effect refers to any bias introduced by our methods of modeling and observing the population. For example, pollsters who contact participants by phone have to determine whether people with phones answer in the same way as those without. Modeling and accounting for selection effects requires detailed background knowledge about the target system as well as the observational program, in order to assess and control for biases. This rapidly becomes quite complex, and has led astrophysicists to develop sophisticated modeling techniques that employ simulations in two distinctive ways. First, the (hypothetical) demographics of the target population of objects needs to be specified. Sometimes these can be treated as the output of a simulation. For example, several large-scale structure simulations evolve forward from an initial state in the early universe (constrained by observations) to yield a distribution of galaxies and other structures at later times. In most cases, however, the simulations do not yield sufficient information about the relevant target population, and so specifying the demographics involves, by necessity, further modeling, physically motivated extrapolations, or inspired guesswork. A second type of simulations model the integrated effects of the telescope, instruments, and observing program design. We will call these observing system simulations, and they model what we should expect to see given our assumptions about the target population. We characterize the cumulative impact of uncertainties as a "selection effect," because failures in either type of simulation lead us astray in treating the actual observations as a fair sample.

As an example, mock "true" galaxy populations, informed by the outputs from physical simulations of galaxy populations evolving over time, can be "observed" using the known properties of an observational survey (specifying details regarding, e.g., detector sensitivities and criteria used to select target objects). The actual output of the observational survey will then be compared to the simulated observations of the "true" population. The model, "true" galaxy populations can be modified until the actual and simulated galaxy catalogues converge. This methodology has enabled an efficient and sophisticated treatment that accounts for cumulative selection effects, as we will describe in more detail in Sect. 12.2 below. This practice leads to simulations being woven into the fabric of galaxy surveys and various other observational programs in both experimental design and data interpretation, and raises questions about the reliability of the results. This role for simulations differs from their use to provide, for example, detailed models of specific types of astrophysical systems, and raises different challenges to assessing reliability. Our argument complements recent work in philosophy that emphasizes the essential role of simulations in astrophysics (Anderl 2018; Jacquart 2020), albeit in different senses, and critically assesses how their reliability can be established (Gueguen 2020).

One novel challenge to reliability is apparent in a different example: evolutionary simulations of mergers of galaxy pairs. These simulations can be stopped and compared to real images to search for "how plausibly" explanations for how a particular observed structure could have formed. In this use of simulations, as with their use in galaxy surveys, the scope of possibilities considered is often informed by observations that are themselves plagued with selection effects. This is a familiar problem, even though it is difficult to account for all potential sources of systematic bias. But there is a second more subtle kind of selection effect that we will emphasize, that arises due to computational constraints. If it is only possible to model a suite of mergers of two galaxies, the possibility that an observed system results from interactions among three or more galaxies may not even be explicitly acknowledged or considered. This computational selection effect limits the space of hypotheses being considered and therefore influences the type of observing programs undertaken. We will consider the ramifications of this kind of limitation, based on a detailed case study, in Sect. 12.3, before stating our conclusions in the final section.

12.2 Selection Effects in Astrophysics

Astronomers often count things. This is typically the first step in the observational study of different types of objects, leading to quantitative measures of a population. This might include organizing objects (e.g., stars) into bins based on, for example, their intrinsic brightnesses. But there are obstacles to getting an accurate count. It is almost always the case that luminous objects—such as the most massive galaxies and the hottest main sequence stars—are rare. So, in a circular section of sky (which represents a cone in volume), the numbers for the most luminous objects are small and have correspondingly large uncertainties from counting statistics. Intrinsically faint objects, such as low mass stars, will be much more numerous, but can only be probed to a much smaller volume before reaching the brightness limit of the survey. If one did not take this observational selection effect into account and correct for the different volumes that are visible for objects of different luminosities, then one would get a very skewed understanding of the true distribution of objects of different luminosities.

This concern was recognized early by Malmquist (1922), and is a well-known example of a selection effect. If distances to objects are known (and thus observed brightnesses corrected to true luminosities), then Malmquist bias can be accounted for with a simple geometric correction for the relative volume at which one could detect objects of a given true luminosity.

The consequences of Malmquist bias are complicated when the numbers of intrinsically fainter objects are greater than the numbers of brighter objects (as is typical), and measurement uncertainties are taken into account. If measurement uncertainties are symmetric about the measured value, and greater numbers exist at fainter fluxes, then objects will preferentially be scattered from fainter into brighter luminosity bins. This effect, known as Eddington bias (Eddington 1913), can also skew the understanding of population demographics if not taken into account.

Malmquist and Eddington bias are not always relevant, but they are two of several selection effects astronomers need to take into account that require alternate methods. For example, for a population study of globular clusters (dense clusters of thousands to millions of stars, all born at the same time) in another galaxy in the nearby universe, all of the clusters are effectively at the same distance from the observer's point of view, and so the volume probed is the same for all intrinsic brightnesses. However, as the detection limit of the survey is approached, the fraction of objects detected drops. The process for recovering the true luminosity distribution of globular clusters from the observed distribution is to perform a completeness study (Whitmore et al. 1999). Specifically, a large number of fake globular clusters with a distribution of brightnesses are randomly added to the observed image of the galaxy, and then the algorithm used to detect and measure the brightness of each source is run on the image which includes the false sources. The input population of false sources is compared to the extracted population. The fraction of detected objects as a function of input brightness is determined so that the observed distribution can be corrected for completeness (see Sect. 2.3 of Gallagher et al. (2001) for an example). The brightness distribution of fake sources does not have to match that of the globular clusters; it is only important that each brightness bin is well-enough sampled (has enough objects) that the uncertainties from counting statistics are small.

This type of simulation to correct for selection effects is fairly straightforward and reliable as long as there are no systematic differences in the distribution of faint versus bright sources. For example, suppose that the most luminous sources are preferentially located in regions with high background light; in this case, the completeness correction would have to account for the negative impact of this on the detection rate. More specifically, the completeness correction to take this effect into account would depend on more than the single parameter of observed brightness. Not recognizing this characteristic of the true globular cluster population could lead to an underestimate of the numbers of brighter objects, and therefore result in a systematic bias.

After decades of study, globular clusters have been well-characterized (Harris 1991). Population demographics are known to depend on such astrophysical properties as cluster ages, metallicities, and the mass and type of galaxy they inhabit. The observability of a particular globular cluster will depend on the wavelength and sensitivity of the observation, the presence of obscuring dust in our Galaxy and the host galaxy, the projected location of the globular cluster within the host galaxy, as well as the globular cluster's brightness. How many of these astrophysical and observational selection effects need to be accounted for in generating a completeness correction will depend on the specifics of the population under study and the characteristics of the observing program. The background knowledge developed over decades of study of globular clusters and the other relevant aspects of astrophysics support reliable estimates of systematic biases.

As mentioned above for the specific case of globular clusters, there are selection effects induced by astrophysics, such as the effects of dust along the line of sight, that affect observations of many systems. A screen of dust between the observer and a star will make the star's light both fainter and redder; these effects are called extinction and reddening. In the Milky Way (and other disk galaxies), dust lies preferentially in the plane of the Galaxy's disk. It is also the case that the most luminous stars are typically in the plane, because they are from a younger population that formed there. In the example of counting stars and binning them based on their intrinsic brightnesses, not accounting for the selection effect caused by dust that differentially affects the most luminous stars would lead one to undercount them. The magnitude of the undercounting would also be sensitive to the color of the images, with blue images being more strongly affected.

Another astrophysical selection effect relevant in extragalactic surveys is a consequence of cosmic variance. This refers to the non-uniformity of the distribution of extragalactic objects that can be detected if one does not sample a large-enough area. Observations of the cosmic microwave background support taking the mass distribution of the universe to be homogeneous and isotropic to a high degree of approximation at early times. Yet the distribution of galaxies only approaches homogeneity at very large scales. Samples collected at smaller scales, using individual galaxies as probes, would be expected to depart from homogeneity. Evaluating samples in different small-area surveys (such as intermediate mass galaxies in the Chandra Deep Field South region, Ravikumar et al. 2007), often reveals significant differences in their distributions.

Modern galaxy surveys include wide-field imaging in many color filters, and subsequent spectroscopic follow-up. Spectroscopy enables obtaining accurate redshifts (essential for calculating distances and therefore luminosities), and determining other galaxy properties such as star-formation rates and the ages of the dominant stellar population. To collect sufficient light for analysis, targets for spectroscopic follow-up must be brighter than the limit of imaging surveys, and different kinds of galaxies are more amenable to spectroscopy. For example, star-forming galaxies are typically blue and have emission lines, the latter make measuring redshifts much easier than for quiescent (non-star-forming) galaxies (generally red) that only have absorption lines. For absorption-line galaxies, the signal-to-noise ratio in the continuum must be higher to detect the features required to measure a redshift. For emission-line galaxies, some redshift ranges—including the "redshift desert" near $z \sim 1.5$—have few bright emission lines in the observed-frame optical wavelength bandpass of most spectroscopic surveys. If we consider each step of this process (measuring the light from multi-color imaging, spectroscopic target selection, and spectral analysis), there are distinct selection effects in detecting and characterizing each particular class of object. For the DEEP2 galaxy survey, Newman et al. (2013) list 7 distinct selection effects for the final sample chosen for spectroscopic follow-up:

1. Galaxy color bias due to the R magnitude limit
2. Loss of bright star-like objects
3. Misclassification of faint stars as galaxies
4. Loss of objects due to missing B or I photometry
5. Loss of small, distant, faint red galaxies

6. Loss of objects at small separations
7. Multiple galaxies masquerading as single galaxies

These can occur because of observing conditions, e.g., bad weather might result in missing data (item 4), the limits of the instrumentation, e.g., the spectrograph cannot observe two objects too close together (item 6), or an inability to accurately identify a galaxy based on how it presents in imaging (items 3 and 7). This list does not even incorporate the subsequent issues that can occur once spectra are obtained, such as not finding sufficient distinguishing features to determine an accurate redshift. Each selection effect will have a differential impact on the detection and characterization of distinct classes of galaxies. Clearly, understanding and accounting for these interconnected selection effects rapidly becomes extremely complicated.

As galaxy surveys have become larger and more sophisticated, the tools to address selection effects have similarly developed. Computer simulations now play an essential role, because the layers of selection effects have become too complex to account for with simple numerical corrections. A specific technique is to use "mock galaxy catalogs"—a model of the true galaxy population, informed by the best understanding of galaxy demographics and evolution—and to forward-model the impact of each observational step (and its associated uncertainties) in a survey and then compare the actual observed data to the simulated observed population (e.g., Coil et al. 2007; Newman et al. 2013). The input population in the mock catalog can be adjusted within a parameter space informed by cosmological simulations until the simulated and observed populations are consistent.

This is a successful solution to the challenge of understanding and then correcting for observational selection effects as long as the input catalogs are a reasonable representation of the true galaxy population. As a new survey pushes into new parameter space (e.g., by imaging at different wavelengths, pushing to fainter fluxes, or covering a larger volume), the possibility of unanticipated objects grows. In this case, the parameter space explored in generating input catalogs can also have selection effects that generate biases; in the most extreme case, galaxies with unexpected properties—unknown unknowns—may simply not be included at all in the mock catalog.

This is particularly clear in cases where it is challenging to determine a reasonable "mock" catalog in order to understand selection effects for objects near the detection threshold. As an illustrative example, consider the efforts to determine accurate redshifts using photometry. Spectroscopic determinations of redshifts are much more accurate, but cannot be feasibly used to measure the redshifts for the number of galaxies used in contemporary surveys, particularly at faint fluxes. Astronomers have turned to easier, but coarser, photometric methods to measure redshift as an alternative. The photometric redshift measurements are then calibrated with the spectroscopic measurements. This requires demographic completeness of the two sets of measurements, so that they are calibrated over galaxy distributions with similar physical properties. This is a major challenge, however, because the properties of the galaxy distributions themselves are uncertain; it is difficult to establish how closely the catalog of galaxies based on spectroscopic observations matches that of photometric observations. There are ongoing efforts to respond to what are called "catastrophic failures" of photometric redshifts (namely, cases where they depart dramatically from spectroscopic estimates), based on new types of observations and refined estimates of the systematic biases these failures induce in determinations of other parameters.

Here it is natural to wonder whether an analog of "experimenter's regress" arises.[1] Collins (1992) claims that there is no way to avoid circularity in identifying correct experimental results: good results are obtained with a good experimental apparatus, and vice versa. Anomalous results can always be rejected as the product of a malfunctioning apparatus. According to Collins, the decision to accept certain experiments and their results is grounded in social interactions in the community and cannot be based solely on epistemic considerations. An analogous "observer's regress" would regard the apparently circular trade-off between assumptions regarding the true population of astrophysical objects and selection effects. What grounds do we have for choosing between the two, particularly for surveys extending into new parameter space at the detection threshold of existing instruments?

Our response is similar in spirit to Franklin (1994)'s rebuttal of Collins: we should not amplify the legitimate challenges with calibrating experiments, or conducting astrophysical surveys, into an impossibility claim. Frontier research faces challenging questions regarding selection effects and how to model them. But historical cases, such as the study of globular clusters described above, reveal that the threat of circularity is only temporary: there are several sufficiently independent lines of evidence that eventually led to a clear choice between attributing a particular result to the true population vs. a selection effect. A culture that embraces open data policies (common practice for many observatories) also means independent teams can tackle the same datasets and apply their own suite of simulations to interpret them. Furthermore, technological advances typically resolve some outstanding uncertainties about the nature of objects on the boundary or beyond what is currently observable; investments in developing future facilities are justified by exactly these sorts of outstanding science questions. While this provides no guarantee that current challenges, such as that associated with redshift measurements, can be resolved in the short term, there is little support for an impossibility claim like Collins's.

We next turn to a different kind of case, an example where the limitations of what is feasible computationally can create a novel type of selection effect.

12.3   Case Study: What Triggers Quasar Activity?

Above we described the types of knowledge, primarily regarding properties of a population of target objects and details of the observational program, that are required to handle selection effects. These aspects of selection effects and sources

---

[1] Thanks to an anonymous reviewer for raising this question.

of systematic bias are well-known in astrophysics, but we will now turn to a type of selection effect that has drawn less attention. We will call this the "problem of uncomputed alternatives" (following Stanford 2006): the neglect of physically plausible scenarios that are, however, computationally intractable. This neglect can lead to designing observational programs that have an unjustifiably narrow scope, neglecting the kind of evidence that could be relevant to assessment of the uncomputed alternative. But by its very nature the uncomputed alternative is difficult to assess because it is computationally inaccessible: there is at present no clear way to set up a clean comparison between observations and the alternative hypotheses. We will illustrate this general issue through a concrete case study.

From observations in the local universe, it appears that every massive galaxy hosts a supermassive black hole at its core (Kormendy and Richstone 1995). These black holes grew primarily as quasars during the epoch known as "cosmic noon" ($z = 1$ to 3) when the universe was approximately a quarter to a half its present age (Soltan 1982; Yu and Tremaine 2002). The question of what triggers quasar activity is an area of active past and current research. Answering that question is challenging, for reasons that intersect.

A natural experiment that could address this question would be to observe the hosts of quasars, to characterize the galaxies they inhabit. This is because the fuel that powers quasars comes from galaxies. Therefore, knowing what kinds of galaxies host quasars—for example star-forming or quiescent, with disk or spheroidal morphologies—would put important constraints on triggering mechanisms. However, this is more easily proposed than accomplished for several reasons. First, a quasar often outshines the light from its host galaxy by factors of up to 1000. Second, quasar host galaxies at the distances commensurate with cosmic noon have small angular extents, on the order of $\sim1''$. From the ground, this is close to the angular resolution of most telescopes (from the smearing of the atmosphere), and thus separating the lower surface brightness host galaxy from the very bright quasar in its center in an image is typically not possible. This challenge of ground-based observations is why studying quasar host galaxies has been a science focus for both the Hubble and Webb Space Telescopes.

The first samples with Hubble imaging of quasar host galaxies showed a range of morphologies, including some indicative of interacting galaxies (e.g., Hutchings and Morris 1995; Bahcall et al. 1997). The varied selection criteria and relatively small sample sizes (a few to 20 objects) make drawing conclusions from the fraction of observed galaxies that showed evidence of mergers challenging. For example, some of the galaxies chosen for Hubble imaging were selected based on evidence from ground-based observations for extended, asymmetric structures, and so it is not surprising that these galaxies were often found to be likely merger remnants (Hutchings et al. 1994). Time on a valuable resource such as Hubble is allocated through a very competitive process, and an observing program that is more likely to yield a positive result (such as a clear detection of interesting structure) is more likely to get chosen.[2]

A second empirical path would be to look at nearby quasars, where these observational challenges can be mitigated because the host galaxies are significantly larger and have higher surface brightnesses. Locally, many luminous quasars are found in 'warm' ultra-luminous infrared galaxies, the highest luminosity galaxies (with $L_{IR} \geq 10^{12}$ $M_\odot$), with infrared properties that indicate higher temperature dust, most plausibly heated by a quasar (as opposed to active star formation) (Sanders et al. 1988). High-resolution Hubble Space Telescope imaging of some of these galaxies revealed signatures of recent galaxy mergers, including tidal features and young, massive star clusters whose formation could be triggered by a merger event (Surace et al. 1998). Mergers of gas-rich disk galaxies are plausible triggers for quasar activity, as the collision of gas clouds can efficiently shed sufficient angular momentum to drive gas towards the gravitational center of the merger remnant, where the supermassive black hole is found.

With the first generation of galaxy-merger simulations that included gas (which can dissipate energy and cool radiatively) and stars (which behave as collisionless particles that only interact gravitationally), the theoretical support for the idea that quasars could be caused by mergers was demonstrated (Barnes and Hernquist 1991). However, it should be recognized that the models themselves did not include supermassive black holes, nor did they have sufficient spatial resolution to follow the gas to the center of the potential well at the scales of the gravitational sphere of influence of a supermassive black hole.

One of the challenges of setting up a galaxy-merger simulation is choosing appropriate initial conditions for the encounter from among a very large parameter space of possibilities. For example, the relative initial positions and velocities for each galaxy, the inclinations of the disks, and the sense of their motions (clockwise or counterclockwise), all impact on the progress of the merger and the final outcome. Furthermore, the structure of the galaxy itself (such as how prominent the central bulge is relative to the disk) impacts the gas flows within the galaxies in the course of merging and thus the amount and timing of induced star formation (Mihos and Hernquist 1996). Since the first galaxy-merger computer simulations of Toomre and Toomre (1972), the touchstones for these merger simulations are often local ultraluminous infrared galaxies, and one measure of success claimed by simulators is to match (at some point in the progression of a merger) well-known examples of merging pairs. Full-blown merger simulations are computationally expensive, and so judicious choices of initial conditions are important. Toomre and Toomre (1972) chose parabolic passages, and were able to come up with reasonable representations of four well-known merging galaxy pairs.

Taken together, both the simulations of galaxy mergers and observations of nearby quasar host galaxies provided a consistent picture whereby a merger of two

---

[2] This illustrates another potential selection effect in astronomy, that of the telescope time allocation committee.

gas-rich disk galaxies could drive gas towards the center of the potential well of the merger remnant and provide the fuel to power a quasar and grow a supermassive black hole. Empirically, this scenario holds up well in the local universe, where both mergers of gas-rich galaxies and quasars are quite rare. At earlier times, galaxies were more numerous and closer together, and quasars were both more common and more luminous. So, does this story, well-supported at low redshift, also hold at $z \sim 2$?

The observational story at higher redshift is complicated, because it is still challenging to separate out the light from quasar host galaxies. Signatures of mergers such as tidal tails and young massive star clusters become significantly harder to resolve spatially. In this case, the role of simulations becomes even more important. From the first generation simulations of Barnes and Hernquist (1991), subsequent researchers made correspondingly more sophisticated merger simulations (e.g., Di Matteo et al. 2005; Hopkins et al. 2005), that supported the original success of mergers accounting for quasar activity at early times. Typically, the initial conditions for galaxy interactions are generated from low-resolution cosmological simulations, and then a higher-resolution simulation is performed to follow the subsequent evolution, with analytic prescriptions for the onset of star formation and black-hole feeding that are below the spatial resolution of the galaxy simulations.

The case for inferring that what happens locally also works at higher redshifts breaks down when we consider the significant evolution of galaxies over billions of years. In particular, at higher redshifts, disk galaxies have a higher fraction of their baryonic mass in gas, and also have dynamically 'hotter' disks, with significant vertical (in addition to primarily rotational) motions. A consequence of these structural properties is that star-forming regions are typically larger because it takes more mass to cause gravitational collapse against gas motions (Elmegreen et al. 2007). Next, gravitational instabilities in the disk gas, such as spiral arms and bars, can happen through secular evolution, without an external trigger.[3] Bars are evidence of radial motions in the gas, and are effective at funneling gas to smaller radii. These factors together mean that a starburst episode coupled (or followed by) quasar activity can plausibly happen without significant dynamical shocks triggered by a merger (Hopkins et al. 2010).

In addition, quasars at $z \sim 1$ are found typically in galaxy group environments (Coil et al. 2007), with several galaxies gravitationally bound to each other. With a handful of galaxies (rather than just two), gravitational interactions become much more complex, and are less likely to lead to a merger of a pair. However, in groups galaxies do interact gravitationally, but the effects—such as low surface brightness tidal features and depletion of cold gas reserves—can be much more subtle than the dramatic impacts of a merger (e.g., Konstantopoulos et al. 2010). These

---

[3] Spatially resolved kinematics of a small sample of star-forming galaxies at $z \sim 1.5$ indicate galaxies with both disk-like and merger-remnant structures, and higher velocity dispersions in star-forming clumps than typically seen at low z (Mieda et al. 2016).

empirical results on secular disk evolution and small galaxy group interactions suggest alternate pathways for triggering quasar activity accompanied by active star formation than the merger of a gas-rich galaxy pair. Such subtle effects would also be challenging to detect beyond the local universe.

One reason that the pair-merger pathway to quasar activity has been so widely accepted is the success of the computer simulations of the physical system. Observations of any single system will necessarily capture only a moment in time, and a collection of observations of different systems has to be put together into a coherent picture to understand evolution over billions of years. Computer simulations thus serve an essential role in filling in the time gaps, and following a single type of system over time.[4] As a recent example, Moreno et al. (2021) investigate the effects of galaxy-pair interactions on star formation within each galaxy (black-hole fueling is not included in the simulations) with a suite of 24 galaxy-pair simulations (varying the initial conditions).

One should also consider, however, which computer simulations are *not* being done. A specific example is a simulation of a small group of galaxies to investigate if (and how) modest and perhaps recurrent gravitational interactions between more than one galaxy could trigger star formation and quasar activity. Practical constraints explain the lack of simulations of this type of system to address the question. First, a single simulation of even three galaxies would be computationally very challenging. Second, such a simulation would also require choosing initial conditions (such as galaxy properties and relative positions and velocities) from a very large parameter space of potential values. Running a large number of simulations to investigate the influence of initial conditions would be computationally extremely expensive. But there are no physical grounds to rule out this kind of interaction. This is an example of an "uncomputed alternative," a reasonable hypothesis that has not been explored because the simulations required are not currently feasible.

One of the plausible explanations for the triggering of quasar activity has thus not been explored using simulations, and therefore is not subject to detailed observational evaluation. This is an example of a novel type of selection effect induced by what is computationally tractable that is limiting the space of hypotheses under consideration.

There are several consequences of a computational selection effect. One is a limitation on the types of observational programs that may be undertaken to test the merger-trigger hypothesis, and also how those data are interpreted. For example, the empirical study of Ellison et al. (2011) considered low-redshift pairs of galaxies to see if evidence for accretion onto a black hole (spectroscopic identification as an active galactic nucleus) was correlated with being classified as a close-separation pair. Though higher multiples (triples or more) were not selected against, the target sample and control sample of isolated galaxies were all chosen from

---

[4] This is a further instance of a role for simulations that Jacquart (2020) emphasizes, namely amplifying astrophysical observations—in this case, moving from isolated instants to an evolutionary trajectory for a type of system.

the Sloan Digital Sky Survey galaxy sample, which has a relatively bright flux limit and (because of instrumental limitations) a known high level of spectral incompleteness for close galaxy separations. The authors are well aware of the potential consequences of these selection effects, but limit their discussion to the evaluation of pair-wise interactions, described as merger candidates, versus the alternate pathway of secular disk evolution to explain black hole fueling.

In another empirical study, Patton et al. (2013) considered observed enhancements in star formation in galaxy pairs and used a suite of 75 merger simulations for comparison. Though their discussion of sample selection of the target sample and control sample of isolated galaxies accounted for local environment (acknowledging that most galaxies are found in groups and clusters), the simulations themselves did not incorporate more than two galaxies. In this case, the use of simulations to reveal the mechanism for the observed increased star formation rate of paired galaxies provides less convincing evidence.

The impact of this "problem of uncomputed alternatives" resembles that of Stanford (2006)'s problem of unconceived alternatives: the force of an eliminative argument in favor of a hypothesis depends on whether all reasonable alternatives have been considered. In our view, the example above illustrates a viable physical mechanism for triggering quasar activity that has not been eliminated, and the case in favor of the predominance of the pair-merger pathway is hence less compelling. (That is not to downplay the importance of the positive case in favor of this proposal: it is based on extrapolating a successful account from low redshift back to the earlier universe. But it does call into question the epistemic support added by the simulation studies.)

There are also two contrasts with Stanford's account worth noting. The assessment of the space of "plausible" competing hypotheses is challenging, and Stanford's historical arguments are intended to illustrate ways in which scientists have routinely failed to consider viable alternatives in the form of radically different theories. This example has a different character: the apparent success of simulations of (relatively speaking) simple cases may lead to an overconfidence in extrapolating to more complex cases, where other causal factors may be in play. In the case we discuss, the alternatives involve different assessments of what physical interactions are relevant to a particular phenomenon, but do not raise questions about the underlying physical theories. The failing is not insufficient exploration of the space of possible theories, but insufficient exploration of how to treat complex situations with existing theory.

But the second contrast is more significant. The failure to include "uncomputed alternatives" undermines an eliminative argument, but it also has a more subtle impact on the interpretation of observations. Analyzing potential selection effects requires a comprehensive understanding of how the properties of a target population interact with the observational program, and any biases that these produce. It is much harder to characterize the impact of the observing programs that are not undertaken because of how the science question is formulated. Specifically, an observing program addressing the question of whether group interactions (without mergers) can trigger quasars at cosmic noon would be fundamentally different than the programs of Ellison et al. (2011) and Patton et al. (2013) described above.

12.4 Conclusion

Astronomers use simulations routinely in order to model selection effects. In cases like large galaxy surveys, inter-related selection effects from a variety of sources, such as details of the instrument and observational program to the astrophysics of the target systems, can no longer be treated through individual numerical corrections. As we have described above, astronomers instead simulate the expected output of an observing program for a "mock" catalog of sources, and use the comparison of these extracted results to actual observations to assess and account for selection effects. We have described a few concrete examples above, with the aim of illustrating this technique in more detail and clarifying the kinds of background knowledge that are needed for it to be reliable. Establishing reliability is particularly difficult when uncertainties regarding selection effects are compounded with uncertainties regarding the population of target objects. Finally, we identified a novel kind of computational selection effect that we called the "problem of uncomputed alternatives." In some cases, physically reasonable proposals simply cannot be followed through computationally, at least at present, to determine their observational signatures. The neglect of these possibilities may lead to the design of observational programs that cannot reveal problems with simpler, albeit incomplete or incorrect, alternative hypotheses.

References

Anderl, S. 2018. Simplicity and simplification in astrophysical modeling. *Philosophy of Science* 85(5): 819–831.

Bahcall, J.N., S. Kirhakos, D.H. Saxe, and D.P. Schneider. 1997. Hubble space telescope images of a sample of 20 nearby luminous quasars. *Astrophysical Journal* 479(2): 642–658.

Barnes, J.E., and L.E. Hernquist. 1991. Fueling starburst galaxies with gas-rich mergers. *Astrophysical Journal Letters* 370: L65–L68.

Boge, F.J. 2021. Why trust a simulation? Models, parameters, and robustness in simulation-infected experiments. *British Journal for the Philosophy of Science*. https://doi.org/10.1086/716542

Coil, A.L., J.F. Hennawi, J.A. Newman, M.C. Cooper, and M. Davis. 2007. The DEEP2 galaxy redshift survey: Clustering of quasars and galaxies at z = 1. *Astrophysical Journal* 654(1): 115–124.

Collins, H. 1992. *Changing order: Replication and induction in scientific practice*. University of Chicago Press.

Di Matteo, T., Springel, V. and Hernquist, L. 2005. Energy input from quasars regulates the growth and activity of black holes and their host galaxies. *Nature* 433(7026): 604–607.

Eddington, A.S. 1913. On a formula for correcting statistics for the effects of a known error of observation. *Monthly Notices of the Royal Astronomical Society* 73: 359–360.

Ellison, S.L., D.R. Patton, J.T. Mendel, and J.M. Scudder. 2011. Galaxy pairs in the Sloan Digital Sky Survey - IV. Interactions trigger active galactic nuclei. *Monthly Notices of the Royal Astronomical Society* 418(3): 2043–2053.

Elmegreen, D.M., B.G. Elmegreen, S. Ravindranath, and D.A. Coe. 2007. Resolved galaxies in the hubble ultra deep field: Star formation in disks at high redshift. *Astrophysical Journal* 658(2): 763–777.

Franklin, A. 1994. How to avoid the experimenters' regress. *Studies in History and Philosophy of Science* 25(3): 463–491.

Gallagher, S.C., J.C. Charlton, S.D. Hunsberger, D. Zaritsky, and B.C. Whitmore. 2001. Hubble space telescope images of Stephan's quintet: Star cluster formation in a compact group environment. *Astronomical Journal* 122(1): 163–181.

Gueguen, M. 2020. On robustness in cosmological simulations. *Philosophy of Science* 87(5): 1197–1208.

Harris, W.E. 1991. Globular cluster systems in galaxies beyond the local group. *Annual Reviews of Astronomy and Astrophysics* 29: 543–579.

Hopkins, P.F., L. Hernquist, T.J. Cox, T. Di Matteo, P. Martini,B. Robertson, and V. Springel. 2005. Black holes in galaxy mergers: Evolution of quasars. *Astrophysical Journal* 630(2): 705–715.

Hopkins, P.F., D. Kereš, C.-P. Ma, and E. Quataert. 2010. When should we treat galaxies as isolated? *Monthly Notices of the Royal Astronomical Society* 401(2): 1131–1140.

Hutchings, J.B., and S.C. Morris. 1995. Imaging of low redshift QSOs with WFPC2. *Astronomical Journal* 109: 1541–1545.

Hutchings, J.B., J. Holtzman, W.B. Sparks, S.C. Morris, R.J. Hanisch, and J. Mo. 1994. HST imaging of quasi-stellar objects with WFPC2. *Astrophysical Journal Letters* 429: L1–L4.

Jacquart, M. 2020. Observations, simulations, and reasoning in astrophysics. *Philosophy of Science* 87(5): 1209–1220.

Konstantopoulos, I.S., S.C. Gallagher, K. Fedotov, P.R. Durrell, A. Heiderman, D.M. Elmegreen, J.C. Charlton, J.E. Hibbard, P. Tzanavaris, R. Chandar, K.E. Johnson, A. Maybhate, A.E. Zabludoff, C. Gronwall, D. Szathmary, A.E. Hornschemeier, J. English, B. Whitmore, C. Mendes de Oliveira, and J.S. Mulchaey. 2010. Galaxy evolution in a complex environment: A multi-wavelength study of HCG 7. *Astrophysical Journal* 723(1): 197–217.

Kormendy, J., and D. Richstone. 1995. Inward bound—The search for supermassive black holes in galactic nuclei. *Annual Reviews of Astronomy and Astrophysics* 33: 581–624.

Malmquist, K.G. 1922. On some relations in stellar statistics. *Meddelanden fran Lunds Astronomiska Observatorium Serie I* 100: 1–52.

Massimi, M., and W. Bhimji. 2015. Computer simulations and experiments: The case of the Higgs boson. *Studies in History and Philosophy of Science Part B: Studies in History and Philosophy of Modern Physics* 51: 71–81.

Mieda, E., S.A. Wright, J.E. Larkin, L. Armus, S. Juneau, S. Salim, and N. Murray. 2016. IROCKS: Spatially resolved kinematics of $z \sim 1$ star-forming galaxies. *Astrophysical Journal* 831(1): 78(37pp).

Mihos, J.C. and L. Hernquist. 1996. Gasdynamics and starbursts in major mergers. *Astrophysical Journal* 464: 641–663

Moreno, J., P. Torrey, S.L. Ellison, D.R. Patton, C. Bottrell, A.F.L. Bluck, M.H. Hani, C.C. Hayward, J.S. Bullock, P.F. Hopkins, and L. Hernquist. 2021. Spatially resolved star formation and fuelling in galaxy interactions. *Monthly Notices of the Royal Astronomical Society* 503(3): 3113–3133.

Morrison, M. 2015. *Reconstructing reality: Models, mathematics, and simulations.* Oxford: Oxford University Press.

Newman, J.A., M.C. Cooper, M. Davis, S.M. Faber, A.L. Coil, P. Guhathakurta, D.C. Koo, A.C. Phillips, C. Conroy, A.A. Dutton, D.P. Finkbeiner, B.F. Gerke, D.J. Rosario, B.J. Weiner, C.N.A. Willmer, R. Yan, J.J. Harker, S.A. Kassin, N.P. Konidaris, K. Lai, D.S. Madgwick, K.G. Noeske, G.D. Wirth, A.J. Connolly, N. Kaiser, E.N. Kirby, B.C. Lemaux, L. Lin, J.M. Lotz, G.A. Luppino, C. Marinoni, D.J. Matthews, A. Metevier, and R.P. Schiavon. 2013. The DEEP2 galaxy redshift survey: Design, observations, data reduction, and redshifts. *Astrophysical Journal Supplement* 208(1): 5(57pp).

Parker, W.S. 2020. Computer simulation, measurement, and data assimilation. *The British Journal for the Philosophy of Science.* 68: 273–304.

Patton, D.R., P. Torrey, S.L. Ellison, J.T. Mendel, and J.M. Scudder. 2013. Galaxy pairs in the Sloan Digital Sky Survey - VI. The orbital extent of enhanced star formation in interacting galaxies. *Monthly Notices of the Royal Astronomical Society* 433(1): L59–L63.

Ravikumar, C.D., M. Puech, H. Flores, D. Proust, F. Hammer, M. Lehnert, A. Rawat, P. Amram, C. Balkowski, D. Burgarella, P. Cassata, C. Cesarsky, A. Cimatti, F. Combes, E. Daddi, H. Dannerbauer, S. di Serego Alighieri, D. Elbaz, B. Guiderdoni, A. Kembhavi, Y.C. Liang, L. Pozzetti, D. Vergani, J. Vernet, H. Wozniak, and X.Z. Zheng. 2007. New spectroscopic redshifts from the CDFS and a test of the cosmological relevance of the GOODS-South field. *Astronomy and Astrophysics* 465(3): 1099–1108.

Sanders, D.B., B.T. Soifer, J.H. Elias, G. Neugebauer, and K. Matthews. 1988. Warm ultraluminous galaxies in the IRAS survey: The transition from galaxy to quasar? *Astrophysical Journal Letters* 328: L35–L39.

Soltan, A. 1982. Masses of quasars. *Monthly Notices of the Royal Astronomical Society* 200: 115–122.

Stanford, P.K. 2006. *Exceeding our grasp: Science, history, and the problem of unconceived alternatives.* Oxford: Oxford University Press.

Surace, J.A., D.B. Sanders, W.D. Vacca, S. Veilleux, and J.M. Mazzarella. 1998. HST/WFPC2 observations of warm ultraluminous infrared galaxies. *Astrophysical Journal* 492(1): 116–136.

Toomre, A. and J. Toomre. 1972. Galactic bridges and tails. *Astrophysical Journal* 178: 623–666.

Whitmore, B.C., Q. Zhang, C. Leitherer, S.M. Fall, F. Schweizer, and B.W. Miller. 1999. The luminosity function of young star clusters in "the Antennae" galaxies (NGC 4038-4039). *Astronomical Journal* 118(4): 1551–1576.

Yu, Q., and S. Tremaine. 2002. Observational constraints on growth of massive black holes. *Monthly Notices of the Royal Astronomical Society* 335(4): 965–976.

Part III
Black Holes

Chapter 13
On the Epistemology of Observational Black Hole Astrophysics

Juliusz Doboszewski and Dennis Lehmkuhl

Abstract We discuss three philosophically interesting epistemic peculiarities of black hole astrophysics: (1) issues concerning whether and in what sense black holes do exist; (2) how to best approach multiplicity of available definitions of black holes; (3) short (i.e., accessible within an individual human lifespan) dynamical timescales present in many of the recent, as well as prospective, observations involving black holes. In each case we argue that the prospects for our epistemic situation are optimistic.

13.1 Introduction

Black holes are philosophically fascinating entities, but in many ways they are also philosophically troubling. Apart from existential questions about spacetime singularities and metaphysical questions about the fundamental theory of quantum gravity, there are epistemological issues to consider. How and what could we ever know about global regions of no escape swallowing every known type of matter? Since we are now entering a golden era of observations of black holes, it is appropriate to consider epistemology of observational black hole astrophysics.[1]

---

[1] We should immediately point out here that this chapter has, by design, a limited scope. Because current empirical evidence does not establish quantum effects related to black holes, we only discuss black holes as seen from the point of view of classical general relativity, and we are only focusing on selected epistemic questions in observational black hole astrophysics. As a consequence, we ignore important issues related to black holes in the foundations of physics, such as the study of singular structure in the black hole interior (Earman 1995), their importance for numerous questions regarding the global structure of spacetime (such as determinism, see Doboszewski (2019), or existence of time machines, see Doboszewski (2022)), or a very closely related issue of the cosmic censorship conjectures (Landsman 2021). We are also setting aside

J. Doboszewski (✉) · D. Lehmkuhl
Institut fur Philosophie, Bonn, Germany

© The Author(s) 2023
N. Mills Boyd et al. (eds.), *Philosophy of Astrophysics*, Synthese Library 472,
https://doi.org/10.1007/978-3-031-26618-8_13

Multiple lines of astrophysical evidence strongly indicate the existence of black holes, and the future of such observations looks bright. Black holes provide the basis for the widely accepted theories of accretion and relativistic jet emission in active galactic nuclei (AGNs). AGNs are observed across most of the electromagnetic spectrum. Jets are measured with X-rays with facilities such as the Chandra X-ray Observatory. High resolution observations of some of them can be done in the optical and infrared part of the spectrum, using bright optical sources such as the star S2 near the center of the supermassive black hole (SMBH) candidate Sagittarius A* in the center of our galaxy,[2,3] and with short wavelength radio interferometry (in particular by recent imaging of multiple sources with the Event Horizon Telescope array and its planned extensions). Most gravitational wave detections with the LIGO-Virgo network of observatories also seem to be generated by collisions involving black holes. Further extensions to the LIGO-Virgo network are under construction, and third generation detectors (such as the Einstein Telescope, Cosmic Explorer, and the space detector LISA) are planned. Furthermore, high redshift evidence concerning formation of supermassive black holes is expected to soon be available from the James Webb Space Telescope.

The number of observations is also growing quickly. To give just two examples: in LIGO-Virgo detections of gravitational waves,[4] the first observational run O1 (in 2015–2016) had 3 events, run O2 (in 2016–2017) 8 events, while runs O3a had 44 and O3b 36 events, for a total of 80 combined in 2019–2020. Some important tests of fundamental physics have already been made with these observations; one example is a strong dis-confirmation of some modified gravity theories, in particular TeVeS, by GW170817.[5] The Earth-spanning EHT network of synchronized telescopes grew from three radio telescopes in 2009 to eight telescopes on six sites in 2017, with further three added in 2018–2020; it has set aside coordinated observational time for a week (typically in early April) every year. EHT images of the M87* (The

---

issues of theory-ladenness, model independence, and robustness—all of which play prominent roles in establishing the reliability of particular lines of evidence for the existence of black holes.

[2] By a common convention the central region of Sagittarius A, M87 galaxy, etc. is denoted with an asterisk.

[3] Black holes come in different sizes, roughly subdivided into the following types. Stellar black holes are observed mostly using gravitational waves, and have masses from $2-5M_\odot$ to $100-150M_\odot$, with currently the highest known being the outcome of the merger event GW190521, of $163.9M_\odot$. Intermediate size black holes are observed through ultraluminous X-ray sources, and have masses ranging from $100M_\odot$ to $1000M_\odot$, perhaps even up to $10^4 M_\odot$. Supermassive black holes are observed in the optical spectrum as well as with radio interferometry have masses of $10^4 M_\odot$ to $10^{10} M_\odot$. And, so far hypothetical, primordial black holes, which might have formed in the very early phase of the universe, and could lie anywhere between 10^{-8} kg to $10^5 M_\odot$.

[4] These passed one of the following thresholds for detection: at least 50% probability of being astrophysical in origin, or have a chance of being a false alarm below 1 for 3 years. For readability we will be omitting confidence intervals throughout this chapter.

[5] For confirmed events, the prefix GW stands for "gravitational wave", with the numbers following it describing day, month, and the last two digits of the year. Gravitational wave astrophysics is discussed in much more detail in Lydia Patton's and Jamee Elder's chapters of this volume.

Event Horizon Telescope Collaboration et al. 2019) and SgrA* (The Event Horizon Telescope Collaboration et al. 2022) were constructed on the basis of data collected in 2017. As a part of its successor, the next generation Event Horizon Telescope, even more stations will be added in the forthcoming years, beginning with five stations in phase 1.

Black holes are in many ways unlike other astrophysical entities, so it is of quite some importance to consider black hole astrophysics' position within astrophysics more generally. Astrophysical tests of the more speculative aspects of black holes, such as the detection of Hawking radiation, remain out of reach for the foreseeable future.[6] But some philosophically interesting observations about the existence of black holes and the character of methodology used in search for them can already be made.

Here we will discuss three questions concerning epistemology:

- are our means of accessing black holes compatible with the belief that black holes exist in the same sense as other physical entities?
- are multiple alternative definitions of a "black hole" detrimental to our overall epistemological situation?
- are observations of black holes limited to effectively static snapshots and other trace-like forms of evidence?

In each case, we provide a cautiously optimistic assessment (in a sense similar to optimism about historical sciences of Currie (2018)) of our overall epistemic situation when it comes to black holes. In Sect. 13.2 we situate black hole astrophysics within considerations about realism, both generally and more specifically within philosophy of astrophysics; these are further exacerbated by the lack of direct access to black holes. However, we argue that the situation is not as problematic as it might seem: if considered jointly with a system coupled to it, there are many directly observable proxies for the geometry of a black hole. In Sect. 13.3 we consider some of the possible reactions to the fact that many different definitions of black holes are available, and argue that relationships between definitions are compatible with there being a substantial common core to the notion of a black hole, mediated by their appropriate behavior in the limiting case of an (idealized) exact solution. Finally, in Sect. 13.4 we point out that dynamical scales in black hole astrophysics are often short (accessible within an individual humans lifespan), and contrast black hole astrophysics with the effectively static snapshot character of many astrophysical lines of evidence, as well as with the view which sees astronomy as analogous to historical sciences. In these regards epistemology of black hole astrophysics is in a considerably better situation than many other branches of astrophysics.

---

[6] See Alex Mathie's chapter in this volume for a discussion of analogue gravity models, which aim at confirming occurrence of these effects by investigating systems similar to black holes and yet available for laboratory manipulation (such as sonic holes in fluids).

13.2 Epistemic Access to Black Holes

Two main issues concerning realism about black holes arise. The first is a general concern about the manipulability of astrophysical entities, the second is related to their indirect observability. If black holes cannot be manipulated, and if they are only indirectly accessible, shouldn't we remain neutral about claims concerning their existence and properties? As for the first clause of the antecedent, we we will argue that the criterion linking manipulability and existence of an entity is too strict, and in any case sufficient lines of evidence are available; as for the second clause, in a substantial sense direct access to black holes is possible (even if not yet realized by human astronomers).

13.2.1 No Interventions on Black Holes

In 1984, Ian Hacking argued that one's belief in the existence of an entity A posited by some theory is justified if and only if A can be used in manipulating and experimenting with some other phenomenon B. He went on to argue that according to this criterion, the existence of most astrophysical entities is doubtful, as they are too far away from us for us to use them in our manipulations. (Hacking's arguments apply to entities outside of the Solar System, as planets within our solar system have been used for gravity assist maneuvers and thus have been used in manipulating other objects, thus fulfilling Hacking's criterion for justified belief in them.) However, in the case of black holes we have good reason to believe that if they exist, then they are so far away from us that using them to manipulate on and experiment with black holes will likely remain beyond human reach, and so they don't fulfill Hacking's criterion for justified belief, as indeed Hacking himself has claimed.[7] It should be noted that the same applies to all stars on the night sky; none of them fulfills Hacking's criterion either, and one might well argue that this speaks against Hacking's criterion rather than against the existence of stars and thus against the possibility of observational astronomy to establish justified belief. Be that as it may; in the following we will argue that even if one accepts Hacking's criterion, the existence of black holes is now much *less* doubtful than it was even just 10 years ago.

First let us note that if a black hole were present anywhere near us, a number of manipulations and experiments using it *would be possible*, and it would thus fulfill Hacking's criterion. These would include extracting energy from a black hole using

---

[7] See Hacking (1989, 561). One could argue that it is an open question whether Hacking's arguments apply to primordial black holes in a similar way, for they might well exist close to us and thus might be amenable to be used in interventions. But despite extensive searches (see Carr et al. 2021 for a recent overview), no trace of those has yet been found. Accordingly, we will ignore primordial black holes in what follows.

a Penrose process (Penrose and Floyd 1971), which would enable us to use that energy in manipulating other objects. It would also be possible to use the black hole to perform gravity assist maneuvers, i.e., to use it in the same way that the planets of the solar system have already been used to speed up a spacecraft in a slingshot maneuver. These manipulations utilising a black hole could be performed by human agents, despite the massive difference in scale between them and the black hole. The outcomes of such interventions can be precisely calculated. Some of these effects are universal general relativistic effects, which only become more apparent in the presence of a strong gravitational field. Some other effects (for example gravitational time dilation or frame dragging) have been experimentally confirmed on Earth, and the corresponding predictions carry over to black holes. Apart from not being readily available for our experimentation, black holes are not special in this regard.

Hacking could admit all this and even be excited about all the things one *could* do with black holes if they were nearby, and yet maintain that the fact remains that they are *not* near enough to do any of these things, so that his criterion for justified belief in their existence is not fulfilled. So let us next look at how far astrophysical objects that are candidates for being black holes actually are beyond human reach.

The location of the black hole nearest to Earth is somewhat uncertain,[8] covering a range between 470 pc to 1530 pc. How far away is this from what humans can reach? After 45 years of travel, Voyager 1 is the human made object farthest from us, at a meager approximately 0.0007 pc. Prospects for any kind of humanity's expedition reaching any of these black hole candidates are, then, even more meager. And so are any experimental interventions, either by using these sources to intervene on something, or on the sources itself. It is practically impossible.

But should we really think of our lack of ability to manipulate things by help of black holes as a fundamental problem, or merely a contingent one? One view is that our location in the cosmos is a highly contingent matter, and thus so is the lack of ability to manipulate with such entities.[9] Drawing conclusions about the existence of some type of physical entities on the basis of a contingent feature would elevate it to

---

[8] To the point where candidates have changed at least three times during the writing of this paper: from HR 6819 and V723 Monocerotis (which seem to be stripped binaries, see Frost et al. 2022 and El-Badry et al. 2022b, respectively) to the gravitational lens which played a role in the microlensing event MOA-2011-BLG-191/OGLE-2011-BLG-0462 (which seems to be an isolated stellar mass black hole of $7.1M_\odot$ (Sahu et al. 2022); this is highly remarkable, because it is the first ever, and so far the only, candidate for an isolated stellar mass black hole), to Gaia BH1 (El-Badry et al. 2022a).

[9] It is not clear whether Earth-like planets and life-as-we-know-it could thrive in the vicinity of a black hole. The so-called black sun hypothesis states that they can. If the hypothesis turns out to be false, then living far away from a supermassive black hole would in some sense be physically necessary for organisms with a biology similar to ours. The jury is still out on this hypothesis. It seems that so-called "blanets", a certain type of exoplanets, could form around some AGNs (Wada et al. 2021). Moreover, blanets might have a temperature (with the gradient provided by the flow of blueshifted flux of cosmic microwave background radiation onto the cold spot of a black hole) within the habitable range (Bakala et al. 2020). On the other hand, arguably (Forbes and Loeb

a privileged epistemic position, and as such would be anthropocentric. Furthermore, human spaceflight is now barely 61 years old. An optimistic outlook on human ability to cooperate would see uniting around a common goal (such as travel to a remote destination) as a possible option. From this perspective inaccessibility might not be an insurmountable difficulty, but a contingent feature of our epistemic position. In any case, it seems like an issue of practice, rather than an issue of principle.[10]

This relates to a point made by Shapere (1993), regarding Hacking's criterion. Remember that Hacking claimed that belief in the existence of A is justified if and only if A can be used in investigating some other phenomenon B. Shapere pointed out (see also Massimi 2004) that the term "use" in Hacking's criterion can be read in two different ways: as "manipulate" and as "employ" or "exploit". Entities posited in astrophysics can rarely be manipulated, but often are employed in mechanistic explanations of various phenomena.[11] Regarding black holes, this is now much more the case than when Hacking first applied his criterion to the question of whether black holes exist. Indeed, such mechanisms have now been probed and tested in various ways in black hole astrophysics. For example, black hole based waveforms have been employed in matching the patterns of gravitational waves detected by LIGO-Virgo. Furthermore, the observed shape of the central brightness depression in the EHT images of the two black hole candidates M87* and SgrA* have provided a good fit for the assumption that the exterior of these objects accords with the Kerr geometry, which in turn strengthens the plausibility that these objects are rotating black holes. (See, however, Bronzwaer and Falcke (2021) and Vincent et al. (2022) for some words of caution: size and shape of the black hole shadow are not unambiguous predictions of GR, but can be recovered from alternative models, and are sensitive not only to geometry of the source, but also to emission models; the photon ring, a strongly lensed thin feature of an image, is such a signature, but has not yet been resolved. This is also of relevance for assessing which of these features can provide direct evidence in the sense discussed in the next section.) Thirdly, the assumption that the respective active galactic nucleus (AGN) is a black hole is currently the only way to explain the bright output of the AGN, which is explained by the hot matter accreting onto a supermassive black hole assumed to be in the center. Finally, light emitted from high redshift quasars (whose high energy output is best explained as being powered by a black hole) has been used by Rauch et al. (2018) in setting up direction of polarization in quantum mechanical tests of

---

2018) XUV irradiation emitted by the gas accreting onto a SMBH might increase loss of planetary atmospheres.

[10] However, this argument is weakened by the fact that we do not have a convincing design for how a spacecraft capable of such a journey could be constructed, even if we had unlimited funding and global cooperation.

[11] This idea also plays well with the view which sees astrophysics as employing natural experiments provided by the universe in a Cosmic Laboratory, cf. Anderl (2016).

Bell inequalities (as an element of an attempt at limiting the so-called freedom of choice loophole).[12]

Thus, at the very least in a passive sense AGNs have already been used in manipulating elements of experiments. It follows that AGNs *do* fulfill Hacking's criterion: there *really are* extremely heavy objects in the center of the M87 galaxy, and in the center of our own Milky Way galaxy. One can still maintain the position that the AGNs in question *may* not be supermassive black holes, despite the fact that this assumption has become ever more fruitful in astrophysics. In other words, the existence of black holes may still be doubtful—but it is now much less doubtful than when Hacking wrote about them in 1984.

13.2.2 Indirect Observability of Black Holes

Hacking's criterion as discussed in the previous subsection required that in order for black holes to exist, we would have to be able to manipulate other objects by help of black holes. A weaker criterion for their existence would be to say that black holes exist if and only if they are observable. This criterion is more in line with van Fraassen than with Hacking, and it brings up the follow-up question of when something counts as observable, and whether it has to be directly or merely indirectly observable.

Recently, Eckart et al. (2017) have argued that "[super-massive black holes] are philosophically interesting entities given that they are only observable by indirect means." Eckart et al. do not define what they mean by "indirect" here, but we can draw on a precise characterization of directness due to Shapere (1982).[13] Shapere considers an entity or a source which undergoes some physical interaction (be it manipulation by a human observer, or some natural process), which leads to the emission of an information-carrying signal, recorded at the detector. According to Shapere's notion, an observation is direct if information received from the source is transmitted without interference to the detector. What constitutes emission, transmission, and interference depends on the theory of the source, the theory of transmission, and the theory of the detector; these, in turn, depend on the particular line of evidence.[14]

Assuming Shapere's notion of direct vs indirect observability, what side do black holes fall on? Eckart et al. (2017) point at the nature of evidence concerning black holes to justify their claims. If an astronomical source is a black hole, there can be

---

[12] The same team has earlier performed similar Bell inequality tests using Milky Way stars, so a similar point could be made about other entities outside the Solar System.

[13] Later elaborated by Franklin (2017); see also Elder (2021) for a recent critical discussion.

[14] One could further make a distinction between a strict notion of directness, where the detector is that of a human sensory system, and a permissive one, which allows for the use of scientific instruments. We will be assuming a permissive notion, as the strict one rules out observation relying on scientific instrumentation.

no emission from the black hole itself; an isolated classical black hole, after all, is a perfect absorber.[15] It is only when that source is coupled with some second system, such as matter in the accretion disk or another black hole, that any signal from the near horizon region can be emitted and detected by a distant observer. Said matter might be used as a proxy for the source itself, but it provides only indirect evidence. The black hole on its own is, then, an in principle unobservable entity. This is in line with Hacking (1989), invoked by Eckart et al. (2017), who notes that "[a] black hole is as theoretical an entity as could be. Moreover, it is in principle unobservable. (…) At best we can interpret various phenomena as being due to the existence of black holes" (561). Evidence for the existence of black holes could then be seen as somehow less certain and less conclusive than the usual empirically collected data, which might be straightforwardly ascribed (through direct observations) to theoretical entities responsible for their production. This argument targets black holes in contrast to other phenomena astrophysics is concerned with, because most other objects are electromagnetic emitters.

Note that some lines of evidence are direct in Shapere's sense. The data collected, for example by LIGO-Virgo, might be a direct detection of gravitational waves (even if arguably only an indirect detection of binary black hole mergers, on the grounds of relying on models of the merger; see Elder 2021). However, interference of radio waves (in the EHT) or an optical signal (in adaptive optics measurements) with the Earth's atmosphere provides interference which could be interpreted as invalidating the "without interference" clause of Shapere's definition. On the other hand, the signal emitted by these sources is present in the data and can be reconstructed; and if it would be the issue of atmospheric noise that invalidates the clause, then the question of indirectness becomes contingent on the location of a telescope. Gravitational waves couple weakly to interstellar matter, and so the "without interference" clause is easier to establish in that case.

What could the signal emitted by a black hole be? If the black hole is truly isolated, then (again, apart from quantum effects such as Hawking radiation or superradiance) the prospects for detecting any signal originating from it are by definition impossible. But once it is coupled to either another black hole, or to hot matter in the accretion disk, the situation changes dramatically. The shape of emissions is sourced by the gravitational field of the black hole. Insofar as the theory describing emission involves the strength and shape of said field, that part of Shapere's notion of directness can be satisfied. This line of argument relies on having a black hole coupled to some other system, and so one could complain that it is that other system which is directly measured, and that the black hole is accessed only indirectly through it. However, one might answer that then everything is an indirect observation: we never observe the table itself but only the light reflected from the table.

---

[15] One limitation of this line of argument is that, arguably, processes such as emission of Hawking radiation, or superradiance mechanisms involving black holes, do constitute a form of emission of energy from a black hole.

In order to judge whether black holes are indeed at best indirectly observable, we have to consider the distinction between the region external to the black hole, the region of no escape inside of the black hole, and a surface separating them, the event horizon. Of these three regions, the interior is merely in-principle observable: only an observer ready to jump inside can observe what happens in there (but not transmit that to the outside).[16] The exterior region, including regions arbitrarily close to the separating surface, *is* epistemically accessible in the same sense as any other region of spacetime.

One might say that no-one had ever claimed that the exterior *outside* a black hole is not directly observable; the question was about whether the black hole *itself* is (directly) observable! But here is the crux that makes black holes special, at least in this respect: if the exterior of a black hole candidate is found to accord with the Kerr geometry, then we can reliably conclude that the object in question is a rotating gravitational source like a star *or* a black hole.[17] If, in addition, the object does not emit any light but is supermassive and sufficiently compact, then arguably the best explanation, even the only available explanation, is that the object in question is a black hole.[18] It's not really different from observing the exterior of a table: you only have to really know how the table looks from the outside to conclude that the object in question is indeed a table.

So, can we make experiments or observations that would tell us that the exterior of a black hole candidate accords to the Kerr geometry and thus is, in all likelihood, the exterior *of a black hole*? Yes: coordinated observers could, for example, shoot lasers towards the black hole candidate and test whether their paths agree with the trajectories of null geodesics of the Kerr geometry. One could also test whether light is on the verge of being trapped in a certain region, how strongly a given region of spacetime lenses light, what the shape of this lensing region is, whether frame dragging effect occurs, and so on.

In less abstract astrophysical situations, luminous matter such as gas or plasma in the accretion disk of an AGN is used for establishing the geometry of the gravitational field. Instead of considering a black hole on its own, one is considering a coupled system of a black hole and luminous matter, which is sufficient to establish exterior geometries that would provide signatures for various kinds of black holes: rotating, charged, those compatible with modified gravity theories, horizon-less black hole mimickers, and so on. Recent work of the EHT measuring the shadow of a black hole is a good example: here the shape of bright emissions from the accretion

---

[16] Again, apart from the possibility that the measurement results leave the interior of a black hole during the semi-classical evaporation process. It is, however, worth pointing out that some physical mechanisms, such as the blueshift heuristic underlying investigations of the cosmic censorship conjecture, do constrain properties of the deep interior on the basis of perturbations of matter in the exterior. See Chesler et al. (2019) and references therein for this line of investigations.

[17] The same could be said if the exterior of the black hole candidate accords with the Schwarzschild geometry or the Reissner-Nordström geometry; however, all current observations are compatible with sources being Kerr, i.e., rotating bodies.

[18] Of course, Stanford's problem of unconceived alternatives always remains.

disk, measured at 230 GHz and 20μas resolution, can be used to rule out some of the shapes incompatible with the source being a black hole. In such situations direct experimental probing of structures and geometry close to the event horizon is possible, given enough time and resources. Even though a black hole is unlike other astrophysical entities,[19] the exterior of a black hole is accessible, and has such a distinctive signature that access to the exterior might be enough to conclude that it is the exterior *of* a black hole.

Finally, something should be said about the localization of a black hole. Hacking claimed that "we cannot with any confidence point to any region of the sky and say, there's one there" (Hacking 1989, 561). Indeed, the concept of a black hole event horizon is a global notion: one would need to know the whole history of spacetime in order to establish that an event horizon exists, and as such it cannot be localized to a finite region of spacetime observed for a short interval of time. In this sense, when taken at face value, it is not an epistemically accessible property of spacetime. But often global spacetime properties should not be taken at a face value: they rather express idealizations about the systems. We consider ourselves far enough from the source that for all practical purposes all light from it has reached us, and so, we can pretend we are observers located at future null infinity (Ellis (2002) even suggests that for a local group of galaxies the appropriate distance is 1.2 Mpc). In this sense a "black hole" as defined by "having an event horizon" *can* be localized (and the situation further improves if some quasi-local notion of a horizon is adopted).

To sum up: following Shapere's criterion, some means of direct access to black holes are possible. But, interestingly, even indirect observations can give us evidence strong enough that we can be quite sure what the object in question actually is; arguably as sure as in many cases of direct observation.

13.3   Interpreting Many Definitions of Black Holes

Issues of epistemic access are further exacerbated by the observation that a 'black hole' is a polysemic term: many definitions of a 'black hole' are available. One could be concerned: what do we even mean when talking about black holes? Do we have sufficient conceptual control over these notions? We will first survey various possible reactions to the occurrence of many definitions, and then argue for cautiously optimistic assessment of the situation: many of the definitions are compatible with each other.

Curiel (2019) recently surveyed some of the definitions of black holes used by practitioners of different sub-communities. These sub-communities include observational and theoretical astrophysicists, classical relativists, mathematical rela-

---

[19] Arguably one can have direct access to the interior of e.g. the Sun by measuring neutrino flux generated within, or simply by entering it with a sufficiently sturdy spacecraft and come out again to tell the tale.

tivists, physicists working on semi-classical gravity, quantum gravity, and analogue gravity. He found at least twelve different ways of defining a "black hole" (see Figure 1 of Curiel (2019)). This includes the characterization as a physical object whose defining feature is that it is simply a very compact object that is incredibly massive, or one that is characterised by an event horizon, or, alternatively another geometric feature: an apparent horizon, or instead a trapped surface; that it is an object featuring a singularity; or instead a region of no escape for low energy modes; or that a black hole is a particular type of engine producing an enormous power output.[20]

13.3.1 Cluster Concepts, Perspectives, and Other Possible Reactions to the Many Definitions of Black Holes

How should we react to this plethora of definitions of black holes? First, we need to note that the fact that different sub-communities operate with these different definitions is crucial. One might say that given that different communities have different purposes, different definitions are not really a problem; a chemist has a different working definition of "molecule" than a quantum physicist. Still, the question remains if the different communities could come to an agreement about the notion of "molecule" or "black hole" that fits all their purposes and that they would accept as the underlying "proper" definition of the term in question—a set of necessary and sufficient conditions for something to be a "black hole" that can be agreed on across all communities. In the case of "black hole", no such agreement on necessary and sufficient conditions has as of yet been found, and it is not clear at all that it ever will be found. Indeed, it is not even clear whether we should hope for such a set to be found, as many definitions for many purposes may well be seen as more flexible and fruitful for the conduct of further research on these objects.

Thus, we see six different options to react to the plethora of definitions of a "black hole" stemming from different sub-communities: (1) the classic hope of an "inner core" to all these definitions, i.e., a set of necessary and sufficient conditions for something to be a black hole; (2) that the different definitions form a Wittgensteinian family; (3) that a "black hole" is a cluster concept; (4) that the different definitions of a "black hole" correspond to different perspectives in the sense of perspectivism; (5) that the different definitions of a "black hole" are so disjoint that one is forced into semantic anti-realism; and (6) a kind of pragmatic pluralism about what a "black hole" signifies. We are going to elaborate on each of these options in the rest of this subsection. It will turn out that the question of which of these options is *actually* the most convincing will turn on how the different definitions of a "black hole"

---

[20] Curiel's list is non-exhaustive; for instance, a quasi-local horizon is one possibility, but there are many inequivalent candidate quasi-local horizon notions; see Booth (2005) for an accessible introduction, including unwelcome features of such notions.

are *actually* related to one another; a question that will be investigated in the next subsection.

But first let us look at the different possibilities. We have already looked at the first option, most familiar from analytical philosophy more generally: it is that there is, in the end, a set of necessary and sufficient conditions for something to be a black hole, and that we just have not yet found this set of conditions. The second option is that the different definitions of a "black hole" form a Wittgensteinian family, i.e., a set where any two family members have *something* in common, but where no trait or property is shared by *all* family members. The third option is that "black hole" is what Baker (2021) called a "cluster concept", i.e., a concept that cannot be captured by necessary and sufficient conditions but instead "can be satisfied in a variety of different ways by different entities falling under" the concept (S279). Baker sees the "best realizer" variant as the most plausible version of a cluster concept view: "only the (ideally unique) structure that best satisfies the criteria of the cluster concept counts as spacetime, even in cases where other structures also meet the criteria to a sufficient degree that they would count as spacetime if they existed alone" (Baker 2021, S290). Under this approach, a list of criteria for being a black hole should be produced (Baker provides just such a list of candidate criteria for a given structure to be a spacetime), and candidate definitions should be compared against it; the one which is a best fit to the criteria becomes the official definition.

At first sight, this looks rather similar to claiming that a given concept forms a "Wittgensteinian family" of definitions, but the idea is actually rather different. In such a cluster of definitions, in contrast to a Wittgensteinian family, there may well be two members of the set that don't have *anything* in common, precisely because the something can fall under the concept in question in "a variety of different ways". If a "black hole" is a cluster concept, then it would be possible to find a set of n conditions of which any $n - m$ conditions (with $n > m$) must be fulfilled in order for something to be a black hole. Of course, the task would be not only to find the set of n conditions but also to justify the number m.

The fourth option one could take in light of the multiple definitions of a "black hole" is perspectivism. Perspectivism (Giere 2010; Massimi 2018) associates the presence of many (possibly inconsistent) scientific models with multiple equally valid perspectives on a phenomenon. Taking many definitions as providing equally valid perspectives or aspects of the same entity may be tempting especially in contexts when these definitions are inconsistent with each other. For example, in non-stationary spherically symmetric spacetimes, the definition of a black hole relying on the presence of an event horizon picks up a different surface from the definition relying on the apparent horizon. (See figure 6 in Senovilla (2013) for a simple illustration of this incompatibility in Vaidya spacetimes.) Similarly, the so-called regular black hole spacetimes with non-singular interiors (see Berry et al. 2021 for an example construction) do not qualify as black holes for definitions relying on the presence of a spacetime singularity. On the other hand, Morrison (2011) argued that inconsistent models signal lack of theoretical understanding of the phenomenon in question. In such cases, she argues (Morrison 2011, 350) that perspectivism about models of the nucleus "amounts to endorsing a claim of the form: Taken as a classical system (...) the nucleus looks like X; as a quantum system it looks like Y, and so on for any given model we choose". Morrison finds this unsatisfactory, because "none of these 'perspectives' can be claimed to 'represent' the nucleus in even a quasi-realistic way since they all contradict each other on fundamental assumptions about dynamics and structure". An analogous point can be made about incompatible definitions of black holes.

The fifth option is a form of semantic anti-realism. Having established that the different definitions of a "black hole" have little in common, one could worry: what could a realist even be a realist about when it comes to black holes? This position has been suggested by Martens (2022) in the context of dark matter. However, arguably the situation of black holes is not so dire: in contrast to dark matter candidates there are consistent estimates for the masses of black holes—this is not the case for dark matter particles, whose mass varies over many orders of magnitude. There are also two commonly used theoretical models of a black hole, given by the Schwarzschild and (subextremal) Kerr geometries for the cases of non-rotating and rotating black holes, respectively. Again, this is not the case for dark matter, which could be accounted for using very different theoretical models, from primordial black holes through axions to entirely new species of particles.

The sixth option is the one that Curiel has argued for. It is a form of pragmatic pluralism: the many definitions of black holes are seen as something positive. Curiel concedes that "there is a rough, nebulous concept of a black hole shared across physics, that one can explicate that idea by articulating a more or less precise definition that captures in a clear way many important features of the nebulous idea, and that this can be done in many different ways, each appropriate for different theoretical, observational, and foundational contexts" (Curiel 2019, 33). He does not see this as a problem, but a virtue. A single precise definition, he argues, would likely be more constraining and less fruitful than the variety of tools provided by the many definitions of a "black hole".

Listing logical possibilities and options one could choose is all well and good, but which of these options we *should* choose will depend on the actual relations between the different black hole definitions. Semantic anti-realism with respect to black holes is only a viable option if the different definitions do indeed have little in common, and whether the definitions are better described as forming a Wittgensteinian family or a cluster concept likewise draws on what precise relationships can actually be found between the different definitions. Thus, we shall look at least at *some* black hole definitions in some detail, and investigate which relations hold between them.

13.3.2 Relationships Between Different Definitions of Black Holes

Curiel's analysis stops short of discussing relationships between more or less precise definitions used in different theoretical contexts. If one does, one finds that some of these definitions are not fully independent of each other; there are subtle relationships between them. The extent and precise nature of these relationships will determine which of the six options discussed in the previous subsection regarding how one could react to the many different definitions of a "black hole" is the most viable one. Here, we can only give a tentative foray into what these relationships are.

Since we are concerned with observational black hole astrophysics (where currently properties of classical black holes are at the frontier of investigations), we will conveniently restrict our attention to some of the definitions most useful in that context. In other words, we want to understand how the definitions of a black hole in terms of it being a compact object, an engine for enormous power output, an object that is characterised by an event horizon or an apparent horizon, or one that is characterised by a singularity are related to each other. By introducing this restriction we are making the task comparatively easy on ourselves: in the context of semi-classical gravity and quantum gravity relationships become more difficult to ascertain.

So what are some of the relationships between the different definitions of a "black hole"? At least two types of relationships can be found.

The first type of relationship obtaining between many definitions of black holes is restricted equivalence: two definitions may be equivalent in a restricted setting (but not in full generality). Consider the definition of a black hole in terms of it possessing an event horizon and the definition in terms of a foliation of spacetime by a sequence of apparent horizons. It turns out that even though in time-dependent spacetimes these definitions do pick out different surfaces, in static cases theses surfaces coincide; see fig. 5 and 6 of Senovilla (2013) for an illustration in the Schwarzschild spacetime and Vaidya spacetimes.[21] So these two notions are provably equivalent in a restricted setting, where the restriction in question is the condition of staticity.

The second possible relationship is one of reliable proxyhood. By this we mean a situation where in a particular theoretical context the fact that one definition holds strongly suggests (though not necessarily in the sense of a logical implication) that some other definition also holds. In that case, one notion of a black hole is a reliable proxy for another notion. Reliable proxies differ from restricted equivalences in two ways. First, the relationship may hold in typical cases (in a sense to be specified in a given context) only. Thus, it would not be appropriate to speak of an equivalence.

---

[21] The stationary case remains open; see Carrasco and Mars (2013) for a summary of results suggesting that the answer will be positive.

Second, one definition may be a proxy for another even in cases where the regions of spacetime picked out by these two definitions fail to coincide. Nevertheless, by learning about the properties that the first definition relies on, we also learn about the properties that the second definition relies on. We will now give two examples of relationships between prominent definitions which illustrate these differences. In the first case, the proxy is located outside the surface of a black hole as characterized by most other definitions; in the second case, the proxy is located inside the surface of a black hole as characterized by some other definitions.

Astrophysical models of accretion and jet launching are usually constructed based on the assumption of a general relativistic background geometry. In this way, models are fitted to an exact solution (typically Schwarzschild or Kerr). The standard active galactic nuclei model strongly suggests that features of the AGNs, so characterizations of black holes in terms of "compact object" and an "engine for enormous power output" can plausibly be associated with the exterior of the object in question being characterised by the Kerr spacetime (and so its geometric structure, including its event horizons and apparent horizons). In this situation "engine for enormous power output" becomes an elliptic expression for "accretion onto a Kerr spacetime with large mass". Whether this proxy remains reliable can change—for instance, if observationally viable accretion models onto Exotic Compact Objects[22] are constructed, "engine for enormous power output" (or "compact object") might no longer be a reliable proxy for a spacetime region with a Kerr geometry. It is also not a scale-invariant characterization of a black hole: stellar mass black holes might not be definable in this way. If the microlensing event MOA-2011-BLG-191/OGLE-2011-BLG-0462 is indeed a black hole, it seems to have effectively zero energy output. In this situation the "engine for enormous power output" is not a universally reliable proxy for the Kerr geometry, despite it being a reliable proxy in the case of supermassive black holes (as long as no well-established alternative models for AGNs are available).

Another example of a reliable proxy are marginally outer trapped surfaces (MOTS). A marginally trapped surface is a closed 2-dimensional surface S such that outward future pointing null vectors have vanishing expansion. If there are many such surfaces, some contained inside others, then an apparent horizon is the outermost one; in other words, it is a MOTS which is not contained in any other MOTS. If the spacetime is asymptotically flat, has an event horizon, and the null energy condition holds, then an apparent horizon is located inside the event horizon (Wald 1984). In that setting, locating an apparent horizon is a reliable proxy for an event horizon. In the 3 + 1 ADM approach to numerical relativity apparent horizons are easier and faster to find than event horizons (Thornburg 2007), because codes for finding MOTS' and apparent horizons can be run during the numerical construction

---

[22] These are a large and very heterogeneous class of objects which have similar masses and sizes as black holes, but do not contain a horizon-like surface; see Cardoso and Pani (2019) for a recent overview of such models. Since by definition ECOs have surfaces instead of event horizons, such models can be constrained by an analysis of their luminosity (Lu et al. 2017).

of the spacetime. In contrast, event horizon finders have to be run as a separate step after the construction. However, as long as the appropriate background conditions are satisfied, a MOTS is located inside an event horizon. A definition of a black hole relying on a MOTS is thus a reliable proxy for the definition in terms of an event horizon.

13.3.3 Consequences of Relationships Between Many Definitions

At least some of the definitions of black holes are related to each other in interesting ways. This strengthens the hope that there might, after all, be a set of necessary and sufficient conditions to be found for something to be a black hole (option 1 from the previous subsection). But it is also entirely consistent with the idea that the different, yet related, definitions of a "black hole" form a Wittgenstein family (option 2) or a cluster concept (option 3). The relationships between definitions seem to weaken the perspectivism account (option 4), yet not rule it out, and also weaken the case for semantic anti-realism (option 5). The case for semantic pluralism (option 6) still stands strong, though the above has raised the question how much of a plurality of definitions there really will be in the end.

We should also note that the relationships between definitions, which we could only point to in this review, typically flow from the empirical and conceptual adequacy of an exact solution of Einstein's field equations, in particular the Kerr and Schwarzschild solutions of the Einstein's field equations. From these, many definitions are further derived, abstracted, or generalized. In this way the exact solution might provide a core concept of a black hole. Many of these definitions are formally or plausibly related under additional auxiliary assumptions (such as stationarity, asymptotic flatness, and the null energy condition), many of which, in turn, express idealizations, such as a system not varying over time, the system being isolated, or neglecting effects due to quantum nature of matter fields.

From this point of view, the plurality of definitions can be seen as resulting from an ongoing process of de-idealization and extension of a concept well understood in a particular limited domain to larger domains. It is then not surprising that many inequivalent definitions of a "black hole" are available. Indeed, one should expect that many definitions will appear: for any highly idealized notion, many ways of de-idealising are available. Depending on the particular context of investigation, different aspects of the object investigated are taken to the be of relevance. In any case, existence of many definitions does not have to constitute a worry for the epistemology of black hole astrophysics.

13.4 Short Dynamical Timescales

Accessible timescales influence available interpretative positions and the assessment of the overall epistemic situation, so it is appropriate to consider them here. Two particular aspects are worth discussing here, in some ways the epistemic situation of black hole astrophysics is richer than in many other areas.

First, astronomy and astrophysics are often seen as analogous to historical sciences such as archaeology, paleontology, or geology: the finite speed of propagation of light implies that the light reaching us from distant sources carries information about events that transpired, in some sense, long ago. Thus, epistemic access to dynamic processes occurring in these sources is limited to their downstream "traces", often sparse and partial, and impoverished in similar ways (see Anderl (2021) for an extended exposition of this view).

Second, in her recent analysis of epistemic roles played by astrophysical simulations, Jacquart has pointed out that astrophysics suffers from the fact that observed sources typically vary very slowly, remaining unchanged over thousands and more years. Access to signals emitted by such sources is effectively confined to an "observational 'snapshot'—a single time-slice of the object under investigation" (Jacquart 2020, p. 4).

Interestingly, a wide range of observations in black hole astrophysics deal with dynamical signals changing on timescales (much) shorter than the average human lifespan. Jacquart concedes that some astrophysical phenomena[23] change during the observation. However, she sees these as "by far the minority", as "[m]ost objects or phenomena of study in astrophysics take place over cosmic time scales of millions of years" (p. 4), which are too large to be observable for humans. Jacquart uses this observation in pointing out an amplifying role of astrophysical simulations, which provide stand-ins for the dynamical evolution of the source. However, notably, in black hole astrophysics short dynamic timescales are present much more commonly than in gravitational wave observations. They are the norm, and snapshots are an exception. This has a further consequence: if observable "traces" carry information about dynamical processes, then the commonly accepted analogy with historical sciences is weakened. Historical traces are in important respects unlike highly dynamical signals carrying information about black holes. We will now discuss some of the examples of the plurality of dynamical timescales present in the current main observational lines of evidence and in their prospective generalizations.

---

[23] Such as supernova explosions and black hole mergers; but one could also point to pulsars, fast radio bursts, gamma ray bursts, etc.

13.4.1   Timescales in Black Hole Astrophysics

Although black holes seem to come in different sizes, the timescale of variability is not a function of the mass of a black hole. It rather relates to the source's immediate environment, distance from Earth to particular sources, and to particular ways of accessing them.

As already noted by Jacquart, current observations of gravitational waves are among the most striking examples of short dynamical timescales in astrophysics. The current generation of gravitational waves detectors is tuned towards very fast transients: those of 2–500 seconds are considered long. Events in LIGO-Virgo observational runs O1-O2 varied from 0.2 seconds (GW150914) to 100 seconds (neutron star-neutron star merger GW170817); the latter, however, is the only long one. But there are many intermediate timescales between these transients and effectively static snapshots.

An increasingly important line of evidence comes from short radio wavelength observations performed using Very Long Baseline Interferometry techniques utilising the Earth-spanning Event Horizon Telescope array. Here, time variability differs between sources and radio frequencies measured. The main targets of these observations are the central object in the Messier 87 galaxy M87* (the subject of the famous first image of a black hole from 2019) and SgrA* in the center of the Milky Way. A number of secondary targets, such as Centaurus A, 3C 279, supermassive black hole binary candidate OJ 287, and others, have also been observed. A natural variability timescale is set by the period of the innermost stable circular orbit, which in turn depends on the mass and spin of the source. In SgrA* this range is between 4 and 30 minutes (The Event Horizon Telescope Collaboration et al. 2022), while for M87* it ranges from 5 to 30 days. This intra-hour variability in the emitted flux leads to the possibility of producing not just single black hole images, but black hole movies, as SgrA* changes its state during a single observing night. However, even here other timescales occur: one example are bright flares, occurring daily, and observable in the near-infrared and X-ray spectrum. Such flares might be interpreted as hotspots generated in the accretion flow (Tiede et al. 2020) and used to map the surrounding spacetime region as a function of the hotspot passing through various near-to-far horizon scales. The M87* variability timescale is of the order of a month. But using the total 2009–2017 data set, the 2019 image as a prior, and under a simplifying assumption that the set of alternatives is limited (to asymmetric ring and a Gaussian), the evolution of the shape of the source over time can be constrained, with the asymmetric ring being preferred (Wielgus et al. 2020). In the case of another source, 3C279, its jet exhibits day-to-day variability (Kim et al. 2020). RadioAstron orbiting VLBI observations from 2014 suggest the presence of helical threads, or filaments, in the jet; this will soon be followed by analysis of 3C279 with the EHT data from 2017 onward, from which time variability of the filaments can be estimated. Finally, OJ 287 is a supermassive black hole binary candidate in which an elliptical orbit of the less massive component takes nearly 11.6 years to complete (Shi et al. 2007). For more than 22 years it has been monitored long term, occasionally with a daily cadence. In the case of this source, periodic flares can be predicted to occur on a particular day; Laine et al. (2020) find[24] that the 2019 flare arrived within 4 hours of the predicted time. In all of these examples multiple short timescales, varying from minutes to years, are accessible to the astronomers.

Timescales accessible within an individual human lifespan are also available in the optical part of the spectrum. One example is Cygnus X-1, where the primary star HDE 226868 is orbiting around an unseen companion, with a period of 5.6 days.[25] Other bright tracers of candidate black holes are utilized; perhaps the most important one is the bright star S2 with an orbit of approximately 16 years. It has been monitored for 26 years (as of data published in 2018) by the ongoing GRAVITY collaboration (Abuter et al. 2018) observations of the center of the Milky Way, and also by the UCLA group (Do et al. 2019) independently observing the same region. The outcomes are consistent with the hypothesis that the central region SgrA* is a single highly concentrated mass. The star S2 is just one of many tracers, and multiple other similar objects are monitored. Multiple observations of objects on such orbits can be made within the lifetime of an individual observer.

Not all prospective observations of black holes are so dynamic: some are likely to be very long and slowly varying, even while supplementing other observations at shorter timescales. One example (following section 7 of Abbott et al. (2016)) are black hole binaries such as GW150914. These binaries emit gravitational waves in the frequency range of space detectors such as (e)LISA, plausibly 0.1–10 mHz; it is scheduled to launch in the 2030s. It takes approximately 1000 years to evolve from 2 to 3 mHz emission to the merger phase. The dynamics of such a system would be a time-varying signal (and so not a static snapshot), which could be monitored during a time interval much longer than an individual human lifespan.

Examples of snapshots can also be expected. Numerous mechanisms for formation and growth of SMBHs have been proposed; these include light seeds ($<10^3 M_\odot$), heavy seeds (10^4–$10^6 M_\odot$), and other intermediate pathways, and it remains an open question what proportion of these mechanisms can explain which proportion of the SMBHs population. Light seeds are likely to be too faint to be seen by the James Webb Space Telescope (JWST), but the possibility that heavy seeds at up to very high redshift $z \sim 15$ might be seen with the JWST (which at the moment of writing started releasing first images) is of relevance here. That evidence is likely to consist of snapshots, but (e)LISA and third generation gravitational wave detectors (Cosmic Explorer and Einstein Telescope) might be able to detect mergers at $z > 10$, and such supplementing evidence would again have a dynamical character. Pointers for a discussion of these mechanisms as well as other possible lines of evidence can be found in Chen et al. (2022).

---

[24] Using the Spitzer Space Telescope rather than VLBI; but dynamics of OJ287 is also observed using VLBI methods, see Sawada-Satoh et al. (2015).

[25] Lack of emission from the companion counts among the lines of evidence for the existence of black holes.

There is a clear sense in which—JWST observations notwithstanding—evidence in observational black hole astrophysics is not confined to effectively static snapshots. It rather concerns a wide range of dynamic processes across different timescales, with the duration of a process typically accessible within an individual human lifespan. In cases of some particular sources (like the SMBH candidate Sagittarius A*) various timescales are accessible simultaneously.

13.4.2   Consequences of Short Dynamical Timescales

Short dynamical timescales provide more information than snapshots. The dynamics of the source accessed through snapshots needs to be inferred from the single trace only. But with short timescales accessible in black hole astrophysics it is not only a single state of the system that can be observed; its change over time can be recorded as well. The character of such downstream traces is different: they are dynamically rich records of the evolution of the black hole and its surroundings. In this way, epistemic access to black holes can be seen as more informative than in many other astrophysical contexts (dominated by effectively static snapshots).

In this way, the analogy with historical sciences is weakened: transient events and other observations associated with black holes in an important sense are unlike trace fossils or geological layers. In some—EHT sources—though not all—transients observed with the current generation of LIGO-Virgo detectors—cases, a dynamically evolving source is available for further sampling with subsequent observations, because the source can be monitored over extended periods of time.

An additional constructive perspective concerns transient observations which cannot be re-sampled at will. Recall that LIGO-Virgo made 91 detections until the end of observational run O3. Out of these 91 detections, black holes seem to be responsible for the vast majority of events: only 2 are classified as neutron star-neutron star collisions, 4 as black hole-neutron star collisions, and a further 2 as involving a black hole and an uncertain object. Plausibly, these proportions will remain similar in the future observational runs. If so, population studies will provide more immediate and more reliable constraints on evolution, production mechanisms, and statistical properties of the population of stellar black holes than on neutron star mergers. From this point of view, black hole astrophysics is in a comparatively better epistemic situation than astrophysics of many less "exotic" entities.

13.5   Conclusions

We have surveyed four problems which are *prima* facie detrimental to the epistemic situation of observations of black holes. The first one concerned lack of manipulability; we diagnosed this as a contingent feature, and pointed out that

AGNs have been used in setting up some experiments. The second concerned alleged lack of direct access; we have argued that under a particular notion direct access is possible (even though not yet realized in practice). The third concerned multiple available definitions of black holes; we have classified possible reactions, proposed two types of relationships between definitions, and suggested the sense in which exact solutions of general relativity might provide a core concept of a black hole. Finally, we have explored the consequences of empirical access to dynamical processes involving black holes. The overall conclusions are optimistic: the future of observations of black holes is bright, and so are prospects for the corresponding philosophical analysis.

Acknowledgments We are grateful to two anonymous reviewers, Nurida Boddenberg, Erik Curiel, Jamee Elder, Sophia Haude, Niels Martens, Jan Potters, Aleksandra Samonek, and Noah Stemeroff for their feedback on previous versions of this article, and to Nora Boyd for her patience. We would also like to thank the Volkswagen Foundation for its support in providing the funds to create the Lichtenberg Group for History and Philosophy of Physics at the University of Bonn.

References

Abbott, B.P., et al. 2016. Astrophysical implications of the binary black hole merger GW150914. *The Astrophysical Journal Letters* 818 (2): L22.

Abuter, R., et al. 2018. Detection of the gravitational redshift in the orbit of the star S2 near the galactic centre massive black hole. *Astronomy & Astrophysics* 615: L15.

Anderl, S. 2016. Astronomy and astrophysics. In *The Oxford handbook of philosophy of science*, ed. P. Humphreys. Oxford University Press.

Anderl, S. 2021. Evidence in astrophysics. In *The Routledge companion to philosophy of physics*, 744–752. Routledge.

Bakala, P., J. Dočekal, and Z. Turoňová. 2020. Habitable zones around almost extremely spinning black holes (black sun revisited). *The Astrophysical Journal* 889(1): 41.

Baker, D.J. 2021. Knox's inertial spacetime functionalism (and a better alternative). *Synthese* 199(2): 277–298.

Berry, T., A. Simpson, and M. Visser. 2021. General class of "quantum deformed" regular black holes. Preprint. arXiv:2102.02471.

Booth, I. 2005. Black-hole boundaries. *Canadian Journal of Physics* 83(11): 1073–1099.

Bronzwaer, T., and H. Falcke. 2021. The nature of black hole shadows. *The Astrophysical Journal* 920(2): 155.

Cardoso, V., and P. Pani. 2019. Testing the nature of dark compact objects: a status report. *Living Reviews in Relativity* 22(1): 1–104.

Carr, B., K. Kohri, Y. Sendouda, and J. Yokoyama. 2021. Constraints on primordial black holes. *Reports on Progress in Physics* 84(11): 116902.

Carrasco, A., and M. Mars. 2013. On uniqueness results for static, asymptotically flat initial data containing mots. In *Black holes: New horizons*, ed. S. Hayward, 55–92. World Scientific.

Chen, H.-Y., A. Ricarte, and F. Pacucci. 2022. Prospects to explore high-redshift black hole formation with multi-band gravitational waves observatories. Preprint. arXiv:2202.04764.

Chesler, P.M., E. Curiel, and R. Narayan. 2019. Numerical evolution of shocks in the interior of Kerr black holes. *Physical Review D* 99(8): 084033.

Curiel, E. 2019. The many definitions of a black hole. *Nature Astronomy* 3(1): 27.

Currie, A. 2018. *Rock, bone, and ruin: An optimist's guide to the historical sciences*. MIT Press.

Do, T., et al. 2019. Relativistic redshift of the star S0-2 orbiting the galactic center supermassive black hole. *Science* 365(6454): 664–668.

Doboszewski, J. 2019. Interpreting cosmic no hair theorems: Is fatalism about the far future of expanding cosmological models unavoidable? *Studies in History and Philosophy of Science Part B: Studies in History and Philosophy of Modern Physics* 66: 170–179.

Doboszewski, J. 2022. Rotating black holes as time machines: An interim report. In *The foundations of spacetime physics*, 133–152. Routledge.

Earman, J. 1995. *Bangs, crunches, whimpers, and shrieks: singularities and acausalities in relativistic spacetimes*. New York: Oxford University Press.

Eckart, A., A. Hüttemann, C. Kiefer, S. Britzen, M. Zajaček, C. Lämmerzahl, M. Stöckler, M. Valencia-S, V. Karas, and M. García-Marín. 2017. The Milky Way's supermassive black hole: How good a case is it? *Foundations of Physics* 47(5):553–624.

El-Badry, K., H.-W. Rix, E. Quataert, A.W. Howard, H. Isaacson, J. Fuller, K. Hawkins, K. Breivik, K.W. Wong, A.C. Rodriguez, et al. 2022a. A sun-like star orbiting a black hole. *Monthly Notices of the Royal Astronomical Society* 518(1): 1057–1085.

El-Badry, K., R. Seeburger, T. Jayasinghe, H.-W. Rix, S. Almada, C. Conroy, A.M. Price-Whelan, and K. Burdge. 2022b. Unicorns and giraffes in the binary zoo: stripped giants with subgiant companions. *Monthly Notices of the Royal Astronomical Society* 512 (4): 5620–5641.

Elder, J. 2021. On the "direct detection" of gravitational waves. *draft*.

Ellis, G.F. 2002. Cosmology and local physics. *New Astronomy Reviews* 46 (11): 645–657.

Forbes, J.C. and A. Loeb. 2018. Evaporation of planetary atmospheres due to XUV illumination by quasars. *Monthly Notices of the Royal Astronomical Society* 479(1): 171–182.

Franklin, A.D. 2017. Is seeing believing?: Observation in physics. *Physics in Perspective* 19(4): 321–423.

Frost, A.J. et al. 2022. HR 6819 is a binary system with no black hole - revisiting the source with infrared interferometry and optical integral field spectroscopy. *A&A* 659: L3.

Giere, R.N. 2010. Scientific perspectivism. In *Scientific perspectivism*. University of Chicago press.

Hacking, I. 1984. Experimentation and scientific realism. In *Science and the quest for reality*, 162–181. Springer.

Hacking, I. 1989. Extragalactic reality: The case of gravitational lensing. *Philosophy of Science* 56(4): 555–581.

Jacquart, M. 2020. Observations, simulations, and reasoning in astrophysics. *Philosophy of Science* 87(5): 1209–1220.

Kim, J.-Y. et al. 2020. Event Horizon Telescope imaging of the archetypal blazar 3C 279 at an extreme 20 microarcsecond resolution. *Astronomy & Astrophysics* 640: A69.

Laine, S. et al. 2020. Spitzer observations of the predicted Eddington Flare from Blazar OJ 287. *The Astrophysical Journal Letters* 894(1): L1.

Landsman, K. 2021. Singularities, black holes, and cosmic censorship: A tribute to Roger Penrose. *Foundations of Physics* 51(2): 1–38.

Lu, W., P. Kumar, and R. Narayan. 2017. Stellar disruption events support the existence of the black hole event horizon. *Monthly Notices of the Royal Astronomical Society* 468(1): 910–919.

Martens, N. 2022. Dark matter realism. *Foundations of Physics* 52(1): 1–19.

Massimi, M. 2004. Non-defensible middle ground for experimental realism: Why we are justified to believe in colored quarks. *Philosophy of Science* 71(1): 36–60.

Massimi, M. 2018. Perspectival modeling. *Philosophy of Science* 85(3): 335–359.

Morrison, M. 2011. One phenomenon, many models: Inconsistency and complementarity. *Studies in History and Philosophy of Science Part A* 42(2): 342–351.

Penrose, R. and R. Floyd. 1971. Extraction of rotational energy from a black hole. *Nature Physical Science* 229(6): 177–179.

Rauch, D., et al. 2018. Cosmic Bell test using random measurement settings from high-redshift quasars. *Physical Review Letters* 121(8): 080403.

Sahu, K.C., et al. 2022. An isolated stellar-mass black hole detected through astrometric microlensing. Preprint. arXiv:2201.13296.

Sawada-Satoh, S., et al. 2015. Apparent inward motion of the parsec-scale jet in the BL Lac object OJ287 during the 2011–2012 γ-ray flares. *Publications of The Korean Astronomical Society* 30(2): 429–432.

Senovilla, J.M. 2013. Trapped surfaces. In *Black holes: New horizons*, 203–234. World Scientific.

Shapere, D. 1982. The concept of observation in science and philosophy. *Philosophy of Science* 49(4): 485–525.

Shapere, D. 1993. Astronomy and antirealism. *Philosophy of Science* 60(1): 134–150.

Shi, W., X. Liu, and H. Song. 2007. A new model for the periodic outbursts of the BL Lac object OJ287. *Astrophysics and Space Science* 310(1): 59–63.

The Event Horizon Telescope Collaboration, et al. 2019. First M87 Event Horizon Telescope results. I. the shadow of the supermassive black hole. *The Astrophysical Journal Letters* 875(1): L1.

The Event Horizon Telescope Collaboration, et al. 2022. First Sagittarius A* Event Horizon Telescope results. I. the shadow of the supermassive black hole in the center of the Milky Way. *The Astrophysical Journal Letters* 930(2): L12.

Thornburg, J. 2007. Event and apparent horizon finders for 3+1 numerical relativity. *Living Reviews in Relativity* 10(1): 1–68.

Tiede, P., H.-Y. Pu, A.E. Broderick, R. Gold, M. Karami, and J.A. Preciado-López. 2020. Spacetime tomography using the Event Horizon Telescope. *The Astrophysical Journal* 892(2): 132.

Vincent, F., S. Gralla, A. Lupsasca, and M. Wielgus. 2022. Images and photon ring signatures of thick disks around black holes. *Astronomy & Astrophysics* 667: A170.

Wada, K., Y. Tsukamoto, and E. Kokubo. 2021. Formation of "blanets" from dust grains around the supermassive black holes in galaxies. *The Astrophysical Journal* 909(1): 96.

Wald, R. 1984. *General relativity*. University of Chicago Press.

Wielgus, M., et al. 2020. Monitoring the morphology of M87* in 2009–2017 with the Event Horizon Telescope. *The Astrophysical Journal* 901(1): 67.

Chapter 14
Black Holes and Analogy

Alex Mathie

Abstract It is generally accepted that science sometimes involves reasoning with analogies. Often, this simply means that analogies between disparate objects of study might be used as heuristics to guide theory development. Contemporary black hole physics, however, deploys analogical reasoning in a way that seems to overreach this traditional heuristic role. In this chapter, I describe two distinct pieces of analogical reasoning that are quite central to the contemporary study of black holes. The first underpins arguments for the existence of astrophysical Hawking radiation, and the second underpins arguments for black holes being 'genuinely' thermodynamical in nature. I argue that while these are distinct analogical arguments, they depend on one another in an interesting way: the success of the second analogical argument presupposes the success of the first. This induces a tension for those who wish to take black hole thermodynamics seriously, but who are sceptical of the evidence provided for astrophysical Hawking radiation by the results of analogue gravity. I consider three ways to resolve this tension, and show that each fails.

14.1 Introduction

To put it flippantly, the trouble with black holes is that they are black, and that they are holes. This makes direct empirical contact with black hole systems an intrinsically troublesome business, since by definition the (classical) black hole necessarily has no optical signature and the fact that its conventional definition—in terms of an event horizon—makes reference to future null infinity means that it has no well-defined location in spacetime. But the experimental astrophysicist's loss is the philosopher's gain: the difficulties that hinder straightforward empirical access to black holes have necessitated the use of less conventional epistemic techniques

---

A. Mathie (✉)
Munich Center for Mathematical Philosophy, Munich, Germany
e-mail: alex.mathie@lmu.de

N. Mills Boyd et al. (eds.), *Philosophy of Astrophysics*, Synthese Library 472,
https://doi.org/10.1007/978-3-031-26618-8_14

by physicists, and it is precisely these difficulties that make the epistemology of black hole physics such a fertile ground for philosophical scrutiny. Some of these techniques are comparatively quotidian: indirect detection of black holes by the observation of their interaction (gravitational or electromagnetic) with ordinary matter, for instance.[1] But others are more philosophically interesting, and perhaps controversial. One example is the use of analogical reasoning in black hole physics.

That there is analogical reasoning in contemporary black hole physics is, by now, an uncontroversial point. Indeed, there is a burgeoning literature on the epistemology of analogue experiments, and analogue gravity (as we shall examine in greater detail below) is a central case study in this literature (Dardashti et al. 2017, 2019; Thébault 2019; Evans and Thébault 2020; Crowther et al. 2021; Field 2022). Similarly, black hole thermodynamics (BHT)—which takes as its departure point the striking analogy between the laws of black hole mechanics and those of ordinary thermodynamics—has been of great interest to physicists since its discovery by Bardeen et al. (1973). Recent years have seen a growing interest in—and dispute over—how we should interpret this unexpected correspondence between two pillars of modern physics by philosophers (Curiel 2014; Wallace 2018, 2019; Wüthrich 2019; Prunkl and Timpson 2019; Dougherty and Callender 2016).

Relatively little attention, however, has been paid to BHT *qua* a piece of analogical reasoning. The literature that does exist focuses almost exclusively on the conceptual difficulties that arise from attempting to interpret various black hole mechanical quantities as genuinely equivalent to their thermodynamical analogues, and the attendant difficulties for the claim that black holes are capable of genuinely thermodynamical behaviour. But it does not explicitly apply the philosophical literature on analogical reasoning *to* the BHT analogy. This leaves two blind spots in the literature on analogical reasoning in black hole physics. First, there is the need for precisely this kind of focused analysis of BHT *qua* a piece of analogical reasoning. Second, although philosophers have scrutinised both analogue gravity and BHT in isolation, scant philosophical work has been done on the connection between them.[2]

And yet such a connection clearly exists: on the standard view, it is Hawking's (1974) prediction of radiative flux from a black hole's event horizon that "removes the blemishes in BHT and transforms it from a suggestive analogy to a full equivalence" (Wallace 2018, p. 60).[3] Because direct empirical detection of the Hawking flux borders on the impossible, however, it is analogue gravity research

---

[1] See Rees (1998) for a helpful—if dated—survey, and Bambi (2018) for a more up-to-date review of recent developments.

[2] The connection between Hawking radiation and BHT is no doubt a subtext to much of the literature on analogue gravity epistemology, but the connection is rarely considered explicitly.

[3] Wallace is certainly not alone here, nor am I suggesting that he is incorrect. I single him out because this is the most lucid—and perhaps most provocative—statement of the role that Hawking radiation is purported to play in BHT. See also various comments to this effect in Wald (1984), Wald (1994), Jacobson (1996), Lüst and Vleeshouwers (2019), and cf. Wüthrich (2019, p. 203): "[U]ntil Stephen Hawking offered a persuasive semi-classical argument that black holes radiate,

that provides the closest thing to empirical evidence for the Hawking effect. Here, then, the two instances of analogical reasoning clearly make contact. I shall argue that their doing so creates an interesting tension for those who are compelled by BHT and yet remain skeptical about analogue gravity epistemology.

This chapter is primarily intended to clarify the *structure* of the instances of analogical reasoning in contemporary black hole physics, and to clarify how they make contact with one another, rather than intended to directly evaluate whether those instances of analogical reasoning succeed. In a sense, then, this chapter plays the role of prelude to both the existing philosophy of physics literature that seeks to defend or downplay the physical significance of the analogue gravity or BHT analogies, and the existing philosophy of science literature that examines the possibility of confirmation via analogue experimentation in black hole physics.

The structure of this piece is as follows: The first half of the paper (Sect. 14.2) provides a survey of the two central instances of analogical reasoning in contemporary black hole physics: analogue gravity and black hole thermodynamics. I begin with some groundwork on analogical reasoning (Sect. 14.2.1), and then survey, and explicitly reconstruct analogical arguments for, analogue gravity (Sect. 14.2.2) and BHT (Sect. 14.2.3). The second half of the paper (Sect. 14.3) examines what the relationship between these two analogical arguments might be, arguing that they are distinct but nevertheless importantly interdependent, and teasing out an important tension that this implies. I then consider three possible ways to resolve this tension and argue that each will fail. In Sect. 14.4 I conclude.

14.2 Two Analogies in Contemporary Black Hole Physics

14.2.1 Analogical Reasoning

Before we examine the ways in which analogies and analogical reasoning are invoked in contemporary black hole physics, we should make explicit exactly what is meant by these terms in the round. In this chapter, I adopt a broadly Hessean picture of analogy, following in particular Keynes (1921), Hesse (1966), and Bartha (2010), since these works build, sequentially, on one another and I take them collectively to constitute the mainstream orthodoxy in the analogical reasoning literature. It is worth noting, however, that plenty of dissenting accounts of analogical reasoning exist.[4] My concern, then, is with specifically *Hessean* analogical reasoning in contemporary black hole physics. Whether the problems

---

and so exhibit thermodynamic behaviour like a body with a temperature, most physicists were not moved by Bekenstein's earlier case for black hole entropy."

[4] See Bartha (2019) for an excellent overview.

I describe in this chapter are remedied or exacerbated by adopting an alternative account of analogy is a question that must be left for another time.[5]

Bartha (2010, p. 1) takes an analogy to be "a comparison between two objects, or systems of objects, that highlights respects in which they are thought to be similar". The conventional way of formalising this idea appeals to the notion of a mapping. Given a source domain, S, and a target domain, T, an analogy is a one-to-one mapping, ϕ, that maps elements of S to elements of T. Here, 'elements' can be construed quite broadly: elements could be properties, relations, objects, or functions depending on the nature of the domains in question (p. 13).

$$a \in S \leftrightarrow_\phi a^* \in T$$

$$b \in S \leftrightarrow_\phi b^* \in T$$

$$c \in S \leftrightarrow_\phi c^* \in T$$

Starred and unstarred elements are 'analogues' of one another under ϕ. From here on I drop the '$\in$' notation and take it as understood that elements on either side of the horizontal relation are from different domains (as a convention, I shall stick to 'source' being left, and 'target' being right).

It is worth making two useful terminological refinements to our account of analogy:

1. Following Keynes (1921), we can distinguish between 'positive', 'negative' and 'neutral' parts of an analogy. For expositional simplicity, let the elements of our source and target domains be propositions. The positive analogy is the set P of propositions in S whose images under ϕ hold in T. The negative analogy is the set N of propositions in S whose images under ϕ do *not* hold in T. The neutral analogy is the set of propositions in S for whom it is not known whether their images under ϕ hold in T, i.e. $S \setminus P \cup N$.

2. Following Hesse (1966), we can make a further distinction between 'horizontal' and 'vertical' relations in an analogy. Horizontal relations are, according to Hesse (1966, p. 59), those concerned with the similarity of counterparts determined by the mapping, ϕ. The horizontal relation is the relation of identity or difference between an element of S and its image under ϕ in T. What makes a and a^* analogues is the fact that ϕ picks them out as identical with respect to,

---

[5] In any case, I draw only lightly on the specifics of the Hessean account, and so it seems plausible that much of what I say here will generalise to other accounts of analogy. One account for which this may not be true, however, is that of John Norton, who eschews the pursuit of "some universal schema that separates the good from the bad analogical inferences" (2021, p. 120) in favour of a case-by-case analysis of whether there are appropriate empirical facts to warrant specific analogical inferences. Formal analogies, which do not take into account these empirical facts, cannot, on Norton's account be sufficient to underpin analogical inference. On those grounds, it would seem that much of what I say in Sects. 14.3.1–14.3.3 will not be compelling to someone that adopts Norton's account of analogy. I thank an anonymous reviewer for this point.

e.g., function, property, mathematical structure, and so on. By contrast, vertical relations, for Hesse are the *causal* relations that obtain between elements of a single domain. For example, how *a* relates to *b*, and so forth.

It is worth noting that this definition sets the bar trivially low for the existence of *analogies* in black hole physics. One can always propose a comparison between any two systems (black holes and thermodynamical systems, for example) that moots certain similarities, and as such one can always trivially assert the existence of analogies more or less anywhere. This makes the question of whether there are analogies in black hole physics somewhat uninteresting. The answer is yes, for the same reason that there are analogies everywhere—we are pattern-seeking beings, and we are apt to propose patterns or similarities wherever we like. Our concern, rather than with analogy *simpliciter*, is with the earnest deployment of *analogical reasoning* in black hole physics.

Analogical reasoning, Bartha suggests, is "any type of thinking that *relies* upon an analogy" (2010, p. 1, my emphasis). In formal terms, this usually comprises conjecturing that an element currently in the neutral analogy belongs in the positive analogy, with the chosen element of the neutral analogy being referred to by Keynes as the 'hypothetical analogy'. An 'analogical argument' is simply an explicit representation, by way of premise and conclusion, of a piece of analogical reasoning, conforming roughly to the general schema below:

General Schema

(i) $P \leftrightarrow_\phi P^*$ [Positive Analogy]

(ii) $Q \leftrightarrow_\phi Q^*$ [Positive Analogy]

(iii) $R \leftrightarrow_\phi R^*$ [Positive Analogy]

$\therefore$ $S \leftrightarrow_\phi S^*$ [Hypothetical Analogy]

The rest of this section applies this schema to the two cases of analogical reasoning in contemporary black hole physics, identifying the salient positive analogies, the types of horizontal relationships involved, and any critical disanalogies that appear to block the conclusion. Section 14.2.2 discusses and formalises the analogical argument that underpins analogue gravity research; Sect. 14.2.3 does the same for BHT.

14.2.2 Analogue Gravity

Analogue gravity research exploits a precise mathematical isomorphism, first discovered by Unruh (1981), between the behaviour of sound waves in a convergent fluid flow, and the behaviour of light in black hole spacetimes. Because our understanding of the propagation of sound in a fluid is so much better than our understanding of quantum field theory in curved spacetime, the primary benefit of analogue gravity is that a system whose microphysics is well understood can be used as an empirical surrogate for a system whose microphysics is incredibly poorly understood (since we expect a complete description to require an adequate theory of quantum gravity), and almost entirely inaccessible. To return to the comment I made at the start of this chapter, analogue gravity's ingenuity stems from the fact that it seems to offer a way to study systems that are not black, and not holes, but are—in some sense—still black holes.

14.2.2.1 The Positive Analogy

Crudely put, the positive analogy between analogue black holes and astrophysical black holes is that in both cases we can describe them well mathematically by "[taking] some sort of 'excitation', travelling on some sort of 'background', and [analysing] its propagation in terms of the tools and methods of differential geometry" (Visser 2013, p. 31). For present purposes, it is not necessary to delve too much further into the technical details of semiclassical or analogue gravity.[6] Rather, we can simply note (as Unruh 1981 originally did) that sound waves in a converging fluid flow can be modelled as the excitations of a minimally-coupled massless scalar field propagating in a $(3 + 1)$-dimensional Lorentzian geometry with the covariant acoustic metric:

$$g_{\mu\nu}^{\text{acoustic}} = \frac{\rho_0}{c_{\text{sound}}} \left[\begin{array}{c:c} -(c_{\text{sound}}^2 - v_0^2) & -v_0^j \\ \hdashline \cdots & \cdots \\ -v_0^i & \delta_{ij} \end{array} \right] \tag{14.1}$$

Which bears a striking resemblance to the Painlevé-Gullstrand form of the Schwarzschild metric:

$$g_{\mu\nu}^{\text{Schwarzschild}} = \left[\begin{array}{c:c} -(c_0^2 - \frac{2GM}{r}) & -\sqrt{\frac{2GM}{r}}\mathbf{r}_j \\ \hdashline \cdots & \cdots \\ -\sqrt{\frac{2GM}{r}}\mathbf{r}_i & \delta_{ij} \end{array} \right] \tag{14.2}$$

Thus, the Schwarzschild spacetime and the converging fluid flow share the same effective geometry, encoded by these isomorphic metrics. Using the tools of quantum field theory of curved spacetime, we approximate quantum gravity by propagating quantum fields across this curved but classical spacetime background. In much the same way, sound waves moving in the fluid medium described by (14.1) can be modelled as scalar fields propagating across this geometry.

---

[6] Readers interested in such details should consult the excellent and comprehensive living review by Barceló et al. (2011), or the philosophical analysis thereof by Thébault (2019, Sects. 11.3.1–11.3.3).

Despite the numerous ways in which analogue black holes remain radically dissimilar to their astrophysical counterparts,[7] the thought is that this underlying mathematical isomorphism is nevertheless strong enough to capture at least *kinematic* similarity between acoustic black holes and their astrophysical counterparts. Thus, the thinking goes, the positive analogy between the mathematics of analogue black holes and astrophysical black holes can support certain inferences about the behaviour of astrophysical black holes from the behaviour of their analogue counterparts.

In principle, there could be many different candidates for what these inferences might be. To put it in Keynes' terms, any property of analogue black holes in the 'neutral analogy' could, in principle, be proposed as the 'hypothetical analogy' (i.e. the property that we are suggesting astrophysical black holes might have, based on the strength of the existing 'positive analogy'). This flexibility is simultaneously analogical reasoning's greatest strength and greatest weakness. But in practice, the inference most commonly made from analogue gravity research has to do with the Hawking effect.

The Hawking effect is the result that at asymptotically late times, particles of a quantum field in a stationary black hole spacetime are radiated from the event horizon out to infinity in precisely the same way as radiation from a perfect blackbody at the *Hawking temperature*, $T = \frac{\kappa}{2\pi}$.[8] But because surface gravity scales inversely with mass, and because black holes are typically very massive objects, the Hawking temperature for even solar mass black holes is minuscule (of the order 10^{-8} K) which makes astrophysical Hawking radiation nigh on impossible to detect empirically. The inference that analogue gravity research is commonly taken to support is that direct observation of the analogue Hawking effect in analogue black holes allows us to infer (at least analogically) that astrophysical Hawking radiation does, indeed, exist.[9]

We might worry that because there remain so many differences between analogue and astrophysical black holes, this inference is a spurious one. There are two ways to allay this concern, which both appeal to the robustness of the Hawking effect. First, we can appeal to the fact that various universality arguments have been given to support the idea that the Hawking effect will be robust under changes in the

---

[7] Many of these dissimilarities will be benign. But others seem plausibly more serious. For example, the analogue gravity case involves *two* relevant metrics: acoustic black holes obey the relevant acoustic metric, which is the one we are interested in, but they also obey—purely by dint of being physically realised systems on Earth—the physical spacetime metric, which for us is approximately Minkowski (Barceló et al. 2011, p. 16). Discussing these kinds of issues is regrettably beyond the scope of this chapter—I mention them only to flag that one should perhaps be wary of waving away 'dissimilarities' too hastily.

[8] I will say slightly more about the technical details of the Hawking effect in Sect. 2.3.1.2, but for a more rigorous treatment, see Wald (1994, ch. 7), and for a more thorough discussion, see Wallace (2018, Sects. 4.1–4.2).

[9] Indeed, the title of Unruh's seminal (1981) paper, *'Experimental black hole evaporation?'* attests to this being the central inference of analogue gravity research.

microphysics of the system in which it appears.[10] It is worth noting, however, that Gryb et al. (2020) argue forcefully that none of the available universality arguments for Hawking radiation are entirely convincing (especially in light of the fact that none provide a fully satisfactory response to the trans-Planckian problem, which we shall see in more detail in Sect. 14.3.2). Field (2022) goes one step further, questioning whether the epistemic situation in analogue gravity is even one that can be ameliorated by universality arguments. On Field's view, "our state of knowledge with respect to Hawking radiation is not currently universality-argument-apt" (2022, p. 25). A safer option, then, is the second: we can note that because the Hawking effect is entirely kinematic and depends only on the existence of a Lorentzian metric and an appropriate horizon (Visser 2003), it would seem to be the kind of thing that *is* determined by the isomorphism we saw above. Thus, analogue gravity advocates suggest, the existence of the Hawking effect in analogue black hole systems would support the existence of the Hawking effect in astrophysical black hole systems. And furthermore, since researchers have demonstrated the existence of both stimulated and spontaneous phononic Hawking radiation from acoustic horizons in fluids (Weinfurtner et al. 2011), and from optical horizons in Bose-Einstein condensates (Steinhauer 2016; de Nova et al. 2019; Drori et al. 2019; Kolobov et al. 2021), the antecedent in this conditional seems to be satisfied.[11] Thus, the results of analogue gravity experiments may be considered to indirectly confirm Hawking's (1974; 1975) prediction of thermalised radiation from astrophysical black holes.

14.2.2.2 Formalisation

Here is a reasonable first pass at formalising the analogical argument for astrophysical Hawking radiation:

Analogue Gravity (Mathematical Similarity)
  (i) Mathematics of analogue black holes $\leftrightarrow_\phi$ Mathematics of astrophysical black holes

---

  ∴ Radiation from analogue black holes $\leftrightarrow_\phi$ Radiation from astrophysical black holes

This is, of course, not deductively valid. No analogical arguments are. But it is also not sufficiently detailed. For a start, we need to be more precise about the nature of ϕ. In what sense is the mathematics of the two systems similar? After all, the mathematics used to describe analogue black holes is not *entirely* similar to the mathematics used to describe astrophysical black holes—gravitation is not fluid flow.[12] So which parts of the mathematics for the two types of systems are we

---

[10] See Gryb et al. (2020, Sect. 4) for a review.

[11] See also Unruh (2014) for an appraisal (and endorsement) of early attempts to measure Hawking radiation in black hole analogues, and Leonhardt (2018) for a dissenting view.

[12] See also my fn. 4, above.

interested in, and in what sense are these parts connected? Whichever mathematical commonality (i) is picking out will have to be appropriately related to the proposed physical commonality in the conclusion.

If Visser is correct that Hawking radiation requires only the existence of a Lorentzian geometry and a suitable horizon, then the relevant parts of the mathematics are the $(3 + 1)$-dimensional Lorentzian metric we saw for both systems in Sect. 14.2.2.1, and a horizon within that geometry (i.e. some region that excitations of the scalar field may enter but not leave). This also takes care of the question of vertical relations: Visser's argument explains why these two parts of the mathematics are sufficient to license an inference to the existence of Hawking radiation. And since we are dealing with mathematical structures, the relevant type of mapping is an isomorphism such that '$\leftrightarrow_\phi$' denotes 'is isomorphic to'. Here is a second pass:

Analogue Gravity (Isomorphic Kinematics)
  (i) Lorentzian geometry for analogue black holes $\leftrightarrow_\phi$ Lorentzian geometry for astrophysical black holes
 (ii) Horizons in analogue black holes $\leftrightarrow_\phi$ Horizons in astrophysical black holes
(iii) Hawking radiation requires only a suitable Lorentzian geometry and the existence of a horizon

  $\therefore$ Radiation from analogue black holes $\leftrightarrow_\phi$ Radiation from astrophysical black holes

But this is too quick, because we can actually read off a deductive argument for astrophysical Hawking radiation by only looking at the right hand side: if astrophysical black holes have a suitable Lorentzian geometry and suitable horizon structure, then the addition of (iii) entails that astrophysical Hawking radiation exists. But if this were the case, there would be no need for analogue gravity! What has gone wrong here is that Visser's argument, i.e. (iii), is fundamentally an argument about differential geometry, not about empirical physics. What I mean here is that Visser's assertion about the 'essential' features of Hawking radiation concerns the *derivation* of the Hawking effect from within a particular theoretical framework—either semiclassical gravity, or hydrodynamics, or condensed matter. It is not—nor does it purport to be—an argument about the existence of Hawking radiation in any empirical sense. This version therefore states Visser's argument too strongly. A third pass:

Analogue Gravity (Derivation from Isomorphic Kinematics)
  (i) Lorentzian geometry for analogue black holes $\leftrightarrow_\phi$ Lorentzian geometry for astrophysical black holes
 (ii) Horizons in analogue black holes $\leftrightarrow_\phi$ Horizons in astrophysical black holes
(iii) The *derivation* of Hawking radiation requires only a suitable Lorentzian geometry and the existence of a horizon

  $\therefore$ Radiation from analogue black holes $\leftrightarrow_\phi$ Radiation from astrophysical black holes

This seems to capture the thinking behind analogue gravity without overreaching it.

14.2.3   Black Hole Thermodynamics

14.2.3.1   The Positive Analogy

BHT hinges upon the recognition that there is a remarkable mathematical similarity between the laws of classical black hole mechanics derived by Bardeen et al. (1973) and the ordinary laws of thermodynamics (Table 14.1).

The formal analogy between (a) surface gravity and thermodynamical temperature, and (b) horizon area and thermodynamical entropy suggests, as Wald (1994, p. 133) puts it, "a close—and, undoubtedly, deep—relationship between the laws of black hole physics and the laws of ordinary thermodynamics". The basis for this suggestion is that a black hole's surface gravity *behaves*, mathematically, as if it were the temperature of a thermodynamical system, and the area of a black hole's event horizon *behaves*, mathematically, as if it were the entropy of a thermodynamical system. This follows from the fact that these quantities occupy structurally equivalent positions in each set of laws. But it does not follow from the formal analogy alone that surface gravity and horizon area *physically are* a black hole's thermodynamical temperature and entropy respectively. Indeed, there are many good reasons to resist this conclusion in the classical regime, with two of the most obvious being precisely those that make empirical access to black holes so difficult in the first place: (i) there are severe problems when it comes to even delineating where a black hole *is*, which makes it difficult to see how to distinguish 'the system' in a thermodynamical context; and (ii) in classical relativity, a black hole is a perfect absorber, and thus, if it has a thermodynamical temperature at all, that temperature could only ever be absolute zero. These are seemingly critical disanalogies between the two domains that prevent the analogy from being taken seriously.[13]

---

[13] These are not the only problems with the analogy between black hole mechanics and ordinary thermodynamics. For instance, there are several deep conceptual puzzles around equating horizon area with entropy—and the concomitant need for a 'generalised second law'—that I set aside here for reasons of space. For a helpful review of these issues, see Curiel (2019b, especially Sects. 5.3–5.4). I also omit: the *prima facie* lack of a microphysical basis for phenomenological BHT; the puzzling fact that the laws of black hole mechanics are theorems in differential geometry, yet the laws of thermodynamics are bulk empirical generalisations and do not admit of analytical proof (Curiel 2014); the fact that in its current guise, BHT has been accused of recovering only a 'pale shadow' of ordinary thermodynamics (Dougherty and Callender 2016). I focus on (i) and (ii) because they seem to pose the most severe challenge to taking BHT seriously, and because the resolution to (ii)—the prediction of Hawking radiation in semiclassical gravity—in particular sets up an interesting tension that provides the basis for discussion in Sect. 14.3.

Table 14.1 The formal analogy between the laws of black hole mechanics and ordinary thermodynamics

	Black hole mechanics	Thermodynamics
Zeroth law	κ constant across the event horizon	T constant throughout the system
First law	$dM = \kappa dA + \Omega dJ + \Phi dQ$	$dU = TdS + pdV + \Omega dJ + \Phi dQ$
Second law	$dA \geq 0$	$dS \geq 0$
Third law	$\kappa \to 0$ not physically realisable in finite steps	$T \to 0$ not physically realisable in finite steps

The fundamental idea of BHT is that the analogy between the laws of black hole mechanics and the laws of ordinary thermodynamics is more than just a formal coincidence *in spite of* these disanalogies. According to BHT, the analogy in Table 14.1 is physically significant, which is to say that it broadly indicates that black holes are thermodynamical systems in the fullest sense of the term. At least part of the case for this physical significance comes from addressing critical disanalogies like those above, and thereby extending the positive analogy such that black holes and thermodynamical systems are more plausibly equivalent in some sense. In the next section, I consider the disanalogies (i) and (ii), and show how the responses to these problems bolster the analogical argument for BHT.

14.2.3.2 The Negative Analogy

System Boundaries

One critical part of the negative analogy between the physics of classical black holes and the physics of ordinary thermodynamical systems is that while ordinary thermodynamical systems are *local* systems that exist in some region of spacetime, classical black holes are not. Classical black holes are typically defined as the region bounded by an event horizon, and the standard definition of an event horizon is the boundary of the causal past of future null infinity (Hawking and Ellis 1973, pp. 311–312; Wald 1984, pp. 299–300).[14] But as Erik Curiel remarks:

> This definition is global in a strong and straightforward sense: the idea that nothing can escape the interior of a black hole once it enters makes implicit reference to all future time—the thing can never escape no matter how long it tries. Thus, in order to know the location of the event horizon in spacetime, one must know the entire structure of the spacetime, from start to finish, so to speak, and all the way out to infinity. As a consequence, no local measurements one can make can ever determine the location of an event horizon. That

---

[14] This is the canonical definition in classical general relativity, though as Curiel (2019a) notes, the precise definition of a black hole varies widely across subdisciplines of physics. I lack the space to do justice to the heterogeneity of these definitions, so I assume the canonical definition for the exposition here.

feature is already objectionable to many physicists on philosophical grounds: one cannot operationalize an event horizon in any standard sense of the term.

(Curiel 2019a, p. 29)

It is extraordinarily difficult to see how global spacetime structures like the event horizon could ever number among the things treated by the theory of thermodynamics, itself a fundamentally operational theory that makes crucial use of the distinction between system and environment. *Prima facie*, this poses a severe challenge for any argument to the effect that we should take BHT seriously.

Any practically-minded physicist would rightly baulk at the idea that the global nature of event horizons in classical relativity somehow prohibits us from believing BHT, however. For example: Astrophysicists routinely speak about the 'location' of black holes (Sagittarius-A* being at the centre of the Milky Way for instance); the LIGO collaboration are credited with detecting black hole mergers for the first time (Abbott et al. 2016); and scientists from the Event Horizon Telescope were lauded for producing the first direct image of the shadow of what is thought to be a $6.5 \times 10^9 M_\odot$ black hole at the centre of the elliptical galaxy M87 (The EHT Collaboration et al. 2019). There is clearly a sense in which these fabulous achievements presuppose that black holes are entities occupying a well-defined region of spacetime—how else would the EHT team have known where to point their telescope? Thus, we might object, that to argue that black holes cannot be local, dynamical entities because their classical definition is non-local is to fixate on a technicality. If anything, the objection would go, this problematic global nature of the event horizon speaks to a deficiency of classical relativity as a tool for describing the universe, not to an intrinsic deficiency of astrophysical black holes that prevents them from being thermodynamical systems.[15] But an objection such as this, that suggests that classical relativity should not be the final arbiter of the physical nature of black holes, also suggests, a fortiori, that an analogy between classical relativity and ordinary thermodynamics is equally deficient as a guide to the physical nature of black holes.

As such, there are good pragmatic arguments against allowing the global nature of the event horizon to block the argument for BHT. But there are also more principled arguments, which draw from the wide variety of locally-definable alternatives to the global event horizon to argue that BHT can be made suitably 'local'. The most well-known of these is the trapping horizon, which takes as

---

[15] Many other physical theories will face similar problems. An adiathermal barrier in ordinary thermodynamics is arguably global in much the same sense: how long and how closely do we have to watch to convince ourselves that a system is 'truly' isolated? Much as we are happy to say that a thermodynamical system can be considered isolated *for practical purposes*, we can say the same for the event horizon. As Ramesh Narayan remarks in a comment quoted in Curiel (2019a, p. 31, Box 1): "for all intents and purposes we *are* at future null infinity with respect to Sagittarius-A*". The ubiquity of strictly global concepts in physics that are nevertheless operationalisable is grist to the practically-minded physicist's mill.

Fig. 14.1 A closed spacelike
surface (dotted) with two
future-directed null vectors.
The surface is *trapped* when
both θ_+ and θ_- are negative
(Color figure online)

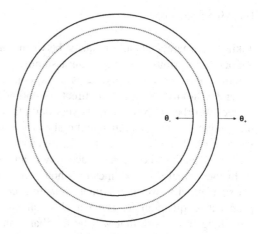

its departure point the notion of a 'trapped surface' (Penrose 1965).[16] A trapped
surface is a closed, spacelike surface with normal, future-directed null vectors that
have an expansion parameters, θ_+ and θ_-, which are both negative (Nielsen 2009).
Loosely put, this means that not just inward but also outward lightlike vectors are
infalling, toward the interior of the trapped surface (Fig. 14.1). A trapping horizon
is an extremal trapped surface such that $\theta_+ = 0$.[17]

Whether BHT can be localised in terms of trapping horizons is a matter of
dispute. Nielsen (2009) identifies them as the forerunner in a sizeable group of
locally-defined alternatives to the event horizon, for a number of reasons whose
technical details go beyond the scope of this paper (pp. 39–40).[18] While the
suitability of the trapping horizon for truly 'localising' BHT has been questioned by
Dougherty and Callender (2016, Sect. 4), others have come to its defence (Wallace
2018, A.2). Insofar as the jury remains out on this question, it seems reasonable
to treat it as a live and promising strategy for reformulating BHT without the
global event horizon. For our purposes, all that matters is that there *are* viable local
alternatives to the event horizon that take the sting out the critical analogy discussed
above.

---

[16] See Nielsen (2009) for a more detailed discussion of why we might want to expel the globally-
defined event horizon from BHT, and for a nice review of a wider range of locally-defined
alternatives to the event horizon.

[17] I gloss over some technical subtleties here, to do with distinguishing trapping horizons from
marginally trapped tubes. See Hayward (1994) for further details.

[18] Furthermore, he notes (p. 36) that in the case of analogue gravity, it is trapping horizons (not
event horizons) that we are dealing with. As such, there are reasons to adopt the trapping horizon as
a local alternative to the event horizon if one wishes to further substantiate the connection between
astrophysical black holes and their analogue counterparts we discussed in Sect. 14.2.2.

Hawking Radiation

Taking the mathematical similarity between surface gravity and thermodynamic temperature seriously would mean assigning a black hole a thermodynamic temperature of $T = \alpha\kappa$, where α is some constant to be determined. In classical general relativity this stands in direct contradiction to the fact that a black hole is a perfect absorber, prevented by its very definition from emitting radiation. The only thermodynamic temperature that it makes sense to assign to a black hole is zero, and yet such black holes generically have nonzero surface gravity. This rules out taking the formal analogy between surface gravity and temperature seriously.

In the semiclassical framework, however, this is no longer the case. In semiclassical gravity, we approximate quantum corrections to classical relativity by propagating quantum fields on a classical but curved spacetime background (a modelling framework that will be familiar from Sect. 14.2.2). And in this framework, as Hawking's (1975) celebrated result shows, black holes emit radiation with a perfect thermal spectrum corresponding to the Hawking temperature (setting $G = c = \hbar = 1$):

$$T_H = \frac{\kappa}{2\pi} \tag{14.3}$$

Which is exactly what the formal analogy suggests.

The precise role of Hawking radiation in bolstering the analogy between black hole physics and ordinary thermodynamics is not always entirely clear, but *prima facie* it achieves at least the following two things. First, it remedies the numerical disagreement between the classical black hole's surface gravity and its temperature, since if black holes are no longer perfect absorbers, they need not have zero temperature. Second, it provides a way to make sense of the notion of thermal contact between black holes. Let two black holes of surface gravity κ_1 and κ_2, respectively, occupy a box that is sufficiently large so as to allow us to neglect their gravitational interaction with one another. Radiation from each will cause the box to reach a temperature $\kappa_1/2\pi > T_{\text{Box}} > \kappa_2/2\pi$, resulting in the movement of heat from the hotter black hole to the colder (Wallace 2018). Thus, Hawking radiation moves two important elements of the negative analogy (numerical disagreement between T and κ; the lack of a well-defined notion of thermal contact between black holes) into the positive analogy.

Two problems hamper Hawking radiation's role in substantiating the BHT analogy, however. The first problem is that for a solar mass black hole, T_H is of the order 10^{-8} K, making it some 100 million times smaller than the 2.7 K cosmic microwave background temperature, and a thousand times smaller than the $\sim 10^{-5}$ K fluctuations in the cosmic microwave background alone (Smoot et al. 1992). The prospect of direct detection of astrophysical Hawking radiation is therefore incredibly poor. As Thébault (2019, pp. 185–186) vividly puts it: "Trying to detect astrophysical Hawking radiation in the night sky is thus like trying to see the heat from an ice cube against the background of an exploding nuclear bomb."

Indeed, as we saw in Sect. 14.2.2, the entire field of analogue gravity research can be understood as a reaction to the astonishingly dim prospects for the direct detection of astrophysical Hawking radiation. The second problem is that there are a battery of fairly severe conceptual problems with Hawking's theoretical derivation. Some of these have been resolved in more modern derivations of the Hawking effect by other means,[19] but some remain unresolved and are therefore cause for concern over the existence of astrophysical Hawking radiation. I discuss both of these problems in more detail in Sect. 14.3, but for now we have enough to sketch a formalisation of the analogical argument for BHT.

14.2.3.3 The Hypothetical Analogy?

Together, local alternatives to the event horizon, the existence of astrophysical Hawking radiation and the generalised second law substantially strengthen the analogical argument for BHT. But towards which conclusion? There are two plausible candidates. On the one hand, the conclusion might be that the formal analogy between the specific mathematics under comparison in Table 14.1 is now endowed with physical significance. On the other hand, the conclusion might be that the formal analogy between black holes and ordinary thermodynamical systems is *complete*—that is, that it extends far beyond what is already encoded in the similarities of Table 14.1, to *all* the physical quantities of black holes and thermodynamical systems. The first conclusion can be understood as attempting to *deepen* the purported connection between black holes and ordinary thermodynamical systems; the second as attempting to *broaden* that connection.

There are reasons both for and against each interpretation. If the connection between black holes and thermodynamical systems is to be deepened by an appeal to 'physical significance', then we owe an account of what this mysterious property might be.[20] But similarly, if the scope of this connection is to be broadened by extending the formal analogy, we soon run into problems. For instance: How are we to account for the missing pdV term in the first law of black hole mechanics (Dolan 2012)? What about the fact that the Hawking temperature generically disagrees with the Tolman temperature, which is the canonical measure of temperature in relativistic thermodynamics, since the two only coincide at future null infinity?[21] And how are we to understand the fact that the laws of black hole mechanics are

---

[19] See the useful discussion and provided by Wallace (2018, Sect. 4.2), and references therein, for more detail.

[20] This is a notoriously vexed question that cuts right to the heart of the relationship between mathematics and physics. Some recent accounts attempt to develop the connection between physical significance and topological stability (Fletcher 2014, 2016), but this seems to miss the mark for our purposes: the formal analogy in Table 14.1 does not seem to be the kind of thing that could be considered 'topologically stable'.

[21] I thank Erik Curiel for useful discussions on this point.

theorems, while the laws of ordinary thermodynamics admit of no analytic proof (Curiel 2014)?

I shall adopt the 'deepening' interpretation, going forward, since the problems with it are problems of clarification rather than problems of contradiction. That is to say: it seems more feasible to provide a satisfactory account of physical significance than it does to provide an exhaustive argument resolving all of the myriad possible mathematical disanalogies that the broadening approach would require.[22]

14.2.3.4 Formalisation

The most coarse-grained version of the analogical argument for BHT is simply to state that the positive analogy between the laws in Table 14.1 suggests that the systems described by these laws are really and truly the same kinds of systems. A schematisation of this coarse-grained argument would then be:

Black Hole Thermodynamics (Nomic Isomorphism)
(i) Laws of black hole mechanics $\leftrightarrow_{\phi_1}$ Laws of ordinary thermodynamics

---

$\therefore$ Black holes $\leftrightarrow_{\phi_2}$ Thermodynamical systems

Modulo some minor discrepancies—notably the absence of a pdV term—ϕ_1 is a horizontal relationship of isomorphism, since we are comparing structurally identical mathematical equations. ϕ_2, following the discussion in Sect. 14.2.3.3, has two plausible interpretations, but the better of the pair is that black hole quantities are *physically*, not just mathematically, equivalent to thermodynamical quantities.[23] But as we have seen, this argument seems nevertheless to fail due to the critical negative analogy between various aspects of black hole physics and various aspects of thermodynamics, namely the global nature of event horizons, and the lack of radiation from classical black holes. As we saw above, however, local alternatives to the event horizon, coupled with the derivation of Hawking radiation in the semiclassical framework allow us to smooth over two of the most pressing disanalogies. The result is a stronger and more detailed analogical argument:

Black Hole Thermodynamics (Nomic Isomorphism + Boundaries and Radiation)
(i) Laws of black hole mechanics $\leftrightarrow_{\phi_1}$ Laws of ordinary thermodynamics
(ii) System boundaries for astrophysical black holes $\leftrightarrow_{\phi_1}$ System boundaries for thermodynamical systems

---

[22] Developing a satisfactory account of physical significance goes far beyond the scope of this paper. I endorse the deepening account here only insofar as it seems more plausible than the broadening account.

[23] Cashing this out with greater precision lies regrettably beyond the scope of the paper. I am sympathetic to the broadly functionalist position espoused by Curiel (2014) and later augmented by Prunkl and Timpson (2019), upon which 'physical equivalence' should be understood as a kind of functional equivalence, and will touch upon this briefly in Sect. 14.3.3.

(iii) Radiation from astrophysical black holes $\leftrightarrow_{\phi_1}$ Radiation from thermodynamical systems

---

$\therefore$ Black holes $\leftrightarrow_{\phi_2}$ Thermodynamical systems

14.3   What Is the Relationship Between Them?

We have arrived at plausible formalisations for the analogical arguments for astrophysical Hawking radiation and for the physical significance of BHT. These arguments are distinct from one another. Primarily, because they appeal to different premises to draw different conclusions. But also because the conclusions they seem to license are of a fundamentally different kind. On the one hand, analogue gravity concerns the ability of one type of physical system to serve as an experimental surrogate for another type of physical system. This is an epistemic matter: it concerns the validity of certain kinds of experiments that scientists might conduct in order to gain indirect access to empirically inaccessible systems. On the other hand, BHT concerns the physical nature of black holes. The upshot of taking BHT seriously is, to quote the physicist Robert Wald, "that the laws of black hole mechanics *truly are* the ordinary laws of thermodynamics applied to a system containing a black hole" (1994, p. 163, my emphasis). Whereas nobody appears to think that analogue gravity evinces some profound and hitherto unappreciated connection between waterfalls and black holes, precisely that kind of connection is being claimed in the case of BHT—in its more extreme form, the claim is that "gravity [...] is a fundamentally thermodynamical phenomenon" (Curiel 2014, p. 3, see also Curiel 2019b, Sect. 5.5). Not only do the two analogical arguments invoked in contemporary black hole physics concern different premises in support of different conclusions, those conclusions also constitute fundamentally different kinds of claim.

It would be too hasty, however, to conclude that the two analogical arguments can be divorced from one another completely. Because a large part of our evidence for astrophysical Hawking radiation comes from the impressive results of analogue gravity research, and because the existence of astrophysical Hawking radiation in turn constitutes such a crucial part of the analogical argument for BHT, there is a clear sense in which these two analogical arguments bear on one another. Indeed, if the analogical argument for Hawking radiation fails, so too must the analogical argument for BHT. One way to see this is to note that one side of the conclusion in the final version of the analogical argument for Hawking radiation appears as one side of premise (iii) in the final version of the analogical argument for BHT. Consequently, the analogical argument for astrophysical Hawking radiation appears as a component of the analogical argument for BHT.

Why is this interesting? Because it appears to tie together, in a very specific way, the epistemic legitimacy of the BHT analogy, the basis of a research programme that is "as widely accepted an idea in theoretical physics as an idea with no direct empirical substantiation can be" (Curiel 2019b), with that of analogue gravity, which some have dismissed as little more than an "amusing feat of engineering [that] won't teach us anything about black holes" (Daniel Harlow, quoted in Wolchover 2016). If the analogical argument for BHT presupposes the success of the analogical argument for astrophysical Hawking radiation, then it is difficult, *prima facie*, to see how these two attitudes can be reconciled. That is, it is hard to see how it is consistent to (a) interpret the connection between black hole mechanics and ordinary thermodynamics as having genuine physical significance while (b) remaining skeptical about the epistemic warrant that analogue gravity gives us for astrophysical Hawking radiation. Yet it is exactly this combination of attitudes that some wish to adopt: Harlow himself, for instance, is actively engaged in research on BHT while remaining openly skeptical about the results of analogue gravity. Similarly, Wüthrich (2019, p. 221) claims that BHT certainly has "some support", but elsewhere concludes starkly that analogue confirmation "does not work" (Crowther et al. 2021, p. 3723).

I can see three plausible strategies for resolving this tension. In the rest of this section, I shall examine each in turn.

14.3.1 Naïve Formalism

I take it to be uncontroversial that there is a clear distinction to be drawn between saying "that is a black hole" while pointing one's finger at the Schwarzschild metric, and saying "that is a black hole" while pointing one's finger into the night sky in the direction of Sagittarius-A*. The former identifies a black hole as a type of mathematical entity, while the latter identifies a black hole as a type of astrophysical entity. Of course, we should expect that there is some connection between the two things (indeed, the use of mathematics in physics presupposes such a connection) but the fact remains that they are distinct: the mathematics is not the physical system it describes.

One way to resolve the tension for those who wish to combine belief in the physical significance of BHT with scepticism about analogue experimentation might be to try to interpret the conclusions of the two analogical arguments as being conclusions about *mathematical*—rather than astrophysical—entities. On this view, the term 'black holes' is to be understood as referring to certain classes of spacetime models in general relativity, rather than as referring to real physical systems in our universe, which may be reasonably well described by certain mathematical models under certain idealisations. Call this strategy 'naïve formalism', in recognition of its wholesale focus on the formalism of relativity.

When the naïve formalist says 'black holes are thermodynamical systems', roughly what they mean is that black holes (*qua* particular solutions to the Einstein field equation) have certain mathematical properties in common with thermodynamical systems.[24] One of these mathematical properties will be that which corresponds to their tendency to produce a thermally distributed radiative flux in certain situations: blackbody radiation in the case of ordinary thermodynamical systems, and Hawking radiation from black holes.[25] But this mathematical property can obtain for black holes—and so can be shared with thermodynamical systems—completely irrespective of any empirical confirmation of astrophysical Hawking radiation. Matters of *mathematical* similarity are to be settled by the mathematics alone.

For the naïve formalist, then, the claim that BHT presupposes the epistemic legitimacy of analogue experimentation fails. This is because if we are only interested in black holes as mathematical entities, we need only be interested in Hawking radiation as a mathematical entity, too. And since Hawking radiation can be derived for black hole spacetimes in semiclassical gravity without any assumptions about whether acoustic black holes can serve as legitimate empirical surrogates for astrophysical black holes, black holes (*qua* mathematical entities) can be said to have the correct mathematical properties so as to be considered 'thermodynamical' without any recourse to analogue gravity. To put the same point slightly differently, this objection would say that claims about the thermodynamical nature of black holes and claims about the existence of Hawking radiation are, at bottom, claims about the mathematical machinery of our physical theories. The same claims could be made about *astrophysical* black holes, and indeed then one would need a story about the empirical warrant for *astrophysical* Hawking radiation. But, the objection might go, this is a separate claim.

There are two arguments against the naïve formalist, however. The first is a straightforward sociological observation: nobody actually seems to interpret BHT this way. Presumably, we are interested in formal properties of mathematical entities only insofar as we expect those properties to manifest in the physical world—that is the essence of 'physical significance'. It should not be surprising that nobody interprets BHT in this extremely narrow sense, since to do so is perhaps to fixate on the mathematics at the expense of the physics. While naïve formalism could plausibly resolve the tension for those willing to adopt it, the cost is an unfamiliar and arguably myopic interpretation of BHT.

A second argument is that naïve formalism is simply too weak to capture the central thesis of BHT. If the central idea of BHT is supposed to be that "black holes are thermodynamical systems *in the fullest sense*" (Wallace 2018, p. 52, emphasis added), it stands to reason that this sense includes not just shared mathematical properties of the two kinds of systems, but shared non-mathematical properties too. Black holes must actually *be* thermodynamical systems (whatever that means),

---

[24] This is still terribly imprecise, but the general idea should be clear enough: the two mathematical entities share certain features that renders them 'equivalent' in the sense that BHT has in mind.

[25] Strictly speaking, in the case of black holes, it is not quite right to attribute this mathematical property to the spacetime geometry alone, since one requires appropriate quantum fields propagating on the geometry to produce this flux. I thank an anonymous reviewer for this point.

not just be mathematically equivalent to them. Thus, we might worry, the naïve formalist fails to actually demonstrate anything beyond formal analogy between black hole mechanics and ordinary thermodynamics. Rather, they smuggle in the physical significance of that analogy by simply moving the goalposts, redefining what physical significance means in flat-footed, mathematical terms. It would, of course, be entirely possible to parry this second objection to naïve formalism by adopting a kind of Pythagoreanism about physics, such that there is nothing to *being* a certain kind of thing over and above having the appropriate mathematical properties. But this kind of radical Pythagoreanism is surely an inordinately heavy price for the naïve formalist to pay.

14.3.2 Sophisticated Formalism

A second strategy is to break with the naïve formalist approach by maintaining that 'black holes' are to be understood as astrophysical entities, and therefore to maintain that astrophysical Hawking radiation must exist in more than just a mathematical sense, but to argue that the theoretical basis for astrophysical Hawking radiation is already strong enough to justify this without the results of analogue gravity. Broadly stated, the animating idea for this second strategy is that the absence of empirical evidence for the thermodynamical behaviour of black holes need not be a severe problem so long as we can fall back on non-empirical evidence for that behaviour, including non-empirical evidence for Hawking radiation.

Call this view 'sophisticated formalism'. Like naïve formalism, it retains a focus on the importance of *theory*, in the sense that it takes a sufficiently strong theoretical basis for astrophysical Hawking radiation to render confirmation of astrophysical Hawking radiation by analogue experimentation redundant. Unlike naïve formalism, however, sophisticated formalism does not limit attention to theory completely. Rather, it maintains that theory can be strong enough to support inferences about the empirical world, even in the absence of experiment. Sophisticated formalism can be seen as a refinement of the ideas of naïve formalism: both identify the importance of theory for decoupling the two analogical arguments, but where the naïve formalist attempted to achieve this by radically reinterpreting BHT as a thesis about mathematical entities, their sophisticated counterpart attempts to achieve this by instead lowering the evidential bar for the existence of astrophysical Hawking radiation.

Something like sophisticated formalism is what underpins pre-analogue gravity attempts to endow BHT with physical significance. This extends to black hole entropy, too: Bekenstein's (1972) arguments for black hole entropy on the basis of information theory, for instance, are one such attempt, and these are (trivially) independent of any analogue gravity results, predating Unruh's seminal work by a whole decade.[26] As Wüthrich (2019, p. 203) notes, to the extent that non-empirical theory confirmation is legitimate, "the fact that thermodynamic behaviour of black holes has not been observed to date may not be a worry as long as we have convincing non-empirical reasons for believing the Bekenstein-Hawking formula". This is an elegant statement of the sophisticated formalist position. However, Wüthrich's own conclusion is that Bekenstein's information-theoretic reasons fail to provide such convincing non-empirical reasons.[27]

Although Bekenstein's original arguments for the physical significance of BHT have failed to convince contemporary philosophers of physics, Wallace (2018) has argued forcefully that the contemporary theoretical arguments for the existence of astrophysical Hawking radiation are, in fact, "very powerful" (p. 61), and have only been strengthened in the 50 years since Hawking's original work. Now, we have at least five independent means of deriving the Hawking effect, each with their own strengths and weaknesses. This consilience is, itself, evidence for the robustness of the Hawking effect, but the theoretical case is not entirely watertight, as we shall see in a moment. It is, however, nevertheless true that the *prima facie* strength of the arguments for the Hawking effect make the prospects for the sophisticated formalist seem initially promising, and this is bolstered further by the sociological observation that many physicists seem to espouse this view (one need only look at the huge literature on the information loss paradox to be persuaded that physicists are sufficiently convinced of the existence of Hawking radiation to be concerned about its consequences for unitarity).

But there are two problems for the sophisticated formalist. The first is that despite the theoretical basis for astrophysical Hawking radiation from black holes being very strong, it is not perfect. For example, it is a matter of some delicacy determining *what* the Hawking temperature should actually be attributed to. Giddings (2016) has argued that Hawking radiation originates not from the black hole horizon but from a region "well outside the horizon" (p. 40).[28] If Giddings is correct, then it is not quite right to say that the *horizon* is the source of the thermalised radiation. This poses a challenge for the sophisticated formalist's claim that the theory unequivocally militates in favour of Hawking radiation *from black holes*, since that radiation in fact comes from a region in the black hole exterior. This is a pedantic point, of course—we might think that the quantum region Giddings describes is

---

[26] I thank an anonymous reviewer for this example.

[27] It is worth noting that Wüthrich appears to equivocate on the broader question of whether these kinds of (what I would call) 'sophisticated formalist' arguments could ever be sufficient to establish the physical significance of BHT. On the one hand, he notes that compelling thought experiments can and do "lend some support to the idea that black holes are thermodynamic in nature" (p. 221), but on the other hand he maintains a few lines later that "only the usual kind of experimental and observational work can establish that black holes are thermodynamic objects". I read Wüthrich as rejecting the possibility of sophisticated formalism, but the fact remains that this kind of attitude prevails among the physics community, for whom recovering the Bekenstein-Hawking formula has "become something of a *sine qua non* for programs of quantum gravity" (Curiel 2019b).

[28] See also Unruh (1977), who makes a similar point.

close enough for the sophisticated formalist's purposes, or that the production of Hawking radiation in that region can nevertheless be causally attributed to the presence of the black hole. But even if these objections succeed, there are other technical challenges beyond that posed by Giddings. The most notorious is the so-called 'trans-Planckian problem', which refers to the fact that the finite wavelengths of the Hawking radiation particles measured at future null infinity, which have been gravitationally redshifted to a drastic extent by their journey away from the black hole, must correspond to particles originating from the black hole with wavelengths arbitrarily shorter than the Planck length. Intuitively, this strongly suggests that the physics giving rise to the Hawking effect near the black hole is beyond Planck scale, i.e. from a regime where we should expect quantum gravitational corrections to relativity to become non-negligible.[29] Polchinski (1995) has argued that the trans-Planckian problem can be finessed by the so-called 'nice slice argument', which roughly says that the Hawking effect can be derived using only 'nice' spacetime slices upon which *only* low-energy (i.e. cis-Planckian) physics is happening. Polchinski's argument is echoed by Wallace (2018), but Gryb et al. (2020, Sect. 2.3) argue that certain assumptions in the nice slice argument make it essentially question-begging. The extent to which the trans-Planckian problem remains an issue for the sophisticated formalist is therefore unclear, but generally, it seems prudent to tread carefully. As Wallace himself rightly notes at the end of his defence of the theoretical basis for the Hawking effect, "as good scientists we should remind ourselves that [the Hawking effect] remains purely theoretical, and that tests of quantum field theory itself in the curved-spacetime regime to date have been much less precise and numerous than in the flat-spacetime regime" (2018, p. 62).

Technical worries about the theoretical basis for Hawking radiation aside, there is a second argument against sophisticated formalism. Namely, that upon closer inspection, the position seems inconsistent. At bottom, to be a sophisticated formalist is to try to rid BHT of any dependence on analogue gravity, and to be motivated to do so by a skepticism about the epistemology of analogue experimentation in general. But this seems necessarily to involve replacing one dubious epistemic technique with another. Is bare theory really more capable than analogue gravity when it comes to providing evidence for the existence of Hawking radiation? In thinking so, the sophisticated formalist seems to be trying to use one hand to set the bar for empirical confirmation so high that analogue experimentation cannot reach it, while using the other hand to remove the bar completely. If theory is strong enough to confirm, on its own, then so too should analogue experimentation be, a fortiori. While it would be consistent to maintain either (a) full-blooded empiricism, such that we have reason to believe that astrophysical Hawking radiation exists if and only if we have suitable empirical evidence of astrophysical Hawking radiation; or (b) or full-blooded anti-empiricism, such that a belief in the existence of astrophysical Hawking radiation can be justified even by non-empirical (e.g.

---

[29] See Jacobson (2005) and references therein for more detail on the trans-Planckian problem.

theoretical or analogical) evidence, the sophisticated formalist tries to be both and arguably achieves neither.[30] It seems plainly inconsistent for the sophisticated formalist to demand that the existence of astrophysical Hawking radiation requires, empirically, *more* than the evidence provided by analogue gravity, but no more than the evidence provided by bare theory alone.

14.3.3 Classicalism

The third strategy is the least fiddly, but the most radical. Rather than entering the treacherous ground of attempting to describe how non-empirical evidence of astrophysical Hawking radiation may nevertheless justify belief in the empirical existence of that radiation, the third strategy de-couples the two analogical arguments by severing all ties between the physical significance of BHT and the existence of astrophysical Hawking radiation. It does so by arguing that *classical* BHT—BHT without, inter alia, Hawking radiation—is already physically significant. Call this strategy 'classicalism'.

The clearest statement of the classicalist position is given in Curiel (2014). For a black hole to be considered a thermodynamical system, Curiel tells us, what surely matters is only that "surface gravity and area couple to ordinary thermodynamical systems in the same way as temperature and entropy, respectively, do", and that counterpart quantities are introduced using into the theory using the same "constructions and arguments" (2014, p. 3). Curiel goes on to argue that since one can, indeed, construct a Carnot cycle operating between a black hole and a thermodynamic fluid, and since κ appears in place of the black hole's 'temperature' when one defines the efficiency of that Carnot cycle, there is little room to doubt that surface gravity really *is* a temperature. On the classicalist view, one need not wait for Hawking radiation to establish the physical significance of the analogy between the two quantities. Insofar as Curiel's arguments succeed, they establish that BHT need not presuppose the epistemic legitimacy of analogue gravity in the way that I have described, simply by virtue of the fact that premise (iii) in the second formalisation of the analogical argument for BHT above is doing no work. The conclusion—the classicalist would have it—follows immediately from (i) and (ii).

There are at least two reasons to think that the classicalist response fails, however. The first is that Curiel's Carnot cycle fails to be reversible in the standard thermodynamical sense: it results in an increase in the black hole's irreducible mass,

---

[30] One example of a consistent anti-empiricist attitude would be considering empirical evidence for astrophysical Hawking radiation to be supererogatory. On this view, empirical confirmation is a bonus, but is not necessary for justified belief in astrophysical Hawking radiation. Such an anti-empiricist might consistently maintain that analogue gravity research fails to constitute such 'bonus' evidence, but that the theoretical arguments alone were sufficient to warrant belief in astrophysical Hawking radiation. They would still have to respond to the technical problems laid out above, however. I thank an anonymous reviewer for pressing me on this point.

and thus an increase in its horizon area, which in classical relativity cannot be reversed on pain of violating the second law of BHT—Hawking's area theorem (Curiel 2014, pp. 16–17). One could reasonably question whether this cycle can be considered sufficiently similar to a thermodynamical Carnot cycle to justify the very strong conclusion of the analogical argument for BHT.[31]

The second is that while the classicalist response may be viable for establishing the physical significance of the analogy between surface gravity and temperature, it seems impossible—in principle—to establish the physical significance of the analogy between horizon area and entropy with the same logic. This is because, as Curiel notes (2014, p. 10, fn. 20) thermodynamic entropy mediates no known physical processes, and so there is, ipso facto, no process one could construct so as to demonstrate that it is mediated by horizon area in the same way it would have been mediated by entropy in the thermodynamical case. Thus, while classicalism may succeed in removing the reliance of BHT on the analogical argument for astrophysical Hawking radiation, it seems to do so at the price of only recovering, at best, a partial equivalence between black holes and thermodynamical systems.

14.4   Conclusion

Contemporary black hole physics is an interesting case study for the epistemology of science because it deploys analogical reasoning in a way that seems to overreach the traditional heuristic role of analogy. There are two distinct ways in which it does so. First, it relies on analogue experimentation and the observation of analogue Hawking radiation in acoustic and optical black holes to justify the *existence* of astrophysical Hawking radiation. Second, it relies on the analogy between the laws of black hole mechanics and the laws of ordinary thermodynamics to justify the claim that black holes are genuinely thermodynamical systems.

The analogical arguments in favour of these two claims are distinct. Each one appeals to different premises in order to draw different conclusions, and those conclusions seems to be fundamentally different in kind. But because the BHT analogy is typically only considered physically significant with the inclusion of Hawking radiation, there is an important sense in which the two analogical arguments are linked. Indeed, the first can be nested inside the other in such a way that the analogical argument for BHT cannot succeed without the analogical argument for the existence of astrophysical Hawking radiation. This connection, I argued in Sect. 14.3, leads to a tension for those who wish to combine an optimism about the physical significance of BHT with a pessimism about the epistemic warrant provided by analogue experiments.

---

[31] Interestingly, the cycle Curiel describes can effectively be *made* reversible when one takes Hawking radiation into account (Prunkl and Timpson 2019), but this rather defeats the point of the classicalist strategy we are considering.

I considered three strategies for resolving this tension, and showed that each one fails. The naïve formalist (Sect. 14.3.1) and the classicalist (Sect. 14.3.3) strategies may succeed in decoupling the two analogical arguments, but they do so at the cost of settling for impoverished versions of BHT: one nakedly mathematical; the other only partially complete. The sophisticated formalist (Sect. 14.3.2), on the other hand, seems to replace the tension between optimism about BHT and skepticism about analogue gravity with another, new tension. The sophisticated formalist is committed to the contradictory idea that analogical evidence is inadequate to the task of justifying the existence of astrophysical Hawking radiation, and yet that same justification may be amply provided by bare theory—something *weaker*, not stronger, than analogical evidence.

If these three strategies exhaust the space of possibilities, then the tension I pointed out at the start of Sect. 14.3 remains: the analogical argument for BHT presupposes the success of the analogical argument for the existence of astrophysical Hawking radiation. Thus, it presupposes the epistemic legitimacy of analogue gravity. It is difficult to see how Harlow's assertion that analogue gravity is no more than an amusing feat of engineering can be reconciled with the belief that black holes are, nevertheless, genuinely thermodynamical systems.

Perhaps the best prospects for resolving this tension come from a retreat to the 'traditional' role of analogy: as heuristic. If there are reasons to believe that these analogical arguments will, in time, be superseded by robust non-analogical arguments, then there are, eo ipso, reasons to believe that this tension will, in time, dissipate. It is far from clear, however, that the analogical arguments discussed here are even capable of being replaced by non-analogical arguments. Insofar as the magnitude of Hawking radiation from an astrophysical black hole and the magnitude of CMB fluctuations are fixed by the nature of physical law, the impossibility of direct detection of Hawking radiation seems fixed with at least nomological necessity. One way around this, which we have not considered in this chapter, would be the discovery of primordial black holes, whose low mass would result in a Hawking temperature sufficiently high to admit direct detection. Until such a discovery, it seems we should make peace with the fact that black hole physics seems destined to continue depending upon analogical reasoning in this unusually strong way.

References

Abbott, B.P., R. Abbott, T.D. Abbott, M.R. Abernathy, F. Acernese, K. Ackley, et al. 2016. Binary black hole mergers in the first advanced LIGO observing run. *Physical Review X* 6: 041015. https://link.aps.org/doi/10.1103/PhysRevX.6.041015.

Bambi, C. 2018. Astrophysical black holes: A compact pedagogical review. *Annalen der Physik* 530: 1700430.

Barceló, C., S. Liberati, and M. Visser. 2011. Analogue gravity. *Living Reviews in Relativity* 14. http://www.livingreviews.org/lrr-2011-3.

Bardeen, J.M., B. Carter, and S.W. Hawking. 1973. The four laws of black hole mechanics. *Communications in Mathematical Physics* 31(2): 161–170.

Bartha, P. 2010. *By parallel reasoning: The construction and evaluation of analogical arguments.* Oxford: Oxford University Press.

Bartha, P. 2019. Analogy and analogical reasoning. In The stanford encyclopedia of philosophy, E.N. Zalta, ed., Spring 2019 edn. Metaphysics Research Lab, Stanford University.

Bekenstein, J.D. 1972. Black holes and the second law. *Lettere al Nuovo Cimento (1971–1985)* 4(15): 737–740. https://doi.org/10.1007/BF02757029.

Crowther, K., N.S. Linnemann, and C. Wüthrich. 2021. What we cannot learn from analogue experiments. *Synthese* 198: 3701–3726. https://doi.org/10.1007/s11229-019-02190-0.

Curiel, E. 2014. Classical black holes are hot. arXiv:1408.3691 [gr-qc].

Curiel, E. (2019a), The many definitions of a black hole. *Nature Astronomy* 3(1): 27–34. http://www.nature.com/articles/s41550-018-0602-1.

Curiel, E. (2019b). Singularities and black holes. In The stanford encyclopedia of philosophy, E.N. Zalta, ed. Spring 2019 edn. Metaphysics Research Lab, Stanford University.

Dardashti, R., K.P.Y. Thébault, and E. Winsberg. 2017. Confirmation via analogue simulation: What dumb holes could tell us about gravity. *The British Journal for the Philosophy of Science* 68: 55–89. https://doi.org/10.1093/bjps/axv010

Dardashti, R., S. Hartmann, Thébault, K. P.Y. & E. Winsberg. 2019. Hawking radiation and analogue experiments: A Bayesian analysis. *Studies in History and Philosophy of Science Part B: Studies in History and Philosophy of Modern Physics* 67: 1–11.

de Nova, J.R.M., K. Golubkov, V.I. Kolobov, and J. Steinhauer. 2019. Observation of thermal Hawking radiation and its temperature in an analogue black hole. *Nature* 569: 688–691.

Dolan, B.P. 2012. Where is the PdV in the first law of black hole thermodynamics? In Open questions in cosmology, G.J. Olmo, ed. Rijeka: IntechOpen, chapter 12.

Dougherty, J., and C. Callender. (2016). Black hole thermodynamics: More than an analogy? In Guide to philosophy of cosmology, B. Loewer, ed. Oxford University Press.

Drori, J., Y. Rosenberg, D. Bermudez, Y. Silberberg, and U. Leonhardt. 2019. Observation of stimulated Hawking radiation in an optical analogue. *Physical Review Letters* 122: 010404.

Evans, P.W., and K.P.Y. Thébault. 2020. On the limits of experimental knowledge. *Philosophical Transactions of the Royal Society A: Mathematical, Physical and Engineering Sciences* 378(2177): 20190235.

Field, G. 2022. Putting theory in its place: The relationship between universality arguments and empirical constraints. *The British Journal for the Philosophy of Science* . https://doi.org/10.1086/718276

Fletcher, S.C. 2014. Similarity and spacetime: Studies in intertheoretic reduction and physical significance, PhD thesis, UC Irvine.

Fletcher, S.C. 2016. Similarity, topology, and physical significance in relativity theory. *British Journal for the Philosophy of Science* 67(2): 365–389.

Giddings, S.B. 2016. Hawking radiation, the Stefan-Boltzmann law, and unitarization. *Physics Letters B* 754: 39–42.

Gryb, S., P. Palacios, and K. Thébault. 2020. On the universality of Hawking radiation. *The British Journal for the Philosophy of Science* 72(3): 809–837.

Hawking, S.W. 1974. Black hole explosions? *Nature* 248: 30–31.

Hawking, S.W. 1975. Particle creation by black holes. *Communications in Mathematical Physics* 43: 199–220.

Hawking, S.W., and G.F.R. Ellis. 1973. *The large scale structure of space-time.* Cambridge: Cambridge University Press.

Hayward, S.A. 1994. General laws of black-hole dynamics. *Physical Review D* 49(12): 6467.

Hesse, M.B. 1966. *Models and analogies in science.* Notre Dame: University of Notre Dame Press.

Jacobson, T. 1996. Introductory lectures on black hole thermodynamics. *Institute for Theoretical Physics University of Utrecht.*

Jacobson, T. 2005. Introduction to quantum fields in curved spacetime and the Hawking effect. In *Lectures on quantum gravity*, A. Gomberoff, and D. Marolf, eds., 39–89. New York: Springer, chapter 2.

Keynes, J.M. 1921. *A treatise on probability*. London: Macmillan.

Kolobov, V.I., K. Golubkov, J.R.M. de Nova, and J. Steinhauer. 2021. Observation of stationary spontaneous Hawking radiation and the time evolution of an analogue black hole. *Nature Physics* 17: 362–367.

Leonhardt, U. 2018. Questioning the recent observation of quantum Hawking radiation. *Annalen der Physik* 530: 1700114.

Lüst, D., and W. Vleeshouwers. 2019. *Black hole information and thermodynamics*. Springer Briefs in Physics. Springer.

Nielsen, A.B. 2009. Black holes and black hole thermodynamics without event horizons. *General Relativity and Gravitation* 41: 1539–1584.

Norton, J.D. 2021. *The material theory of induction*, BSPS Open.

Penrose, R. 1965. Gravitational collapse and space-time singularities. *Physical Review Letters* 14: 57.

Polchinski, J. 1995. String theory and black hole complementarity. arXiv preprint hep-th/9507094.

Prunkl, C. E.A., and C.G. Timpson. 2019. Black hole entropy is thermodynamic entropy. arXiv:1903.06276 [physics.hist-ph].

Rees, M.J. 1998. Astrophysical evidence for black holes. In *Black holes and relativistic stars*, R.M. Wald, ed., 79–102. Chicago: The University of Chicago Press, chapter 4.

Smoot, G.F., C.L. Bennett, A. Kogut, E.L. Wright, J. Aymon, N.W. Boggess, E.S. Cheng, G. De Amici, S. Gulkis, M.G. Hauser, G. Hinshaw, P.D. Jackson, M. Janssen, E. Kaita, T. Kelsall, P. Keegstra, C. Lineweaver, K. Loewenstein, P. Lubin, J. Mather, S.S. Meyer, S.H. Moseley, T. Murdock, L. Rokke, R.F. Silverberg, L. Tenorio, R. Weiss, and D.T. Wilkinson. 1992. Structure in the *COBE* differential microwave radiometer first-year maps. *The Astrophysical Journal* 396: L1–L5.

Steinhauer, J. 2016. Observation of quantum Hawking radiation and its entanglement in an analogue black hole. *Nature Physics* 12: 959–965.

The EHT Collaboration et al. 2019. First M87 event horizon telescope results. IV. Imaging the central supermassive black hole. *The Astrophysical Journal Letters* 875: 4. https://iopscience.iop.org/article/10.3847/2041-8213/ab0e85

Thébault, K. 2019. What can we learn from analogue experiments? In *Why trust a theory? Epistemology of fundamental physics*, R. Dardashti, R. Dawid, and K. Thébault, eds., 184–201. Cambridge University Press, Cambridge, chapter 11.

Unruh, W.G. 1977. Origin of the particles in black hole evaporation. *Physical Review D* 15: 365.

Unruh, W.G. 1981. Experimental black-hole evaporation? *Physical Review Letters* 46 (21): 1351–1353.

Unruh, W.G. 2014. Has Hawking radiation been measured? *Foundations of Physics* 44: 532–545.

Visser, M. 2003. Essential and inessential features of Hawking radiation. *International Journal of Modern Physics* D12: 649–661.

Visser, M. 2013. Survey of analogue spacetimes. In *Analogue gravity phenomenology: Analogue spacetimes and horizons, from theory to experiment*. Lecture Notes in Physics, D. Faccio, F. Belgiorno, S. Cacciatori, V. Gorini, S. Liberati, and U. Moschella, eds., vol. 870, 31–50. Springer, chapter 2.

Wald, R.M. 1984. *General relativity*. Chicago: The University of Chicago Press.

Wald, R.M. 1994. *Quantum field theory in curved spacetime and black hole thermodynamics*. Chicago: University of Chicago Press.

Wallace, D. 2018. The case for black hole thermodynamics, Part I: Phenomenological thermodynamics. *Studies in History and Philosophy of Science Part B: Studies in History and Philosophy of Modern Physics* 64: 52–67.

Wallace, D. 2019. The case for black hole thermodynamics, Part II: Statistical mechanics. *Studies in History and Philosophy of Science Part B: Studies in History and Philosophy of Modern Physics* 66: 103–117.

Weinfurtner, S., E.W. Tedford, M.C.J. Penrice, W.G. Unruh, and G.A. Lawrence. 2011. Measurement of stimulated Hawking emission in an analogue black hole system. *Physical Review Letters* 106: 021302.

Wolchover, N. 2016. What sonic black holes say about real ones. *Quanta Magazine*. https://www.quantamagazine.org/what-sonic-black-holes-say-about-real-ones-20161108/.

Wüthrich, C. 2019. Are black holes about information? In *Why trust a theory? Epistemology of fundamental physics*, R. Dardashti, R. Dawid, and K. Thébault, eds., 202–223. Cambridge: Cambridge University Press, chapter 11.

Chapter 15
Extragalactic Reality Revisited: Astrophysics and Entity Realism

Simon Allzén

Abstract Astrophysics is a scientific field with a rich ontology of individual processes and general phenomena that occur in our universe. Despite its central role in our understanding of the physics of the universe, astrophysics has largely been ignored in the debate on scientific realism. As a notable exception, Hacking (Philos Sci 56(4):555–581, 1989) argues that the lack of experiments in astrophysics forces us to be anti-realist with respect to the entities which astrophysics claim inhabit the universe. In this paper, I investigate the viability of astrophysical realism about black holes, given other formulations of entity realism, specifically Cartwright's (How the Laws of Physics Lie. Oxford University Press, 1983), and Chakravartty's (A Metaphysics for Scientific Realism: Knowing the Unobervable. Cambridge University Press, 2007) versions of entity realism. I argue that on these accounts of entity realism, you cannot be a realist with respect to black holes, and likewise, if you want to be a realist about black holes, you cannot be an entity realist of these particular strands.

15.1 Introduction

Astrophysics is a scientific field with a rich methodological profile: it uses explanatory causal inferences, astronomical observation, complex modeling, data analysis, and simulations in order to generate theories about the individual processes and general phenomena that occur in our universe (Anderl 2015; Jacquart 2020). Scientific realism is a philosophical doctrine that seeks to carve out the specific conditions under which we may rationally believe that a scientific theory is true, or when its objects are real. Usually, realists are taken to hold that there is a mind independent world which terms in our best scientific theories successfully refer to, and that we can come to know what that world is really like. Astrophysics is a field

S. Allzén (✉)
Department of Philosophy, Stockholm University, Stockholm, Sweden
e-mail: simon.allzen@philosophy.su.se

© The Author(s) 2023
N. Mills Boyd et al. (eds.), *Philosophy of Astrophysics*, Synthese Library 472,
https://doi.org/10.1007/978-3-031-26618-8_15

in science which contain theories and claims about the nature of various processes and phenomena in the universe. The question is if the epistemological practices in astrophysics satisfy the realist criteria, and so, if we should be realist with respect to the entities which astrophysics take to inhabit the universe. Surprisingly, given the scope of astrophysics, few realists have engaged with this question.[1] A notable exception is Hacking's "Extragalactic Reality: The Case of Gravitational Lensing" in which his brand of realism—entity realism—regarding astrophysics as a whole is deemed unattainable:

> Astrophysics is almost the only human domain where we have profound, intricate knowledge, and in which we can be no more than what van Fraassen calls constructive empiricists. (Hacking 1989, 578)

Although Hacking's skeptic conclusion about astrophysical realism has been challenged (Shapere 1993; Sandell 2010; Anderl 2015), much remains to be said about the specific relation between entity realism and astrophysics. Hacking's entity realism premises belief in a certain entity on the possibility of causally manipulating that entity, which explains why he excludes both theoretical truth and realism about the majority of objects and processes found in astrophysics (as well as in cosmology and astronomy). Entity realism in this form, then, may be taken to exclude realism about astrophysical objects. The question arises as to what degree Hacking's astrophysical anti-realism can be taken to represent the broader entity realist project in the astrophysical context.

Like Hacking, Cartwright (1983) has advocated a form of entity realism which emphasizes the role played by causality in homing in on the proper objects of realism: the entities. For her, however, the connection between causation and realism is not modeled on the manipulation of entities by experimentalists. Instead, *causal explanation* is the epistemic route to realism. Causal explanations, she argues, only make sense if we take the causes described by the explanations to be real. In this sense, she permits ontology based on an *inference to the most likely cause*. That is, if we want to take the causal explanations offered by science seriously, we have to believe in the entities to which they refer. Or as Cartwright herself puts it: "In causal explanations truth is essential to explanatory success." (1983, 10) Prima facie, her view of realism as premised on causal explanations allows for a more permissive epistemology and consequently a richer ontology. Whether or not accepting causal inferences is sufficient to output realism about astrophysical entities is nevertheless opaque.

Yet another kind of entity realist account is *semi-realism*, defended by Chakravartty (2007). The epistemic aim in semi-realism is, like Cartwright's version, more ambitious than Hacking's entity realism. It introduces a spectrum

---

[1] Although adjacent questions have been somewhat explored, for example cosmological realism (Merritt 2021), dark matter realism (Jacquart 2021; Allzén 2021; Martens 2022), String Theory realism (Dawid 2007, 2013), observation and simulation (Jacquart 2020), experimental limits in astrophysics (Evans and Thébault 2020), and simulation and modeling (Guala 2002; Morgan 2005; Parker 2009; Parke 2014).

of causal connection that correlates with degrees of belief. To this end, Chakravartty offers an epistemic distinction between *detection properties*, defined as "the causal properties one knows, or in other words, the properties in whose existence one most reasonably believes on the basis of our causal contact with the world." (Chakravartty 2007, 47), and *auxiliary properties*, defined as the properties which are attributed to objects by a theory. In this framework, auxiliary properties can become detection properties once new experiments and technology facilitates causal contact with them. This enables semi-realism to be firmly realist about empirically confirmed unobservables, and agnostic about unobservables posited for explanatory reasons.

The current paper addresses the viability of entity realism in the case of black holes. Are the epistemic and methodological tools available to the astrophysicist sufficient to generate rational beliefs about the existence and properties of astrophysical black holes, and if so, can this result be recovered in entity realism?[2] Studies of black holes involve many instances of methodological practices found in astrophysics, and there is a fairly wide consensus about their existence. This allows for a comparison between the epistemic justification astrophysicists have for the existence and properties of black holes, and the ontologically committing causal reasoning of the considered forms of entity realism.

15.2 Entity Realism

Scientific realists believe that our best scientific theories can be taken at face value: their terms refer to a mind independent reality, and we can come to know what that reality is like by consulting science. In the early 1980s Laudan (1981) showed that many of our best scientific theories in the past were, *pace* realism, false. Laudan's historical gambit—the so called 'pessimistic meta-induction' (PMI)—targets the fact that scientific realists postulate a connection between empirical success and truth. By breaking this connection, Laudan showed that not only do we have reason to believe that past science was false, but, by induction, our current best science is as well. If there is a connection, we have inductive reasons to think that it is between empirical success and falsehood. Any realist that aims to be taken seriously had to find a way to deal with PMI. One of the strategic revisions to the realist position was to reduce its scope. Perhaps, realists thought, theoretical terms in past successful theories were empty, but the *entities* to which those terms were intended to refer may well have existed nonetheless. If so, that would mean that, under certain specified conditions, it is rational to believe that the 'corpuscles' that J.J. Thomson experimented with in his cathode-ray tubes and the electrons that

---

[2] The idea of letting the particulars of scientific epistemology inform the standards according to which realism is viable is not unanimously accepted in the realist debate. Usually, realists take a principled approach to such standards and then decide on that basis if some particular scientific epistemology merits realism. This issue may be taken to arise as an upshot of the paper's current focus, so it will be alluded to in the concluding remarks.

are essential to the operation of an electron microscope are the same ontological entity, one which is constant through theoretical changes and advancements. It is the entities, not the theory, that realism ought to target, hence *entity realism*. This move is thought to bypass PMI because it does not commit the realist to the truth of any specific theoretical model, predictively successful though it might be, and so does not suffer from being forced to accept the truth of incorrect but predictively successful theories. Prima facie, entity realism sounds like a plausible route for astrophysical realism, given that much of astrophysical theory investigates the nature of entities and processes in the universe. Decoupling realist commitment from theoretical descriptions renders a more robust ontology, and an epistemically safer route to a defensible realism. There is, however, more than one way in which to design the selection criteria for an entity's eligibility for realist commitment. In order to evaluate the specific relation between entity realism and black holes, we will first need to review a representative sample of these different criteria.

15.2.1 Hacking's Manipulationist Account

One of the founders of, and primary advocates for, entity realism, Hacking (1983) suggests taking the *manipulation* of entities to be central to realist commitment:

> Experimenting on an entity does not commit you to believing that it exists. Only manipulating an entity, in order to experiment on something else, need do that. (Hacking 1983, 263)

In order to manipulate an entity, scientists must first establish a certain level of causal connection to it. The causal connection enables scientists to extract some of the causal properties of the entity in order to build devices that can manipulate it. The core premise for realism outlined by Hacking offers a significantly smaller but epistemically safer set of things to be realist about: we may not be licensed to believe in the truth of the Standard Model of particle physics or the theory of electromagnetism, but we are licensed to believe in the reality of the electron and some of its causal properties. Hacking is in a sense employing a methodological approach to realism: since experimentation by manipulation of electrons does not require a full theory of the nature of the electron, philosophers can take a leaf from the experimenter's book and be realist with respect to entities which function, to us, as tools. However, as Hacking himself points out in "Extragalactic Reality: The Case of Gravitational Lensing" (1989, 578), his manipulationist account of entity realism is not the route to astrophysical realism simply because we cannot manipulate astrophysical entities in the way necessary for his realist criteria to kick in. This result is striking because it renders an anti-realism about basically the whole universe, given that the manipulationist premise sets a boundary of accessibility that does not extend to objects outside of our solar system. It is perhaps possible to call Hacking a *qualitative* realist about the stuff in the universe, given that there is a sufficient level of local interaction with the kind of entities that comprise the universe globally. This qualitative realism, however, requires a reductionist programme where astrophysical macro objects can be reduced to their component parts, which are such that we can use them as tools. Alas, this route begs the question against Hacking's own realism, given its reliance on the fact that astrophysical theory is correct about the constitution of macro objects.

15.2.2 Cartwright's Causal-Explanatory Account

Despite being cautious regarding scientific realists' aim of recovering truth in science, Cartwright appears to share at least some of their optimistic spirit:

> I think that van Fraassen and Duhem eliminate more than they should. It is apparent from earlier essays that I share their anti-realism about theoretical laws. On the other hand, I believe in theoretical entities, and that is my main topic in this essay. (Cartwright 1983, 89)[3]

For Cartwright, like for Hacking, the core of a tenable scientific realism is causality and entities. What sets her account apart from Hacking is that she gravitates towards causal explanation, not manipulation, as the locus of causal interest. Causal explanations require a cause as an explanandum, which in turn strongly implies some entity or process that is the real world instantiation of the explanandum. Such explanations are in some sense isomorphic to the world in a way that other forms of explanation just aren't: "In causal explanations truth is essential to explanatory success" (1983, 10). Cartwright might seem to invoke an inference that is merely an instance, or a special case, of inference to the best explanation, but she argues that in causal explanations, truth is an internal part of the explanandum, whereas in other explananda truth is an external addition.[4] The argument is that while inference to the best explanation can be used with explanations that lack this external addition, thereby generating incorrect inferences to theoretical claims, inference to the most likely cause always involves an inference to a causing entity or object, the existence of which is not dependent on theories about it:

> I infer to the most probable cause, and that cause is a specific item, what we call a theoretical entity. But note that the electron is not an entity of any particular theory. In a related context van Fraassen asks if it is the Bohr electron, the Rutherford electron, the Lorenz electron or what. The answer is, it is the electron, about which we have a large number of incomplete and sometimes conflicting theories. (Cartwright 1983, 92)

Again, we can see that this form of realism is aiming at designing principles for realism with theory-invariance of some sort built in, at least with respect to physical theory. It also aims to provide a natural connection between entities and causal explanations.

---

[3] Cartwright is referring to the constructive empiricism of van Fraassen (1980) and the instrumentalism of Duhem (1991). Both views shun a realism about theory and unobservable entities.

[4] An argument to this effect can be found in Psillos (2008).

15.2.3 Chakravartty's Semi-realism

Chakravartty's semi-realism is yet another attempt to protect realism against objections like the pessimistic meta-induction, underdetermination by data, and challenges to inference to the best explanation. His specific view aims to take the idea of selective scepticism—to not accept predictively successful theories wholesale—and pair it with the dictum that "a realist's degree of belief should reflect one's degree of causal contact, with mastery and manipulation at one end of the spectrum, and mere detection and weaker speculation at the other" (Chakravartty 2007, 47). It is clear that causality again plays the main role, setting the parameters for rational belief and mapping realist commitment about properties or entities to the level of causal contact we have with them. Chakravartty fleshes out his semi-realism by distinguishing between auxiliary properties and detection properties, where only the latter are candidates for rational belief. Auxiliary and detection properties are described, and distinguished, as follows:

> An auxiliary property is one attributed by a theory, but regarding which one has insufficient grounds, on the basis of our detections, to determine its status. (Chakravartty 2007, 47)

And;

> The realist requires a practical means of demarcating detection properties (and the structures associated with them) from auxiliary properties. Here is a suggestion. Detection properties are connected via causal processes to our instruments and other means of detection. (Chakravartty 2007, 48)

Causality does much (if not all) of the heavy lifting in order to provide an epistemically safe connection between the detection properties of scientific objects and us. Knowledge about these properties, and their relations, are then thought to constitute knowledge about concrete structures of the world—objects and entities— which then furnishes the ontology of particulars in semi-realism (Chakravartty 2007, 64).

15.3 Astrophysical Black Holes

It is an understatement to say that attempting to provide a universally accepted definition of a black hole is hard. As Curiel (2019) shows, there are more than a few candidate definitions, where each field harbors a definition which suits their specific methodological needs, and in addition, many of them are inconsistent. The astrophysical picture of a black hole is centered around the notion that black holes are *objects* with properties, for example mass (and/or charge, spin, etc.), which can be connected with observational data. A couple of quotes from Curiel (2019) can, if not provide a precise definition, give a sense of the focal point for the conceptual understanding of an astrophysical black hole:

> A black hole is a compact body of mass greater than four solar masses – the physicists have shown us there is nothing else it can be. – Ramesh Narayan, astrophysicist (active galactic nuclei, accretion disk flow) (Curiel 2019, 30)

> [I]n practice we don't really care whether an object is 'precisely' a black hole. It is enough to know that it acts approximately like a black hole for some finite amount of time. . . . [This is] something that we can observe and test. – Don Marolf, theoretical physicist (semi-classical gravity, string theory, holography) (Curiel 2019, 31)

> Today 'black hole' means those objects we see in the sky, like for example Sagittarius A*. – Carlo Rovelli, theoretical physicist (classical general relativity, loop quantum gravity, cosmology, foundations of quantum mechanics) (Curiel 2019, 31)[5]

The definition(s) here clearly take a black hole to be an astrophysical system—a three dimensional object which persists through time and participates in dynamical behavior, such as black hole mergers or in binary systems—which is within the boundary of empirical study. This is the rough definition of a black hole that will be assumed in relation to the issues considered in this paper. Assuming this view means that (some of) the properties of a black hole can be accessed and studied, at least in principle.[6] Whether this in-principle epistemic access to black holes allows us to be realists about them in the philosophical sense, however, remains to be seen.

15.3.1 Discovery of Black Holes

The first black hole ever discovered is called Cygnus X-1. This discovery was not serendipitous, given that black holes would be virtually impossible to find if you don't know what to look for. The preceding work that made the discovery possible was both theoretical (Schwartzschild's solution of Einstein's field equations of GR in 1916) as well as empirical (the discovery of neutron stars in the 1960s). When the Uhuru X-Ray satellite in 1970 found an intensely flickering X-Ray source (later discovered to be part of a binary system) with a high mass in a small region, the once theoretical possibility of black holes took a leap towards becoming a reality.[7] Importantly, the methodology involved in this discovery involves an inference, theoretical background assumptions, and observational astronomy.

---

[5] Sagittarius A* refers to the supermassive black hole at the centre of the Milky Way.

[6] Phenomena that occur in the interior of a black hole are in principle not accessible, given that the interior marks a causal boundary—an event horizon—which means that black holes are only partially in-principle accessible.

[7] In a panel discussion on the existence of black holes, physicist Werner Israel recalls being ridiculed for believing in the mere conceptual possibility of black holes existing: "the Director of the Institute remarked, 'Werner is going to be with us for a year. We should all talk to him and try to cure him of these silly notions he has about the possibility of black holes'" (Collmar et al. 1998, 487).

If we take the discovery of Cygnus X-1 to mark the first time the concept of a black hole was coupled with empirical evidence, we get some idea of the particular epistemology that is employed when detecting black holes. Given the rather strange nature of spacetime regions associated with black holes, the corresponding epistemology has its unique set of challenges:

> How would we know if there were a black hole? The fundamental obstacle to direct detection is, of course, blackness: a black hole will not itself give off any radiation [...]. But black holes will feature extremely strong gravitational fields, so we can hope to detect them indirectly by observing matter being influenced by these fields. As matter falls into a black hole, it will heat up and emit X-rays, which we can detect with satellite observatories. A large number of black-hole candidates have been detected by this method, and the case for real black holes in our universe is extremely strong. (Carroll 2019, 235)

Already, we may note that astronomical observation, both in the visible and X-ray range, is of crucial importance to obtain the data needed to make inferences about likely causes for the dynamical behavior of matter surrounding a specific region of spacetime. But to get a more fine-grained, and hopefully clearer, understanding of black hole epistemology, it will be useful to devote some space to the discovery and reasoning that supported the existence of black holes. Much of the following will be based on Celotti et al. (1999).

15.3.1.1 Stellar Black Holes

As already mentioned, Cygnus X-1 was the first observable source that was coupled with the theoretical understanding of a stellar black hole.[8] The earliest observations that detected discrete X-ray sources outside of our solar system were made in the early 1960s, using X-ray detectors which operated outside of the atmosphere. Nearly a decade later, up to twenty different X-ray sources had been identified this way. Optical observations later determined that there was a star-counterpart to one of the most intense X-ray sources, leading researchers to infer that given that the star could not itself be the source of the X-rays, the source was most likely very hot gas. The gas could only be that hot if it was being accreted from the optical star on to a compact undetected nearby binary object. In the following decade, the data improved with the launch of the X-ray satellite Uhuru, which enabled scientists to conclude that the X-ray source was in fact part of a binary system, most likely a black hole (Rothschild et al. 1974). In the mid 1980s, a detailed analysis of the Cygnus X-1 binary system combined over 55 astronomical observations, concluding that:

> Our results indicate that the mass of the X-ray source is much greater than the neutron star limit, which further strengthens its black hole candidacy. (Gies and Bolton 1986, 387)

---

[8] Taking the mass range of stellar black holes to be $\approx 5M_\odot - 100M_\odot$.

As we can see, this result, though based on observational data, rests on an important piece of reasoning from eliminating alternative possibilities. The only candidate objects compact enough to generate the observed phenomena were neutron stars and stellar black holes, which was precisely the underdetermination that Gies and Bolton (1986) were trying to break. Given that the mass limit for neutron stars was uncertain, Cygnus X-1 and other signature X-ray sources like it suffered from underdetermination: the data retrieved from X-ray sources was consistent with them being neutron stars. This uncertainty was a consequence of the fact that neutron stars are so dense that the equations of state for material go well beyond known nuclear physics, and therefore beyond well confirmed and understood physics for which there is experimental data. Whatever equations of state one determines are appropriate for neutron stars in turn determines the maximum masses they can have. Celotti et al. (1999) describes how Rhoades Jr and Ruffini (1974), based on better known low density equations of state, derived a fixed upper limit on the maximum mass of neutron stars: $M_{max} \simeq 3.2 M_\odot$. Based on this limit, one can estimate the likelihood of a compact object being a stellar black hole or a neutron star based on its mass. It is this upper limit that feeds the inference that the Cygnus X-1 X-ray source is *not* a neutron star, but a stellar black hole (this reasoning is well reflected in the above quote from Ramesh Narayan). The advancements of X-ray detection coupled with optical observations, models of neutron stars, and modeling of accretion flow are clearly methods needed when inferring the existence of a stellar black hole, all of which rely on a solid understanding of basic physical principles. It's interesting to note that the fact that Cygnus X-1 was part of a binary system turned out to be prototypical for discoveries of stellar black holes since "All the known stellar-mass black holes are members of X-ray binaries" (Frampton 2016, 1).

15.3.1.2 Supermassive Black Holes

If the detection, observation, and modeling of neutron stars are significant for the epistemology of stellar black holes, the same is true for quasars and supermassive black holes (SMBHs) which are black holes with masses $\geq 10^5 M_\odot$. Quasars, short for 'quasi-stellar radio sources', are, as the name suggests, a source of immense radiation, far exceeding the luminosity of the Milky Way.[9] In 1964 Edwin Salpeter and Yakov Zel'doviĉ proposed that the mechanism responsible for the radiation of quasars was accretion of gas onto a SMBH, and in 1971 Lyndon-Bell and Rees suggested that our own Milky Way may host a SMBH in its centre. The most compelling candidate objects for SMBHs then, reside in the centre of galaxies. The initial inference made by Salpeter, Zel'doviĉ, Lyndon-Bell, and Rees was one built on the observation that some massive compact object produced extreme levels of

---

[9] Quasars are now often referred to as active galactic nuclei (AGN), since the abbreviation 'quasar' turned out to be misleading.

radiation in the centre of (many) galaxies. Again, the upper limit of mass for neutron stars was essential for eliminating alternatives, and the modeling of accretion around black holes provided a consistency test with known data. Interestingly, scarcity of alternative explanations for the radiation seems to have played a significant part in the acceptance of SMBHs:[10]

> Accretion onto a black hole was at that point the widely accepted model, to be sure, but the seemingly exotic nature of black holes left many astrophysicists with unease; there was, however, no other plausible candidate known. With upper possible mass limits on neutron stars worked out in the 1970s, and more and more observational evidence coming in through the 1980s that the objects at the centre of quasars had to be more massive than that, and compressed into an extremely small volume, more and more doubters were won over as theoretical models of no other kind of system could so well account for it all. (Curiel 2019, 28)

> The main characteristic feature of the AGN phenomenon is the inferred compactness of the sources: luminosities of the order of 10^{46} erg s^{-1} (more than 10^{12} times the luminosity of the Sun) are produced from regions less than a light year across ($\sim 10^{18}$ cm). [...] The most extreme constraint on the compactness comes from the high-energy (X-ray) radiation. [...] This high energy radiation, together with other spectral characteristics, including line emission from gas moving at speeds of thousands of km s^{-1}, cannot be satisfactorily ascribed to any stellar-related (quasi-thermal) process. (Celotti et al. 1999, A13)

Though many in the scientific community were convinced by the strong theoretical reasoning, whatever doubt that remained dissipated with the later infrared observations and data-analysis which determined the density of the compact radio source, prompting the authors to state that "There is no stable configuration of normal stars, stellar remnants or substellar entities at that density" (Genzel et al. 1997, 219), referring to the SMBH Sagittarius A* in the Milky Way. The confidence in this conclusion is in part built on the observed orbital motions of stars in Sag A*, which requires modeling using stellar dynamics. Stellar dynamics is the description of systems containing $N \gg 10$ point masses where the mutual gravitational interaction of the point masses dictate their orbital motion, a description which is sensitive to modeling assumptions: "stars [...] behave basically like point masses in ballistic motion" (Celotti et al. 1999, A15).[11] As with its stellar counterpart, observation, inference, modeling, and eliminative reasoning all appear intrinsically coupled with SMBH epistemology.

---

[10] One may note that on some accounts, the lack of alternatives may amount to confirmation. See Dawid et al. (2015), Dawid (2016) for the probabilistic strength of such an argument.

[11] See Celotti et al. (1999) for a full survey of the astrophysical evidence, and (Murdin 2001) for the equations and concepts involved in stellar dynamics.

15.4 Black Hole Realism?

15.4.1 Cartwright

Can the plurality of methodology displayed in astrophysics be analyzed in terms of causal explanations or detection properties so as go generate realism about black holes? Prima facie, this question is opaque at best, given the variety and complexity of astrophysical epistemology. One of the factors that muddies the waters is the application and use of background theory. One may plausibly claim that an inference to the most likely cause is at work when entertaining causes for extreme gravitational fields and their effects on surrounding systems which—for Cartwright—should entail being realist with respect to the black hole as an entity. However, as we have seen this inference is not only guided by, but dependent on, a multitude of background theories including general relativity; stellar dynamics; optics; accretion flow; et.c. One particularly salient aspect of the inference was that one could rule out neutron stars as a cause based on an upper mass limit, a limit which was determined using further theory:

> On the basis of Einstein's theory of relativity, the principle of causality, and Le Chatelier's principle, it is here established that the maximum mass of the equilibrium configuration of a neutron star cannot be larger than $3.2M_\odot$. (Rhoades Jr and Ruffini 1974, 324).[12]

To avoid any confusion, the 'principle of causality' is used in order to set limits on values in the equations of state so that it does not violate the speed of light. This seems to me to be a minimal requirement for something to count as a causal explanation, but not sufficient in order to categorize the upper mass limit for neutron stars as the kind of causal explanation that would merit realism for Cartwright:

> [W]hen do we have reasonable grounds for counting a causal account acceptable? The fact that the causal hypotheses are part of a generally satisfactory explanatory theory is not enough, since success at organizing, predicting, and classifying is never an argument for truth. Here, as I have been stressing, the idea of direct experimental testing is crucial. (Cartwright 1983, 98-9)

If direct experimental testing is crucial for truth or existence to emerge in Cartwright's account, then the existence of astrophysical black holes as inferred based on eliminating neutron stars as causes is beyond the limit of her entity realism. The lack of experiments was precisely the feature that led Hacking to the conclusion that we ought to be constructive empiricists about astrophysics. We cannot perform direct experimental tests on black holes, and the inference that guides reasoning in this case is so clearly coupled with the upper mass limit for neutron stars, as well as eliminative reasoning. Scientists cannot devise a direct experimental test for

---

[12] Kalogera and Baym (1996) later used the same method to update the maximum mass of neutron stars to $2.9M_\odot$.

the mass limit,[13] and the eliminative reasoning can only be construed as a causal explanation in the most minimal sense. Indeed, it is unclear that the entity realist can even allow for a distinction of neutron stars and black holes at all on the basis of deriving an upper mass limit for neutron stars. The reason is that the distinction only makes sense on the basis of theory—GR. Applying Cartwright's stance on electrons, objects must *somehow* be theory-invariant to be eligible:

> [T]he electron is not an entity of any particular theory. In a related context van Fraassen asks if it is the Bohr electron, the Rutherford electron, the Lorenz electron or what. The answer is, it is the electron, about which we have a large number of incomplete and sometimes conflicting theories. (Cartwright 1983, 92)

Most descriptions of black holes, as well as the mass limit for neutron stars, are intrinsically linked to GR which limits the case for a Cartwright style entity realism about astrophysical black holes, unless direct experimental testing is available.[14] This last caveat may however be exploited by the entity realist by referring to multi-messenger astronomy.

15.4.1.1 Multi-Messenger Astronomy

The advent of gravitational wave astronomy has made it possible to cross check detection of dynamical events like black hole or neutron star mergers. The basic idea is that gravitational signals received in gravitational wave observatories (LIGO, VIRGO and KAGRA) provide the basis for an assessment of what kind of event, and what kind of objects, are the cause of the signals. One may then direct electromagnetic telescopes to the location in order to receive electromagnetic signals from the same event. The types of hypothesized events that are violent enough to create detectable gravitational waves are black hole mergers, neutron star mergers, and black hole neutron star mergers. The entity realist could then claim that this method can be used to decouple the concepts of neutron stars and black holes. The claim is grounded in the fact that the prediction of neutron star merger gravitational signals by GR can be corroborated by following up with electromagnetic observations in the entire EM spectrum (gamma-ray, X-ray, ultraviolet, optical, infrared, and radio wave). This novel kind of observation in multiple regimes was first deployed in the neutron star merger GW170817A on August 17, 2017. Gravitational waves were detected at the two US LIGO locations (coupled with a weaker "blindspot" signal at Virgo) followed by a brief gamma-ray burst detection in the Fermi space telescope seconds later. The GW signal detected

---

[13] "[...] the EOS at $\rho \gtrsim \rho_0$ cannot be reproduced in laboratory, and it cannot be calculated exactly because of the lack of the precise relativistic many-body theory of strongly interacting particles. Instead of the exact theory, there are many theoretical models. The reliability of these models decreases with growing ρ" (Haensel et al. 2007, 14).

[14] There are exceptions: see Kehagias and Sfetsos (2009) for solutions to black holes in non-relativistic gravity.

in LIGO and Virgo was not the short "chirp" associated with GW detection of a black hole merger, but a 100 second long signal. The difference of the signals coupled with the electromagnetic counterpart—the gamma-ray burst—were telling signs of a neutron star binary merger. The detections triggered scientists to do a follow up observation with the Hubble telescope to localize the source of the gamma-ray burst: a bright object in NGC 4993, a lenticular galaxy some 130 million light years away. The particularly striking part of GW170817 is the amount of data gathered by the following EM observations of the object. Over 70 observatories and telescopes were directed at the object, which radiated in all the frequencies of the EM spectrum. Had the binary system been a black hole merger, no such radiation would have been expected.

So, can the entity realist use this event, the first ever detected by multi-messenger techniques, in order to decouple neutron stars from black holes? Perhaps not. While the event may be used in order to allow for the existence of neutron stars (and other astrophysics, like the production of heavy elements like gold and platinum), the issue still boils down to *eliminative reasoning*. Since multi messenger astronomy cannot be used in order to directly detect black hole mergers (since they don't radiate), the only way to infer their existence is to eliminate the possibility that objects detected by gravitational waves are neutron stars. Even in such a well observed event as GW170817, this is a non-trivial matter:

> Gravitational-wave observations alone are able to measure the masses of the two objects and set a lower limit on their compactness, but the results presented here do not exclude objects more compact than neutron stars such as quark stars, black holes, or more exotic objects. The detection of GRB 170817A and subsequent electromagnetic emission demonstrates the presence of matter. (Abbott et al. 2017, 161101-2)

Given that the maximum mass estimates for neutron stars are uncertain and deeply theory driven, the existence of black holes are inferred because there are no other alternatives consistent with background theory, i.e. GR. While this inference is fine as an inference to the best (only?) explanation, it lacks the experimental flavor of causal inference that is central to Cartwright's account.

15.4.2 Chakravartty

For Chakravartty, the issue is whether black holes are "connected via causal processes to our instruments and other means of detection" (2007, 48).[15] X-rays, in the sense of being radiation, may fulfill this sort of relation, but that the detected X-ray *sources* are the product of accretion, either in the X-ray binary case for stellar black holes or in the AGN phenomena for SMBHs, is not detectable in the relevant sense. This is to say in the sense that we detect some phenomena over and

---

[15] In more recent work, Chakravartty (2017) develops his account further and connects it to metaphysical inference and dispositional realism, but the core of his 2007 remains intact.

above the radiation itself. That would be an additional, interpretative, step which requires modeling and theory informed inference. It would be a further step still to say that the X-ray sources should be coupled, again in the semi-realist sense of connected to our instruments, with black holes. The chain of inferences here may be taken to go from detection of X-ray radiation to accretion to black holes, where the only candidate step in the chain pertaining to the causal relation presented by Chakravartty is the first. Prima facie, black hole detection is not well suited to take place in the kind of realist account on offer. However, since semi-realists primarily speak of *properties*, rather than *objects* (even though the latter are coupled with the former), we may switch the target system of realism from black holes qua object and instead focus on its associated properties in order to see if those can be recovered in semi-realism. To do this would better reflect the purpose and metaphysical spirit of semi-realism. In such an analysis, it makes sense to use Chakravartty's spectrum of strength of causal interactions mapping to degrees of belief as a basis for determining the level of commitment that a semi-realist should have towards the properties of black holes. Here, Chakravartty provides a brief statement of the connection:

> In addition to a negative charge, [...] scientists associated many different properties with electrons. Enter semirealism, first and foremost a realism about well-detected properties. This refinement illuminates certain discriminations that are otherwise glossed over: they all believed in negative charge, and certain relations involving negative charge and particulars having it, but many of the other properties they associated with these particulars changed dramatically over the years as subatomic physics developed. And since on this view the realist understands properties in terms of dispositions for relations, there is no question of separating a knowledge of one from a knowledge of the other. A knowledge of entities and their relations is intimately connected here.(Chakravartty 2007, 58-9)

The charitable sentiment may be that while knowledge of entities and their relations cannot be separated, black hole realism may still be recovered if their properties in some sense can stand in a suitable causal relation to our instruments. However, the candidate properties of black holes most likely to be measurable—spin, mass, and charge—are not measurable in the way that Chakravartty needs them to be. Mass estimates use the dynamics of objects in the gravitational field of a black hole to derive a value, and spin is measured by using the hot X-ray gas at the heart of accretion disks. Both methods are dependent on theory in a way unsuitable to satisfy the causal connection condition, at least in way that would license realist commitment. Recall that "the greater the extent to which one seems able to interact with something—at best, manipulating it so as to bring about desired outcomes—the greater the warrant for one's belief in it" (Chakravartty 2007, 59).

Another property of black holes which is strongly endorsed by scientists is Hawking radiation, the eponymous thermodynamic glow theorized by Stephen Hawking (and Jacob Bekenstein). What, for present purposes, is most interesting about Hawking radiation is the level of acceptance it has despite the fact that it is decoupled from any empirical testing:[16]

> [Black Hole Thermodynamics] itself relies almost entirely on theoretical arguments, and its most celebrated result—Hawking's argument that black holes emit radiation—has no direct empirical support and little prospect of getting any. (Wallace 2018, 52)

Wallace argues that despite its disconnect with empirical data, there are good reasons to believe that black holes are thermodynamic systems. For semi-realism, however, this line of evidential reasoning regarding astrophysical black holes will fall far from the mark of realism, given its reliance on theoretical argument. The detection of Hawking radiation, by virtue of its extreme redshift, is not particularly likely to happen, so will be located at the very speculative end of Chakravartty's spectrum of causal contact cum belief (if eligible at all). The epistemological practices of astrophysics appears to greatly outstrip the semi-realist position, leading the latter to an anti-realism about a well established class of astrophysical objects—black holes—and their properties.

15.5 Concluding Remarks

For scientific realism, one of the core questions is what we can be realist about. Different varieties of realism have constructed different criteria for how we can arrive at an answer for this question. The debate over these criteria has for the most part consisted in anti-realists presenting counter examples to proposed accounts, to which realists have responded in kind. Realists have focused on recovering the right verdict with respect to cases either in history of science or in specific scientific areas, for example in particle physics. Curiously, they have neglected astrophysics, cosmology, and astronomy (Hacking excepted). Curious, since these fields jointly encompass the quantitatively (and arguably qualitatively) dominant part of our universe. An unforeseen consequence of this neglect is that the realist criteria have been shaped to square with a specific set of cases, and their extension to astrophysics was far from obvious. Here, I have attempted to ameliorate the opaqueness of this extension, arriving at the conclusion that the criteria for realism forwarded by entity realists are not a promising route for astrophysical realism. Perhaps this result is

---

[16] There may be other epistemic paths to knowledge about Hawking radiation, although it is unclear to what extent it would amount to detection. One class of such paths are analogue experiments with dumb holes in which certain black hole properties, in particular Hawking radiation, are disclosed or inferred by their analogue counterpart: "Our first core claim is that whether a theory regarding certain phenomena can be well supported or established by experiment is not constrained by the requirement that the target system displaying these phenomena be manipulable or accessible, either in principle or practice." (Evans and Thébault 2020, 2) This claim would be able to provide support for realism beyond causal detection as specified by semi-realism, but would of course also violate or alter its conceptual core.

a bullet realists think is worth biting, as Hacking thought. If it is not, realists may have to consider a formulation of their realist criteria based on the contemporary epistemic practices of science.

References

Abbott, B.P., R. Abbott, T. Abbott, F. Acernese, K. Ackley, C. Adams, T. Adams, P. Addesso, R. Adhikari, V.B. Adya, et al. 2017. Gw170817: Observation of gravitational waves from a binary neutron star inspiral. *Physical Review Letters* 119(16): 161101.

Allzén, S. 2021. Scientific realism and empirical confirmation: A puzzle. *Studies in History and Philosophy of Science Part A* 90: 153–159.

Anderl, S. 2015. Astronomy and astrophysics in the philosophy of science. Preprint. arXiv:1510.03284.

Carroll, S.M. 2019. *Spacetime and geometry*. Cambridge University Press.

Cartwright, N. 1983. *How the laws of physics lie*. Oxford University Press.

Celotti, A., J.C. Miller, and D.W. Sciama. 1999. Astrophysical evidence for the existence of black holes. *Classical and Quantum Gravity* 16(12A): A3.

Chakravartty, A. 2007. *A metaphysics for scientific realism: Knowing the unobservable*. Cambridge University Press.

Chakravartty, A. 2017. *Scientific ontology: Integrating naturalized metaphysics and voluntarist epistemology*. Oxford University Press.

Collmar, W., N. Straumann, S.K. Chakrabarti, G. 't Hooft, E. Seidel, and W. Israel. 1998. Panel discussion: The definitive proofs of the existence of black holes. In *Black holes: Theory and observation*, F.W. Hehl, C. Kiefer, and R.J. Metzler, eds., 481–489. Berlin, Heidelberg: Springer Berlin Heidelberg.

Curiel, E. 2019. The many definitions of a black hole. *Nature Astronomy* 3(1): 27–34.

Dawid, R. 2007. Scientific realism in the age of string theory. *Physics and Philosophy* ID: 11.

Dawid, R. 2013. *String theory and the scientific method*. Cambridge University Press.

Dawid, R. 2016. Modelling non-empirical confirmation. In *Models and inferences in science*, 191–205. Springer.

Dawid, R., S. Hartmann, and J. Sprenger. 2015. The no alternatives argument. *The British Journal for the Philosophy of Science* 66(1): 213–234.

Duhem, P.M.M. (1914/1991). *The aim and structure of physical theory*, volume 13. Princeton University Press.

Evans, P.W., and K.P. Thébault. 2020. On the limits of experimental knowledge. *Philosophical Transactions of the Royal Society A* 378(2177): 20190235.

Frampton, P.H. 2016. The primordial black hole mass range. *Modern Physics Letters A* 31(12): 1650064.

Genzel, R., A. Eckart, T. Ott, and F. Eisenhauer. 1997. On the nature of the dark mass in the centre of the milky way. *Monthly Notices of the Royal Astronomical Society* 291(1): 219–234.

Gies, D., and C. Bolton. 1986. The optical spectrum of hde 226868= cygnus x-1. ii spectrophotometry and mass estimates. *The Astrophysical Journal* 304: 371–393.

Guala, F. 2002. Models, simulations, and experiments. In *Model-based reasoning*, 59–74. Springer.

Hacking, I. 1983. *Representing and intervening: Introductory topics in the philosophy of natural science*. Cambridge University Press.

Hacking, I. 1989. Extragalactic reality: The case of gravitational lensing. *Philosophy of Science* 56(4): 555–581.

Haensel, P., A.Y. Potekhin, and D.G. Yakovlev. 2007. *Neutron stars 1*. Springer.

Jacquart, M. 2020. Observations, simulations, and reasoning in astrophysics. *Philosophy of Science* 87(5): 1209–1220.

Jacquart, M. 2021. Dark matter and dark energy. In *The Routledge companion to philosophy of physics*, 731–743. Routledge.

Kalogera, V., and G. Baym. 1996. The maximum mass of a neutron star. *The Astrophysical Journal* 470(1): L61.

Kehagias, A., and K. Sfetsos. 2009. The black hole and FRW geometries of non-relativistic gravity. *Physics Letters B* 678(1): 123–126.

Laudan, L. 1981. A confutation of convergent realism. *Philosophy of Science* 48(1): 19–49.

Martens, N. 2022. Dark matter realism. *Foundations of Physics* 52(1): 1–19.

Merritt, D. 2021. Cosmological realism. *Studies in History and Philosophy of Science Part A* 88: 193–208.

Morgan, M.S. 2005. Experiments versus models: New phenomena, inference and surprise. *Journal of Economic Methodology* 12(2): 317–329.

Murdin, P. 2001. *Encyclopedia of astronomy & astrophysics*. CRC Press.

Parke, E.C. 2014. Experiments, simulations, and epistemic privilege. *Philosophy of Science* 81(4): 516–536.

Parker, W.S. 2009. Does matter really matter? Computer simulations, experiments, and materiality. *Synthese* 169(3): 483–496.

Psillos, S. 2008. Cartwright's realist toil: From entities to capacities. In *Nancy Cartwright's philosophy of science*, 167–194. Routledge.

Rhoades Jr, C.E., and R. Ruffini. 1974. Maximum mass of a neutron star. *Physical Review Letters* 32(6): 324.

Rothschild, R., E. Boldt, S. Holt, and P. Serlemitsos. 1974. Millisecond temporal structure in cygnus x-1. *The Astrophysical Journal* 189: L13.

Sandell, M. 2010. Astronomy and experimentation. *Techné: Research in Philosophy and Technology* 14(3): 252–269.

Shapere, D. 1993. Astronomy and antirealism. *Philosophy of Science* 60(1): 134–150.

Van Fraassen, B.C. 1980. *The scientific image*. Oxford University Press.

Wallace, D. 2018. The case for black hole thermodynamics part i: Phenomenological thermodynamics. *Studies in History and Philosophy of Science Part B: Studies in History and Philosophy of Modern Physics* 64: 52–67.

Part IV
Concluding Thoughts

Chapter 16
Reflections by a Theoretical Astrophysicist

Kevin Heng

Abstract A theoretical astrophysicist discusses the principles and rules-of-thumb underlying the construction of models and simulations from the perspective of an active practitioner, where it is emphasised that they are designed to address specific scientific questions. That models are valid only within a restricted space of parameters and degenerate combinations of parameter values produce the same observable outcome are features, and not bugs, of competent practice that fit naturally within a Bayesian framework of inference. Idealisations within a model or simulation are strongly tied to the questions they are designed to address and the precision at which they are confronted by data. If the practitioner visualises a hierarchy of models of varying sophistication (which is standard practice in astrophysics and climate science), then de-idealisation becomes an irrelevant concept. Opportunities

I am a tenured full professor of theoretical astrophysics. During an exploratory phase of my career (2003—2011), I published peer-reviewed papers on supernova remnants (Heng et al. 2006; Heng and McCray 2007), gamma-ray burst afterglows (Heng et al. 2007, 2008b), X-ray astronomy (Heng et al. 2008a), accretion onto neutron stars (Heng and Spitkovsky 2008), gravitational lensing (Heng and Keeton 2009), planet formation (Heng and Kenyon 2010) and debris disks (Heng and Tremaine 2010). However, the only topic I consider myself to be an expert on is the study of the atmospheres of exoplanets (Heng 2017), which requires mastery of radiative transfer, fluid dynamics, atmospheric chemistry and Bayesian inference. In my work on astrophysics, I construct models (e.g., Heng et al. 2021) and perform simulations (e.g., Heng et al. 2011). I have also constructed infectious disease models (for COVID) in collaboration with epidemiologists and medical doctors (Chowdhury et al. 2020; Heng and Althaus 2020). At the time of publication of this volume, I have published about 170 peer-reviewed papers and 1 textbook, which have garnered over 10,000 citations (h-index of 58; according to Google Scholar). Phenomenology, which is the confrontation of models with data, is an essential part of my research profile and interests.

K. Heng (✉)
Faculty of Physics, Ludwig Maximilian University, Munich, Germany

Department of Physics, Astronomy & Astrophysics Group, University of Warwick, Coventry, UK

ARTORG Center for Biomedical Engineering Research, University of Bern, Bern, Switzerland
e-mail: Kevin.Heng@physik.lmu.de

© The Author(s) 2023

N. Mills Boyd et al. (eds.), *Philosophy of Astrophysics*, Synthese Library 472,
https://doi.org/10.1007/978-3-031-26618-8_16

for future collaborations between astrophysicists and philosophers of science are suggested.

Models are designed by the practitioner to answer specific scientific questions. There is no such thing as a "universal model" that is able to emulate the natural world in all of its detail and provide an answer to every question the practitioner asks. Even if one could, in principle, write down some universal wave function of the Universe, this is useless for the practising astrophysicist seeking to understand the natural world by confronting models with data—it is computationally intractable if one wishes to understand complex, non-linear systems, where the interplay between various components of the system is often the most interesting outcome. If buying a larger computer were the solution to understanding these complex systems, we would have solved biology and economics by now. One example of a spectacular, failed attempt at constructing an emulation is the billion-euro Human Brain Project, which attempted to replace laboratory experiments aimed at studying the human brain with all-encompassing computer simulations. Another fundamental obstacle with emulations, even if we could construct them, is that they merely produce *correlations*. To transform these correlations into statements on cause-and-effect requires theoretical understanding.

One of the first skills a competent theorist learns is to ask when her or his model breaks. What are the assumptions made? What is the physical regime beyond which the model simply becomes invalid? What are the scientific questions one may (or may not) ask of the model? It is the job of the theorist to be keenly aware of these caveats. To give concrete examples: if one's scientific question is, "What is the structure of the water molecule," then one solves the Schrödinger equation. If one's scientific question is instead, "What is the behaviour of waves within a body of water," then one solves the Navier-Stokes equation. For the latter, it is understood that one cannot ask questions on length scales shorter than the mean free path of collisions between water molecules—or on time scales shorter than the collisional time between water molecules. A similar reasoning applies to why one is able to simulate the behaviour of dark matter on large scales—without knowing what dark matter actually is. This is because, in these simulations, one is forbidden from actually querying the nature of dark matter—this is an *input*, rather than output, of the simulation. Another rule-of-thumb that all competent theoretical astrophysicists who run simulations know well is: one gets out what one puts in. Or to put it more colloquially: garbage in, garbage out. This rule-of-thumb bears some resemblance to what philosophers of science term "robustness analysis".

While it is tempting to separate the practice of science from the science itself, the *skill level* of the practitioner is an aspect that philosophers of science cannot ignore. Not all theorists or modellers should be placed on the same footing. For example, questioning what philosophers term the "theory-ladenness" of an observation, which one is interpreting, is a *skill* that is honed over years of practice. There are time-honoured "best practices" in astronomy and astrophysics that do not always make their way into the peer-reviewed literature. Only by interacting with practitioners will philosophers of science uncover them. On short time scales, sensationalism and frivolity may enter our peer-reviewed literature. On longer time scales, our peer-reviewed literature has the tendency to self-correct; practitioners are fairly conservative about what we term "standard" (methods or approaches).

It is instructive to elucidate the *intention* of the practitioner when constructing simulations. Not all simulations are constructed with the same goals. In the grandest sense, one would like to simulate the full temporal and spatial evolution of some system or phenomenon. But sometimes the goals are more modest. The practitioner starts with studying the system on paper and ponders how various physical (or chemical) effects interact with one another. If all of these effects have comparable time scales (or length scales), it implies that they exert comparable influences on the outcome. One is then solving for a complex steady state produced by the interplay between different physical effects, which are often highly non-linear. Simulating the long-term climate of a planet is one such example. If sufficient empirical data are present, one may also incorporate them as initial or boundary conditions in order to predict the future, short-term behaviour of a system. Weather prediction simulations are such an example. In astrophysics, a common goal is to study trends in the predicted observables and how they depend on varying the various input parameters—what the philosophers call "intervention", which we simply term a parameter study or sweep. As simulations are often computationally expensive, few practitioners would claim that any suite of simulations being computed is complete. Rather, the goal is to elucidate trends and (hopefully) understand the underlying physical mechanisms.

As a practitioner of simulations, I consider the "Verification and Validation" framework to be an unattainable dream. Verification has the ideal that one should compare the simulations against all possible analytical solutions in order to establish their accuracy. The fundamental obstacle is that non-linear analytical solutions are rare, e.g., the solution for solitons. Unfortunately, one often runs a simulation precisely because one is interested in the non-linear outcome! If one adheres to this ideal of verification, no simulation will ever be fully verified—and hence such an ideal is irrelevant to the practitioner (and will thus be ignored in practice). Rather, the practitioner often speaks of benchmarking, where one agrees on an imperfect test that multiple practitioners should attempt to reproduce. Agreement simply implies consistency—but there is a possibility that these practitioners could have all *consistently* obtained the *wrong* answer. By contrast, when a practitioner uses the term "validation" it means that one is comparing the simulation to an absolute ground truth—either provided by data or mathematics. In astrophysics, these ground truths are hard to come by. One example of validation is the Held-Suarez test for producing a simple climate state of Earth (without seasons), which was motivated by climate scientists wishing to verify the consistency of simulation codes operated by different laboratories (Held and Suarez 1994).

As a professional maker of models, I find the debate about "fictions" to be puzzling. All governing equations of physics involve approximations—even if one is unaware of them being built into the equations. Rather than visualise a universal model, it is much more useful to think of a *hierarchy* of models of varying sophistication, which is standard practice in climate science (Held 2005) and used widely in astrophysics. Each model in the hierarchy incorporates a different set of simplifying assumptions designed to answer specific questions. If one's scientific question is to understand the evolution of stars over cosmic time scales, then approximations such as spherical symmetry are not unreasonable. However, if one's scientific question is to understand the density structure of stars by studying how sound waves propagate across them over comparatively shorter time scales, then more elaborate models need to be constructed. The question is not whether stars are perfectly spherical—they certainly are not. The real question is: what is the magnitude of the correction to spherical symmetry and how does this affect the accuracy of one's answer for addressing a specific scientific question? Simplicity is intentionally built into these models, because it allows one to more cleanly identify cause and effect, rather than simply recording correlated outcomes in a simulation.

While Einstein's equations of relativity supercede Newton's equations in principle, it is sufficient to solve the latter if one wishes to understand the orbits of exoplanets. While the theoretical foundation of thermodynamics is provided by statistical mechanics, implementing thermodynamics in one's model or simulation is often sufficiently accurate for the scientific question being asked. In the previous example given, it would be unnecessary (and infeasible) to simulate large-scale fluid behaviour by numerically solving the Schrödinger equation. Models are not constructed in an absolute sense. In addition to addressing specific scientific questions, they are constructed to facilitate effective comparison with data—at the quality and precision available at that time. In other words, one cannot discuss models without also discussing the associated errors in comparison to data. Speaking in generalities without quantitative estimates of the approximations and tying them to the specific scientific question being addressed is not useful for the practising theoretical astrophysicist. If one approaches modelling from the perspective of a model hierarchy, what philosophers of science term "de-idealisation" is simply irrelevant, because each member of the hierarchy employs a different degree of idealisation.

In the use of similarity arguments to justify how terrestrial experiments may mimic celestial systems, one should note that similarity may be broken by introducing physical effects that encode intrinsic length scales. To use the well-known Rayleigh-Taylor instability as an example, if one introduces surface tension to the calculation of the fluid then a minimum length scale for features in the flow appears. If one introduces gravity, then a maximum length scale appears. Similarity only appears when one is asking a scientific question that is justified by treating the system purely as a fluid, but radiation, chemistry and other effects exert non-negligible influences in real astrophysical systems.

In confronting models with data, the modern approach is to use Bayesian inference. When multiple combinations of parameter values yield the same observable outcome, this is known as a model degeneracy. Degeneracies are a *feature*—and not a bug—of models. The formal way of quantifying degeneracies is to compute the joint posterior distributions between parameters—a standard feature of Bayesian inference. Testing if the data may be explained by *families* of models and penalising models that are too complex for the quality and precision of data available is a natural outcome of Bayesian model comparison (Trotta 2008). In other words, Bayesian model comparison is the practitioner's *quantitative* method for implementing Occam's Razor. Combining the use of Bayesian model comparison with the construction of a model hierarchy is how modern astrophysics approaches problem solving and the confrontation of models with data. Another feature of Bayesian inference is the specification of prior distributions, which reflect one's state of knowledge of the system or phenomenon at that point in time. A skilled theorist is keenly aware of when the answer to a scientific question is *prior-dominated*—again, one gets out what one puts in.

What is the over-arching goal of the theoretical astrophysicist? Certainly, Nature has laws and our models need to abide by them. The construction of models always has unification as a goal—if I observe N phenomena and I need N classes of models to describe them, then I have failed. Our goal is to advance our understanding of Nature on celestial scales—whether by the use of theory, simulation, observation or experiment. The most useful models are the ones we can falsify using data, because they teach us important lessons about the system we are studying. The sparseness of data in astronomy for any single object is not a bug, but a feature—it is a reality of astronomical data that we have to live with. This requires us to adjust our thinking: instead of asking intricate questions about a single object, we often have to ask questions of the *ensemble* of objects. Instead of tracking a single object or system across time, we have to contend with studying an ensemble of objects at a specific point in time—akin to an astronomical version of the ergodic principle. Such a property distinguishes astrophysics from the rest of physics. I would argue that questions of the ensemble are no less interesting or fundamental, e.g., what fraction of stars host exoplanets and civilisations? A potentially fruitful future direction for astrophysicists and philosophers of science to collaborate on is to combine ensemble thinking and model hierarchy building with thinking deeply about the detection versus auxiliary properties of a system or phenomenon.

Some fundamental issues are missing from the debate about simulations that are often dismissed by philosophers of science as belonging to the realm of practice or implementation. To set up any simulation, the governing equations of physics, which describe continuous phenomena, need to be discretised before they are written into computer code. The very act of discretisation introduces challenges that are ubiquitous to computer simulations, such as an artificial, unphysical form of dissipation that cannot be specified from first principles. Such numerical "hyper-parameters" severely impact the predictive power of simulations, further casting doubt on the analogy between simulations and experiments. Furthermore, simulations often suffer from a "dynamic range" problem—Nature has infinite resolution, but in order to run any simulation within one's lifetime one has to specify minimum and maximum length scales of the simulated system. The practitioner can never implement an *emulation*, where all relevant length and time scales are captured in the simulation. There is often crucial physics (e.g., turbulence) occurring below the smallest length scale simulated: so-called "sub-grid physics". It is not uncommon to have simulated outcomes being driven by one's prescription of sub-grid physics. One example that affects the study of brown dwarfs, exoplanets, climate science, etc, is how clouds form on small length scales. The problems of dynamic range and numerical hyper-parameters are widely debated by practitioners and are relevant to the debate on the epistemic value of computer simulations.

I would like to end with an unsolved problem in physics (and astrophysics) that I find fascinating, but to date has not received much attention from philosophers of science. It concerns our incomplete understanding of turbulence, which is considered to be an important subfield of modern astrophysics. The Nobel laureate and physicist Werner Heisenberg once allegedly remarked, "When I meet God, I am going to ask him two questions: Why relativity? Why turbulence? I really believe he will have an answer for the first." The fascinating thing about turbulence is that we have all of the tools and data at our disposal: we have the Navier-Stokes equation, the ability to perform laboratory experiments, astronomical observations of turbulence on a dazzling range of length scales and all of the computational power to simulate it in computers. Yet, despite decades of research, we do not have a complete theory of turbulence. If we did, then we would be able to exactly calculate the threshold Reynolds number for any flow to transition from being laminar to turbulent—and calculate the variation in this dimensionless fluid number as the geometry and boundary conditions of the system change. We would also be able to understand why some turbulent flows are intermittent. Currently, the determination of these phenomena remains an engineering exercise. Studying why we are unable to understand turbulence will potentially yield valuable insights for philosophers of science on the epistemic value of theory, simulation, observation and experiment—and how these different approaches need one another in order to advance our understanding.

References

Chowdhury, R., et al. 2020. Dynamic Interventions to Control COVID-19 Pandemic: A Multi-variate Prediction Modelling Study Comparing 16 Worldwide Countries. *European Journal of Epidemiology* 35: 389.

Held, I.M. 2005. The Gap Between Simulation and Understanding in Climate Modeling. *Bulletin of the American Meteorological Society* 86: 1609.

Held, I.M., and M.J. Suarez. 1994. A Proposal for the Intercomparison of the Dynamical Cores of Atmospheric General Circulation Models. *Bulletin of the American Meteorological Society* 75: 1825.

Heng, K. 2017. *Exoplanetary Atmospheres: Theoretical Concepts & Foundations* Princeton: Princeton University Press.

Heng, K., and C.L. Althaus. 2020. The Approximately Universal Shapes of Epidemic Curves in the Susceptible-Exposed-Infectious-Recovered (SEIR) Model. *Scientific Reports, 10*, 19365

Heng, K., and C.R. Keeton. 2009. Planetesimal Disk Microlensing. *Astrophysical Journal* 707: 621.

Heng, K., and S.J. Kenyon. 2010. Vortices as Nurseries for Planetesimal Formation in Protoplanetary Discs. *Monthly Notices of the Royal Astronomical Society* 408: 1476.

Heng, K., and R. McCray. 2007. Balmer-Dominated Shocks Revisited. *Astrophysical Journal* 654: 923.

Heng, K., and A. Spitkovsky. 2008. Magnetohydrodynamic Shallow Water Waves: Linear Analysis. *Astrophysical Journal* 703: 1819.

Heng, K., and S. Tremaine. 2010. Long-Lived Planetesimal Discs. *Monthly Notices of the Royal Astronomical Society* 401: 867.

Heng, K., et al. 2006. Evolution of the Reverse Shock Emission from SNR 1987A. *Astrophysical Journal* 644: 959.

Heng, K., D. Lazzati, and R. Perna. 2007. Dust Echoes from the Ambient Medium of Gamma-Ray Bursts. *Astrophysical Journal* 662: 1119.

Heng, K., et al. 2008a. Probing Elemental Abundances in SNR 1987A Using XMM-Newton. *Astrophysical Journal* 676: 361.

———. 2008b. A Direct Measurement of the Dust Extinction Curve in an Intermediate-Redshift Galaxy. *Astrophysical Journal* 681: 1116.

Heng, K., K. Menou, and P.J. Phillipps. 2011. Atmospheric Circulation of Tidally Locked Exoplanets: A Suite of Benchmark Tests for Dynamical Solvers. *Monthly Notices of the Royal Astronomical Society* 413: 2380.

Heng, K., B.M. Morris, and D. Kitzmann. 2021. Closed-Form ab Initio Solutions of Geometric Albedos and Reflected Light Phase Curves of Exoplanets. *Nature Astronomy* 5: 1001.

Trotta, R. 2008. Bayes in the Sky: Bayesian Inference and Model Selection in Cosmology. *Contemporary Physics* 49: 71.

Chapter 17
Annotated Bibliography

Cameron C. Yetman ⓘ

Abstract The following annotated bibliography contains a reasonably complete survey of contemporary work in the philosophy of astrophysics. Spanning approximately 40 years from the early 1980s to the present day, the bibliography should help researchers entering the field to acquaint themselves with its major texts, while providing an opportunity for philosophers already working on astrophysics to expand their knowledge base and engage with unfamiliar material.

17.1 Introduction

The bibliography is divided into seven sections. The first section (17.2) covers methodological issues in astrophysics: how do astrophysicists make observations, interpret data, and solve problems that arise in the process? The section includes case studies on gravitational waves, astroparticle physics, dark matter, extra-galactic objects, and others.

The second and largest section (17.3) covers topics related to astrophysical modelling and computer simulations: their epistemic value, their limits, and their application to major problems in the field. This section contains case studies on galactic modelling, analogue experiments, cosmological simulations, and code comparisons.

The third section (17.4) concerns perhaps the oldest debate within the contemporary philosophy of astrophysics, namely between astrophysical realists and anti-realists, which was initiated by the work of Ian Hacking in the 1980s. The

---

The following text is a revision and expansion of an annotated bibliography the author compiled in 2021 for Drs. Pauline Barmby and Francesca Vidotto, Rotman Institute of Philosophy.

C. C. Yetman (✉)
Department of Philosophy, University of Toronto, Toronto, Canada
e-mail: cameron.yetman@mail.utoronto.ca; https://philpeople.org/profiles/cameron-yetman

N. Mills Boyd et al. (eds.), *Philosophy of Astrophysics*, Synthese Library 472,
https://doi.org/10.1007/978-3-031-26618-8_17

section contains case studies on astroparticle physics, gravitational lensing, dark matter, as well as stellar physics and classification.

The fourth section (17.5) covers the relationships between astrophysical theory, observation, confirmation, and more. This section contains case studies on singularities, general relativity, dark matter, interstellar interlopers, and gravitational waves.

The fifth section (17.6) is somewhat tangential to mainstream work in the philosophy of astrophysics, but nevertheless contains a number of articles of which philosophers should be aware and with which they should be prepared to engage. The section covers issues in the sociology of scientific knowledge (SSK), as well as other social issues related to astrophysics. The section contains case studies on gravitational waves, the Hubble Space Telescope, astronomy and high-energy physics, the Herschel Space Observatory, "star-crushing", the Gemini Telescopes, Pluto, and the use of visualizations.

The sixth section (17.7) contains works on typicality, the anthropic principle, and extra-terrestrial life. Most of the existing literature on typicality has an explicitly cosmological focus, but philosophers of astrophysics may offer fresh perspectives to these debates. The articles were chosen due to their potential interest for philosophers in this field, though few have explicit astrophysical content. Section six, then, serves as an invitation for philosophers of astrophysics to explore a field largely untouched by those with their knowledge and skillset.

The seventh and final section (17.8), compiled by Siska De Baerdemaeker, explores recent work related to dark matter and MOND (Modified Newtonian Dynamics) on both astrophysical and cosmological scales. This section is by no means a comprehensive overview of the philosophical literature on MOND, but the entries included have been chosen for their specific relevance to the philosophy of astrophysics.

At the end of each section, there is a list of articles which deal with the section's theme, but whose primary theme warranted placing them somewhere else.

The reader will notice a number of articles which focus on cosmology or astronomy, rather than astrophysics. These articles were chosen in virtue of their potential applicability to problems in the philosophy of astrophysics, as judged by myself in discussion with other philosophers in the field. Every effort was made to avoid inflating the bibliography beyond its natural bounds, but some material from adjacent fields was necessary to provide a comprehensive overview of the state and future of the philosophy of astrophysics.

Especially given the relatively small size and recent vintage of this field (there are only 87 entries in this bibliography, 66 of which are from 2010 onwards, and 32 since 2020) the articles in this volume constitute a significant and timely addition.

17.2 Methodologies in Astrophysics

Anderl, S. (2016). Astronomy and astrophysics. In P. Humphreys (Ed.), *The Oxford Handbook of Philosophy of Science* (Vol. 1). Oxford University Press. https://doi.org/10.1093/oxfordhb/9780199368815.013.45.

A comprehensive, readable introduction to the main debates in philosophy of astronomy and astrophysics, this article offers a great starting point for those new to the field. Anderl (*Frankfurter Allgemeine Zeitung;* Institut de Planétologie et d'Astrophysique de Grenoble) argues that astrophysics is not vulnerable to Ian Hacking's charge of antirealism due to its unique methodology, which incorporates aspects of both the historical and experimental sciences (including the "cosmic laboratory"), as well as simulations, models, and analyses of large amounts of data.

Cleland, C. E. (2002). Methodological and epistemic differences between historical science and experimental science. *Philosophy of Science, 69*(3), 447–451. https://doi.org/10.1086/342455.

A useful introduction to the distinction between historical and experimental sciences – a distinction central to debates over the reliability of astrophysical findings. Cleland (CU Boulder) contends that the different kinds of evidential reasoning practiced by experimental and historical scientists are underwritten by an objective feature of nature, namely, the time asymmetry of causation between present and past events, and present and future events. Historical sciences exploit information about the present-past events, while experimental science exploits information about present-future events. Thus, each type of science is doing something different, and neither is more objective or rational than the other.

De Baerdemaeker, S. (2021). Method-driven experiments and the search for dark matter. *Philosophy of Science, 88*(1), 124–144. https://doi.org/10.1086/710055.

Given target X, how do scientists argue that their method(s) will be effective in probing X? De Baerdemaeker (Stockholm University) discerns two "logics" of method choice, namely "target-driven" and "method-driven", and argues that scientists employ the latter in situations where previous knowledge about the target system is sparse or unreliable, as illustrated by dark matter production and detection experiments. However, the use of method-driven logic poses difficulties for the employment of traditional robustness arguments due to the assumptions involved in using this logic.

Elder, J. (2020). *The Epistemology of Gravitational-Wave Astrophysics.* Ph.D. dissertation. University of Notre Dame. https://curate.nd.edu/show/3f462517k8t.

The first comprehensive study in the epistemology of gravitational wave (GW) astrophysics, Elder (Black Hole Initiative) discusses the distinction between "direct" and "indirect" observations of gravitational waves, raises a circularity problem facing model-dependent observations (and explains how it is mitigated by GW astronomers), and elaborates on the virtues of multi-messenger astrophysics for creating more robust dependency relations between sources and traces of data, among other topics.

Elder, J. (2022). On the "direct detection" of gravitational waves [unpublished manuscript]. https://www.jameeelder.com/uploads/1/2/1/6/121663585/elder__2021__direct_detection_du_cha%CC%82telet.pdf.

The authors of the LIGO-Virgo collaboration's "discovery paper" for the binary black hole merger GW150914 claim to have made a "direct detection" of gravity waves and a "direct observation" of the merger. Elder (Black Hole Initiative) seeks to disambiguate the meaning of terms like "direct", "indirect", "observation" and "measurement" in a way which is both philosophically adequate and true to how scientists use these terms. Elder argues that the LIGO-Virgo team can only be said to have *indirectly* detected a binary black hole merger due to their reliance on model-based inferences, thereby raising some important epistemic challenges that gravitational wave astrophysicists must overcome.

Falkenburg, B. (2014). On the contributions of astroparticle physics to cosmology. *Studies in History and Philosophy of Science Part B: Studies in History and Philosophy of Modern Physics, 46*, 97–108. https://doi.org/10.1016/j.shpsb.2013.10.004.

Although cosmology proceeds top-down (from theory to data and from large-scale to small scale) and astroparticle physics proceeds bottom-up (from detection of particles to theorizing about their cosmic sources), Falkenburg (TU Dortmund) argues that these disciplines pursue complementary strategies of scientific explanation while aiming at theoretical unification – a fact inadequately captured by contemporary philosophical accounts of scientific explanation and realism. Given this, Falkenburg urges the philosophical community to pay greater attention to astroparticle physics and the way in which it contributes to the empirical basis of cosmology.

Hudson, R. G. (2007). Annual modulation experiments, galactic models and WIMPs. *Studies in History and Philosophy of Science Part B: Studies in History and Philosophy of Modern Physics, 38*(1), 97–119. https://doi.org/10.1016/j.shpsb.2006.05.002.

Groups studying WIMPs have generated apparently incompatible data. Hudson (University of Saskatchewan) argues that this data is only incompatible given certain ancillary assumptions involved in data processing, and that we can reconcile the discordant results into an empirically adequate model *à la* van Fraasen (we cannot be realists about this model).

Hudson, R. G. (2009). The methodological strategy of robustness in the context of experimental WIMP research. *Foundations of Physics, 39*(2), 174–193. https://doi.org/10.1007/s10701-009-9271-3.

Although central to the methodologies of sciences like psychology, robustness is not valued as highly among astroparticle physicists, who often pursue alternative strategies such as "model-independence" in assuring the reliability of their results. Hudson (University of Saskatchewan) contends that in these experimental contexts, robustness may be pragmatically fruitful (it may give us multiple lines of support to fall back on in response to countervailing evidence) while adding no epistemic value.

Hudson, R. G. (2013). Dark matter and dark energy. In R. Hudson, *Seeing Things: The Philosophy of Reliable Observation.* https://doi.org/10.1093/acprof:oso/9780199303281.001.0001.

Using 2006 observations of the Bullet Cluster and mid- to late-1990s observations of Type 1a supernovae as his case studies, Hudson (University of Saskatchewan) argues that robustness reasoning does not play a significant justificatory role in astrophysical theorizing about dark matter or dark energy. Instead, Hudson contends that astrophysicists in these contexts employ the epistemically meritorious methodological strategy of "targeted testing", wherein multiple techniques are used to address an observational question (*à la* robustness) but where alternate techniques are aimed at a specific "strategic goal". For Hudson, mere convergence of results should *not* be considered epistemically significant in the absence of this targeted approach, despite how some astrophysicists have reflectively justified their conclusions.

Meskhidze, H. (2021). Can machine learning provide understanding? How cosmologists use machine learning to understand observations of the universe. *Erkenntnis.* https://doi.org/10.1007/s10670-021-00434-5.

Can cosmological "black-box" machine leaning algorithms provide genuine scientific understanding? Meskhidze (UC Irvine) distinguishes between black-boxes themselves and black-boxing as a methodology – what she calls the "method of ignoration" – and argues that machine learning algorithms can deliver scientific understanding when they are used as part of this "method of ignoration" to investigate emergent statistical relations in the simulations within which they are employed. More broadly, Meskhidze contends that the epistemic value of machine learning algorithms is heavily context-dependent.

Salmon, W. C. (1998). Quasars, causality, and geometry: A scientific controversy that did not occur. In W. Salmon, *Causality and Explanation.* Oxford University Press. https://doi.org/10.1093/0195108647.003.0026.

Astrophysicists have argued on the basis of a "causal argument" that the rapid variability in the brightness of quasars requires that their sources be extremely compact. Salmon (d. 2001, form. University of Pittsburgh) identifies the "cΔt size criterion" – according to which the region of brightness-variation cannot be larger than the distance light travels in its time of variation – as a crucial premise in this causal argument. Salmon claims that scientists have treated this criterion (or at least have often appeared to treat it) as a law of nature derived from special relativity, but that in fact it is "egregiously fallacious". If the criterion has any use at all, it is as a plausibility principle for fixing

Bayesian priors when attempting to construct quasar models, and *not* as a physical requirement which such models must satisfy.

Shapere, D. (1982). The concept of observation in science and philosophy. *Philosophy of Science, 49*(4), 485–525. https://doi.org/10.1086/289075.

A classic and wide-ranging discussion of observation and inference in science which uses the detection of solar neutrinos as its primary case study. Shapere (d. 2016, form. Wake Forest University) contends that philosophical skepticism regarding the use of the term "observation" in astrophysics (for instance, in the claim that solar neutrinos allow us to "observe" the sun's interior) and other domains is unwarranted, especially since even ordinary and uncontroversial cases of observation involve inference and filtering through one's beliefs and background context.

Valore, P., Dainotti, M. G., & Kopczyński, O. (2020). Ontological categorizations and selection biases in cosmology: The case of extra galactic objects. *Foundations of Science.* https://doi.org/10.1007/s10699-020-09699-5.

Using Gamma Ray Bursts (GRBs) as a case study, the authors argue that philosophical analysis of ontological categorizations in astrophysics can help illuminate the limits and distortions of our scientific methods, as well as the theoretical and metaphysical presuppositions which undergird them and our understanding of reality as a whole.

Weinstein, G. (2021). Coincidence and reproducibility in the EHT black hole experiment. *Studies in History and Philosophy of Science Part A, 85*, 63–78. https://doi.org/10.1016/j.shpsa.2020.09.007.

Weinstein (University of Haifa) analyzes the Event Horizon Telescope (EHT) black hole experiment in light of philosophical themes from Ian Hacking, Nancy Cartwright, and Peter Galison. The author argues that EHT scientists employed an "argument from coincidence" in order to establish trust in their results, but that this method is problematic when used for this purpose.

Wilson, K. (2021). The case of the missing satellites. *Synthese, 198*(S21), 1–21 https://doi.org/10.1007/s11229-017-1509-6.

Wilson (University of Melbourne) provides an overview of the missing satellites problem in galactic astrophysics and analyzes how researchers have attempted to solve the problem. According to Wilson, these researchers have "blackboxed" their simulations by treating them as self-contained worlds in which simulated phenomena are epistemically significant, and they have blended these simulated results with real-world observations in generating their solution to the problem. This process of blending can make simulated worlds not merely possible, but *plausible.*

For further articles relevant to this category, see Boyd 2018, Curiel 2019, De Baerdemaeker and Boyd 2020, Gueguen 2020, Gueguen 2021, Massimi 2018, and Meskhidze 2017.

17.3 Models and Simulations

Anderl, S. (2018). Simplicity and simplification in astrophysical modeling. *Philosophy of Science, 85*(5), 819–831. https://doi.org/10.1086/699696.

> Should astrophysical models strive to be "complete" (i.e., to capture all the details of the available data), or simple? Anderl (*Frankfurter Allgemeine Zeitung;* Institut de Planétologie et d'Astrophysique de Grenoble) argues that in many cases, simplicity is a valuable representational ideal because simple models facilitate (1) faster, more comprehensive exploration of the parameter space, and (2) internal validation of a model and the concomitant use of "physical intuition" which is so important for good model building.

Bailer-Jones, D. M. (2000). Modelling extended extragalactic radio sources. *Studies in History and Philosophy of Modern Physics, 31*(1), 49–74. https://doi.org/ 10.1016/S1344-2198(99)00028-3.

> This article discusses practical and epistemological issues associated with scientific modeling of novel phenomena, using extended extragalactic radio sources (EERSs) as a case study. Bailer-Jones (d. 2006, form. University of Heidelberg) argues that models are ways of representing the causal mechanisms behind poorly understood phenomena ("representation [caus.mech.]"), and that they also serve as conventional means of representing the unity of explanations of such mechanisms ("representation [conv.]"). Although models are "a central form of knowledge about empirical phenomena" (69), they can rarely be taken to constitute a definitive statement of what the world is really like; their epistemological status is thus quite complicated.

Boyd, N. M. (2015). Are astrophysical models permanently underdetermined? [Unpublished manuscript]. http://jamesowenweatherall.com/wp-content/ uploads/2014/10/Boyd_SoCal_060615.pdf.

> Against Hacking (1989) and Ruphy (2011), Boyd (Siena College) argues that we ought to be more optimistic about the prospects of breaking underdetermination in representation-driven astrophysical modeling. Using case studies from research into supernovae, dark matter, structure formation, and gamma ray bursts, Boyd articulates a framework according to which models with identifiable distinguishing features can be evaluated separately in light of new empirical evidence. In other words, Boyd argues that the underdetermination of astrophysical models is more often transient than permanent, and that the epistemic status of such models therefore remains significant.

Crowther, K., Linnemann, N. S., & Wüthrich, C. (2021). What we cannot learn from analogue experiments. *Synthese, 198*(S16), 3701–3726. https://doi.org/10.1007/s11229-019-02190-0.

Contrary to Dardashti et al. (2017; 2019), Thébault (2019), and Evans and Thébault (2020), Crowther et al. argue that analogue experiments used to investigate inaccessible target phenomena (for instance, fluid "dumb holes" used to investigate astrophysical black holes) are no more confirmatory than analogical arguments – which is to say, hardly confirmatory at all. More specifically, the authors argue that analogue experiments cannot confirm whether a particular inaccessible phenomenon (such as Hawking radiation) actually exists, and they criticize their opponents for unjustifiably assuming the physical adequacy of analogue modelling frameworks, thereby begging the question. Despite this, the authors admit that analogue experiments can be useful scientific tools for exploring the relevant modeling framework and for demonstrating robustness of the phenomena of which they are designed to be analogues.

Dardashti, R., Thébault, K. P. Y., & Winsberg, E. (2017). Confirmation via analogue simulation: What dumb holes could tell us about gravity. *The British Journal for the Philosophy of Science, 68*(1), 55–89. https://doi.org/10.1093/bjps/axv010.

Using Hawking radiation as a case study, the authors argue that analogue models of inaccessible astrophysical phenomena can be used to confirm predictions about such phenomena given (1) a robust syntactic isomorphism between the modelling frameworks of the analogue and the target systems, (2) diverse analogue realizations of the phenomena under study, and (3) valid universality arguments.

Dardashti, R., Hartmann, S., Thébault, K. P. Y., & Winsberg, E. (2019). Hawking radiation and analogue experiments: A Bayesian analysis. *Studies in History and Philosophy of Science Part B: Studies in History and Philosophy of Modern Physics, 67*, 1–11. https://doi.org/10.1016/j.shpsb.2019.04.004.

Extending Dardashti et al. (2017)'s discussion of universality arguments, the authors provide a quantitative Bayesian model for investigating the inferential structure and confirmatory power of analogue black hole experiments. Their formal model shows how to link evidence about analogue systems to target systems, accounts for the confirmatory relevance of "saturation" (when multiple types of analogues are used to probe the same targets), and shows that the more confident we are about the physics underlying a particular analogue, the less it can teach us about the target system.

Evans, P. W., & Thébault, K. P. Y. (2020). On the limits of experimental knowledge. *Philosophical Transactions of the Royal Society A: Mathematical, Physical and Engineering Sciences, 378*(2177), 20190235. https://doi.org/10.1098/rsta.2019.0235.

Using stellar nucleosynthesis and Hawking radiation as case studies, Evans (University of Queensland) and Thébault (University of Bristol) analyze how scientific models and experiments (both conventional and analogue) justify inductive inferences about unmanipulable and/or inaccessible target systems. The paper is framed as a response to inductive skeptics who doubt the possibility of gaining inductive knowledge. The authors argue that scientists can use inductive triangulation – the validation of one mode of inductive reasoning via independent modes of inductive reasoning – to justify claims about unmanipulable and/or inaccessible target systems and to assuage reasonable doubt about inductive knowledge.

Field, G. (2021a). Putting theory in its place: The relationship between universality arguments and empirical constraints. *The British Journal for the Philosophy of Science*. https://doi.org/10.1086/718276.

Field (Cambridge University) argues that universality arguments such as those discussed in Dardashti et al. 2017 and 2019 cannot fill the empirical gap between analogue black hole experiments and their target systems unless at least one of the following conditions is met: (1) we know that the micro-physics of the two systems are relevantly similar, or (2) we can empirically access the macro-behavior of the systems. These conditions help clarify the confirmatory status of analogue black hole experiments, while emphasizing the need for empirical evidence in determining this status.

Field, G. (2021b). The latest frontier in analogue gravity: New roles for analogue experiments [Unpublished manuscript]. http://philsci-archive.pitt.edu/20365/.

In this preprint, Field (Cambridge University) offers an interpretation of the role of analogue black hole experiments which is at odds with conventional interpretations thereof (such as those due to Crowther et al. 2021, Dardashti et al. 2017 & 2019, Evans and Thébault 2020, and Thébault 2019). According to Field, analogue black hole experiments are valuable not only (or primarily) for their ability to confirm the existence or characterize the behavior of some inaccessible target phenomenon (usually Hawking radiation), but also for the way they can be used – and increasingly *are* being used – to *directly* detect instances of more *general* gravitational phenomena (in this case, the "Hawking process"), to explore the intrinsically interesting behavior of the analogue systems themselves, and to investigate the robustness of predicted phenomena which may contribute to a two-way knowledge flow between analogue and target systems. The paper also contains a helpful discussion of the history of analogue black hole experiments, and explains how old experiments may be reinterpreted using the author's framework

Gueguen, M. (2020). On robustness in cosmological simulations. *Philosophy of Science, 87*(5), 1197–1208. https://doi.org/10.1086/710839.

Scientists use numerical simulations to determine the mass distribution of dark matter halos. Numerical results are normally taken to be confirmatory when they are robust, i.e., when they resist some degree of fluctuation in the values of certain underlying parameters, as typically explored in "convergence studies". However, Gueguen (Institute of Physics of Rennes 1) argues that robustness analysis in the form of convergence studies fails to exclude numerical artifacts, and that in fact convergence can *result* from artifacts; we need a better criterion for determining the trustworthiness of our simulations.

Gueguen, M. (2021). A tension within code comparisons [unpublished manuscript]. http://philsci-archive.pitt.edu/19227/.

While convergence studies like those discussed in Gueguen (2020) are meant to test for the "internal robustness" of astrophysical simulations, code comparisons (which look for shared results across different simulation codes) appear to test for "external robustness". However, Gueguen (Institute of Physics of Rennes 1) argues that the presence of shared results across different astrophysical simulations has little epistemic significance in practice, and that even in principle (with a perfectly constructed ensemble of codes to compare), the requirement that the codes bear on comparable targets is inevitably in tension with the requirement that they differ with respect to their components. Thus, code comparisons cannot help us decide whether to trust a given simulation.

Jacquart, M. (2020). Observations, simulations, and reasoning in astrophysics. *Philosophy of Science, 87*(5), 1209–1220. https://doi.org/10.1086/710544.

Using collisional ring galaxies as a case study, Jacquart (University of Cincinnati) argues that computer simulations in astrophysics play three epistemic roles: (1) hypothesis testing (eg., testing possible explanations for how a galaxy could form a ring-shape), (2) exploring possibility space (eg., to establish the parameter boundaries in which ring galaxy-formation occurs), and (3) amplifying observations (i.e., using the simulation to develop a context in which to interpret observational data).

Jebeile, J. (2017). Computer simulation, experiment, and novelty. *International Studies in the Philosophy of Science, 31*(4), 379–395. https://doi.org/10.1080/02698595.2019.1565205.

Can computer simulations provide genuinely new knowledge? Using the "dark" galaxy called VirgoHI21 as a test case, Jebeile (University of Bern) argues affirmatively that although only concrete experiments can confound scientists and refute theories, simulations can still provide new knowledge *qua* knowledge obtained for the first time which adds to existing knowledge (this is the "first time" criterion of novelty). Importantly, the ability of simulations to generate new knowledge does *not* depend on features that they share with experiments.

Jebeile, J., & Kennedy, A. G. (2015). Explaining with models: The role of idealizations. *International Studies in the Philosophy of Science, 29*(4), 383–392. https://doi.org/10.1080/02698595.2015.1195143.

On the typical representationalist view of model explanation, idealized models are less explanatory than de-idealized models. Using galactic simulations as a case study, Jebeile (University of Bern) and Kennedy (Florida Atlantic University) contend that de-idealization is not always *in itself* explanatorily beneficial; sometimes, comparisons between idealized and de-idealized models allow researchers to extract important explanatory information not available in the de-idealized model alone. Furthermore, the authors argue that model explanation ought to be understood not as a product or feature of models, but as a user-dependent *activity*.

Massimi, M. (2018a). Perspectival modeling. *Philosophy of Science, 85*, 335–359. https://doi.org/10.1086/697745.

It is intuitive that using a plurality of models to represent one target system stifles the quest for scientific realism. However, Massimi (University of Edinburgh) argues that the problem of inconsistent models can be solved if we reconceptualize the role of models as representing not actual or fictional states of affairs, but *possibilities* in a possibility space. If we do so, using and testing a plurality of models can help narrow this space, which is inherently valuable for the realist goal of achieving true or approximately true theories. This article provides an interesting rebuttal to the model anti-realism of Ruphy (2011), and represents a potentially fruitful framework for interpreting inconsistent astrophysical models.

Meskhidze, H. (2017). Simulationist's regress in laboratory astrophysics [unpublished manuscript].

Extending the idea of the "experimenter's regress" from Collins (1985) and of the "simulationist's regress" from Gelfert (2011),[1] Meskhidze (UC Irvine) argues that the widespread use of modular models and bootstrapping methods in astrophysics renders the field susceptible to irresolvable situations of regress. The paper also includes interesting discussions of internal, external, and construct validity.

Reutlinger, A., Hangleiter, D., and Hartmann, S. (2018). Understanding (with) toy models. *The British Journal for the Philosophy of Science, 69*(4), 1069–1099. https://doi.org/10.1093/bjps/axx005.

Can simplified and idealized scientific models ("toy models") provide genuine understanding? In this paper, the authors divide such models into two types –

---

[1] Gelfert, A. (2011). Scientific models, simulation, and the experimenter's regress. In P. Humphreys & C. Imbert (Eds.), *Models, Simulations, and Representations* (pp. 145–167). Routledge.

those which are *embedded* within an empirically well-confirmed framework theory, and those which are *autonomous* from any such framework – and argue that the former can provide "how-actually" understanding and the latter "how-possibly" understanding. Given that astrophysical models are sometimes quite simplified and idealized, this article provides a framework for understanding the epistemic role of such models.

Ruphy, S. (2011). Limits to modeling: Balancing ambition and outcome in astrophysics and cosmology. *Simulation & Gaming: An Interdisciplinary Journal, 42*, 177–194. https://doi.org/10.1177/1046878108319640.

Ruphy (École normale supérieure – PSL) argues that in galactic astrophysics, there are often numerous empirically adequate submodels available for researchers to choose from, and the choice of a particular submodel at a given stage constrains the range of available submodels at later stages. This renders models path-dependent and contingent. Combined with the plasticity and stability of such models, these features can lead to persistent incompatible model pluralism, which thwarts the goal of accurately representing the world; accordingly, we should be anti-realists about galactic models.

Thébault, K. (2019). What can we learn from analogue experiments? In R. Dardashti, R. Dawid, and K. Thébault (Eds.), *Why Trust a Theory? Epistemology of Fundamental Physics* (pp. 184–201). Cambridge University Press. https://doi.org/10.1017/9781108671224.014.

Analogue experiments, for example the use of fluid models to investigate Hawking radiation, can provide us with evidence of the same confirmatory type (and plausibly even of the same confirmatory degree) as conventional experiments. This is because, according to Thébault (University of Bristol) we can externally validate analogue black holes and thus take them to stand in for their astrophysical cousins.

For further articles relevant to this category, see Anderl 2016, Elder 2020, Elder 2021/2, Elder 2022, Meskhidze 2021, Salmon 1998, Suárez 2013, Sundberg 2010, Sundberg 2012, and Wilson 2021.

17.4   Realism and Antirealism

Falkenburg, B. (2012). Pragmatic unification, observation and realism in astroparticle physics. *Journal for General Philosophy of Science, 43*(2), 327–345. https://doi.org/10.1007/s10838-012-9193-1.

This article discusses how the historical and contemporary practices of astroparticle physicists evince a commitment to scientific realism. Falkenburg (TU Dortmund) argues that scientists working in astroparticle physics employ various strategies of pragmatic unification and theories of observation which can only be explained in realist terms, and thus that a commitment to realism is necessary for the coherence of the discipline. See Gava (2019) for a constructive empiricist response to Falkenburg.

Gava, A. (2019). Astroparticle physics, a constructive empiricist account. *Science & Philosophy, 7*(1), 21–40. https://doi.org/10.23756/sp.v7i1.450.

A direct response to Falkenburg (2012), Gava (Paraná State University) contests Falkenburg's claim that the theory and practice of astroparticle physics are unintelligible except from a realist perspective. Instead, Gava argues that astroparticle physicists' realist-sounding claims can be recast in an antirealist light, without doing injustice to the science itself (similar to arguments made by Bas van Fraasen in relation to other disciplines).

Hacking, I. (1982). Experimentation and scientific realism. *Philosophical Topics, 13*(1), 71–87.

This article contains an early formulation of Hacking's "argument from engineering", according to which the reality of unobservable entities in experimental physics (and science more generally) is guaranteed by our ability to manipulate the entities' causal powers in order to generate new phenomena – i.e., to *interfere* with nature. Hacking (University of Toronto, emeritus) elaborates on this argument in his classic book, *Representing and Intervening* (1983), and uses it to explicitly advocate for antirealism about astrophysical entities in (1989).

Hacking, I. (1989). Extragalactic reality: The case of gravitational lensing. *Philosophy of Science, 56*(4), 555–581. https://doi.org/10.1086/289514.

The *locus classicus* for contemporary astrophysical antirealism. Despite being quite confident that gravitational lens systems exist in certain regions of the sky, Hacking (University of Toronto, emeritus) argues that our inability to manipulate those systems or observe them directly, combined with the fact that they are usually explained using different, incompatible, and literally false models, shows that we can only be constructive empiricists – rather than realists – about them.

Martens, N. C. M. (2022). Dark matter realism. *Foundations of Physics, 52*(1), 16. https://doi.org/10.1007/s10701-021-00524-y.

Given the current lack of empirical evidence regarding the nature of dark matter, Martens (University of Bonn) argues that we ought to be anti-realists about it, at least for now. He advocates for a form of "semantic" anti-realism in light of the thinness and vacuousness of the concept of dark matter, but leaves open the possibility that further discoveries will thicken the concept and thereby discredit his anti-realist stance.

Rockmann, J. (1998). Gravitational lensing and Hacking's extragalactic irreality. *International Studies in the Philosophy of Science, 12*(2), 151–164. https://doi.org/10.1080/02698599808573589.

In this critical response to Hacking's astrophysical antirealism, Rockmann (Deutsche Lufthansa AG) offers a realist interpretation of gravitational lenses which is grounded in their observability, in astrophysical common cause arguments, and in "home truths".

Ruphy, S. (2010). Are stellar kinds natural kinds? A challenging newcomer in the monism/pluralism and realism/antirealism debates. *Philosophy of Science, 77*(5), 1109–1120. https://doi.org/10.1086/656544.

Breaking new ground in the debate between natural kind monists/pluralists and realists/antirealists, Ruphy (École normale supérieure – PSL) argues that monism and realism about stellar kinds are both untenable. Furthermore, essentialism (the view that members of natural kinds share essential properties) and structuralism (the view which defines kind membership in terms of structural properties) can come apart, despite usually being presented as a package deal.

Sandell, M. (2010). Astronomy and experimentation. *Techne, 14*(3), 252–269. https://doi.org/10.5840/techne201014325.

Sandell (Discover Hawaii Science) argues that since Ian Hacking's experimental realism requires that unobservables be used in the production of "real" (as opposed to artefactual) experimental data, and since real experimental data are produced by something extra-instrumental, Hacking needs independent justification for realism in order for his own version of realism to work. Furthermore, even if Hacking's view was correct, astronomy would still count as an experimental science because astronomers *do* manipulate the causal powers of the objects they study.

Shapere, D. (1993). Astronomy and antirealism. *Philosophy of Science, 60*(1), 134–150. https://doi.org/10.1086/289722.

A wide-ranging critique of Ian Hacking's experimental realism and conception of science, Shapere (d. 2016, form. Wake Forest University) claims that astronomy is as much of a science as any other, and that Hacking's antirealism depends on an overly static understanding of science. Furthermore, Shapere argues that Hacking's 1989 article on gravitational lenses cherry picks its data, interprets these data too narrowly, and falsely concludes that the use of incompatible models renders realistic treatment of astrophysical phenomena impossible.

Suárez, M. (2013). Fictions, conditionals, and stellar astrophysics. *International Studies in the Philosophy of Science, 27*(3), 235–252. https://doi.org/10.1080/02698595.2013.825499.

Using models of stellar structure as a case study, Suárez (Complutense University of Madrid) contends that the main assumptions of such models are best understood as useful fictions, but that scientists can nevertheless maintain a realist agenda by (1) treating such assumptions as background knowledge required for the generation of "fictional conditionals", or (2) treating such assumptions as components of the antecedents of these conditionals and employing a non-truth-functional semantics for them.

For further articles relevant to this category, see Anderl 2016, Boyd 2015, Hudson 2007, Massimi 2018, and Ruphy 2011.

17.5 Theories and Testing

Boyd, N. M. (2018). *Scientific Progress at the Boundaries of Experience.* Ph.D. dissertation. University of Pittsburgh. http://d-scholarship.pitt.edu/id/eprint/33843.

In this PhD dissertation, Boyd (Siena College) articulates a new empiricist philosophy of science and a non-internalist conception of scientific progress according to which the accumulating corpus of empirical data available to us (despite its theory-ladenness) constrains viable theories and constitutes growing knowledge about the world. Although Boyd offers a general account of scientific progress, her discussion is largely furnished with examples from the observational sciences, especially astrophysics and cosmology. Case studies include Arecibo telescope data, Babylonian astronomical tables, dark energy, and cosmic inflation. For the published version of the dissertation's third chapter, see Boyd, N. (2018). Evidence enriched. *Philosophy of Science* 85, 403–421. https://doi.org/10.1086/697747.

De Baerdemaeker, S., & Boyd, N. M. (2020). Jump ship, shift gears, or just keep on chugging: Assessing the responses to tensions between theory and evidence in contemporary cosmology. *Studies in History and Philosophy of Science Part B: Studies in History and Philosophy of Modern Physics, 72,* 205–216. https://doi.org/10.1016/j.shpsb.2020.08.002.

When comparing predictions from the ΛCDM model with high-resolution astronomical observations, we face three dark matter-related "small-scale challenges": the Missing Satellites problem, the Too Big to Fail problem, and the Cusp/Core problem. De Baerdemaeker (Stockholm University) and Boyd (Siena College) note three potential responses scientists can take to these problems, namely to jump ship (i.e., abandon ΛCDM for something like MOND), to switch gears (i.e., modify ΛCDM with something like warm dark matter), or to keep on chugging (i.e., focus on improving ΛCDM simulations by incorporating known baryonic physics). Based on the heuristics of epistemic conservatism and individuating causal factors, the authors argue that scientists ought to keep on chugging, and they conclude by outlining potential future scenarios in dark matter research.

Elder, J. (2023). Black hole coalescence: Observation and model validation. In L. Patton and E. Curiel (Eds.), *Working Towards Solutions in Fluid Dynamics and Astrophysics: What the Equations Don't Say*. Springer. ISBN-13: 9783031256851.

The models of binary black hole mergers used by researchers at the LIGO-Virgo collaboration are vital for connecting high-level gravitational theory with the observational data produced by the instruments, thereby granting empirical access to gravitational waves and their sources. However, recalling Collins' (1985) "experimenter's regress", Elder (Black Hole Initiative) suggests that these models pose an epistemic circularity problem insofar as they are used to validate the observations, while the accuracy of the observations depends upon the validity of the models. LIGO-Virgo scientists attempt to circumvent this circularity using a variety of tests, including the "residuals test" and the "IMR consistency test".

Horvath, J. E. (2009). Dark matter, dark energy and modern cosmology: The case for a Kuhnian paradigm shift. *Cosmos and History: The Journal of Natural and Social Philosophy, 5*(2), 287–303. https://www.cosmosandhistory.org/index.php/journal/article/view/161.

Horvath (University of São Paulo) argues that current debates over dark matter and dark energy are marked by features characteristic of pre-paradigm shift science, including attempts to isolate and characterize the problematic explanandum, the flourishing of philosophical/methodological analysis, the accelerating proliferation of proposed alternatives, and a sense of despair and discomfort within the community.

Kosso, P. (2013). Evidence of dark matter, and the interpretive role of general relativity. *Studies in History and Philosophy of Science Part B: Studies in History and Philosophy of Modern Physics, 44*(2), 143–147. https://doi.org/10.1016/j.shpsb.2012.11.005.

Kosso (Northern Arizona University) offers a lucid and accessible discussion of the theory and history of dark matter, with special attention paid to the question of whether it is possible to detect dark matter independently of general relativity (GR). Using the Bullet Cluster as his primary case study, Kosso contends that the part of GR employed in detecting dark matter through gravitational lensing (namely, the Einstein Equivalence Principle) is common to all metric theories of gravity. Given that all viable theories of gravity are metric, any such theory can be employed when investigating dark matter lenses – the specifics of GR or any other theory are only required to determine the *amount* of dark matter present. Thus, contrary to the "dark matter double-bind" proposed by Vanderburgh (2003; 2005), Kosso claims that dark matter can be detected without assuming the truth of GR. See Sus (2014) and Vanderburgh (2014b) for responses.

Martens, N. C. M., & Lehmkuhl, D. (2020a). Dark matter = modified gravity? Scrutinising the spacetime–matter distinction through the modified gravity/dark matter lens. *Studies in History and Philosophy of Science Part B: Studies in History and Philosophy of Modern Physics, 72*, 237–250. https://doi.org/ 10.1016/j.shpsb.2020.08.003.

Most proposed solutions to the dark matter problem are either matter-based (eg., WIMPS) or gravity-based (eg., MOND). These types of solutions are typically represented as conceptually distinct, owing to a deeper distinction between matter and spacetime. In this paper, Martens and Lehmkuhl (both University of Bonn) argue that a strict matter-spacetime distinction is untenable, and likewise for the distinction between matter vs. gravity-based solutions to the dark matter problem. Their analysis draws heavily from the recent literature on superfluid dark matter, the scalar field φ of which they interpret both as a kind of dark matter, and as a modification of gravity. This paper constitutes the first part of a pair of articles, the second being Martens and Lehmkuhl (2020b).

Martens, N. C. M., & Lehmkuhl, D. (2020b). Cartography of the space of theories: An interpretational chart for fields that are both (dark) matter and spacetime. *Studies in History and Philosophy of Science Part B: Studies in History and Philosophy of Modern Physics, 72*, 217–236. https://doi.org/10.1016/ j.shpsb.2020.08.004.

Following up from Martens and Lehmkuhl (2020a), the authors advance a "cartographic" taxonomy of interpretations for "Janus-faced" theories like superfluid dark matter (according to which a single scalar field is both a dark matter field and a modification of gravity, in certain contexts). Their taxonomy contains three classes of interpretations with nine subclasses, and they argue that four such subclasses remain viable ways of understanding superfluid dark matter. See p. 231 for their chart of interpretations.

Matarese, V. (2022). 'Oumuamua and meta-empirical confirmation. *Foundations of Physics, 52*(4). https://doi.org/10.1007/s10701-022-00587-5.

Astrophysicist Abraham Loeb has suggested that the interstellar interloper 1I/2017 'Oumuamua is a piece of alien technology. To empirically confirm or confute his hypothesis would require significant expenditure of financial and intellectual resources – for instance by sending a probe to 'Oumuamua, as proposed by *Project Lyra*. How can we be sure that Loeb's hypothesis is viable and thus worth pursuing at all? Matarese (University of Bern) argues that we should use a meta-empirical framework to answer this question, one which provides information about the capacity of Loeb's hypothesis to adequately represent potential future empirical data. Furthermore, Matarese contends that meta-empirical confirmation does not violate the empiricist spirit since it can be fruitfully applied even in empirically grounded research contexts such as this one.

Patton, L. (2020). Expanding theory testing in general relativity: LIGO and parametrized theories. *Studies in History and Philosophy of Science Part B: Studies in History and Philosophy of Modern Physics, 69*, 142–153. https://doi.org/10.1016/j.shpsb.2020.01.001.

Using LIGO as a case study, this paper explains how parametrized theories – specifically the parametrized post-Einsteinian (ppE) framework – can allow for more and better tests of General Relativity (GR). Patton (Virginia Tech) argues that formal reasoning on the theoretical structure of GR can broaden its empirical reach by removing barriers to empirical testing that have been encoded into the theory's formal structure (and into existing testing frameworks, such as those used in the creation and interpretation of LIGO results).

Sus, A. (2014). Dark matter, the Equivalence Principle and modified gravity. *Studies in History and Philosophy of Science Part B: Studies in History and Philosophy of Modern Physics, 45*, 66–71. https://doi.org/10.1016/j.shpsb.2013.12.005.

In this critical response to Kosso (2013), Sus (University of Valladolid) argues that although all viable alternative theories of gravity satisfy the Einstein Equivalence Principle (EEP), Kosso is wrong to think that gravitational lensing (the primary source of evidence for dark matter in the case of Bullet Cluster observations) is a direct consequence of the EEP. Specifically, Sus claims that different metric theories of gravity (including MONDian alternatives like TeVeS) may countenance different conclusions concerning the location and physical properties of the lensing matter. Sus also accuses Kosso of being unclear about whether he takes his argument to support the very basic conclusion that gravitational lensing provides evidence for matter which cannot be luminously detected, or for the more controversial claim that this matter is non-baryonic. See also Sus's interesting discussion of direct vs. indirect evidence.

Vanderburgh, W. L. (2001). *Dark Matters in Contemporary Astrophysics: A Case Study in Theory Choice and Evidential Reasoning*. Ph.D. Dissertation. Western University. https://philpapers.org/rec/VANDMI-4.

Vanderburgh's (CSU San Bernardino) PhD dissertation covers the foundations of the dynamical dark matter problem in twentieth century astrophysics, raises the "dark matter double bind" as an in-principle difficulty we must face when solving the problem, and attempts to identify and evaluate patterns of inference involved in evidential arguments for candidate solutions thereof.

Vanderburgh, W. L. (2003). The dark matter double bind: Astrophysical aspects of the evidential warrant for general relativity. *Philosophy of Science, 70*(4), 812–832. https://doi.org/10.1086/378866.

Is our confidence in the applicability of general relativity (GR) to galactic and supra-galactic scales warranted, given currently available tests of GR? Vanderburgh (CSU San Bernardino) answers in the negative, noting that in order to evaluate the empirical adequacy of competing theories of gravitation at galactic scales, the mass distribution of test galaxies must first be known; however, because of the well-known discrepancy between dynamical mass and luminosity mass, we cannot feel confident in our measurements of mass distribution. In order to infer the distribution, we must assume a gravitational law (whether GR, MOND, Weyl gravity, or something else), but this is illegitimate given that the validity of our gravitational laws is precisely what is being tested. This is the "dark matter double bind".

Vanderburgh, W. L. (2005). The methodological value of coincidences: Further remarks on dark matter and the astrophysical warrant for general relativity. *Philosophy of Science, 72*(5), 1324–1335. https://doi.org/10.1086/508971.

A follow-up to his (2003), Vanderburgh (CSU San Bernardino) addresses the question of whether apparent agreement between four ways of measuring the masses of galaxies and larger structures – namely through rotation curves, the Virial Theorem, observed X-ray emissions, and gravitational lensing – gives us strong evidential warrant for the applicability of general relativity at those scales. At least compared to its rivals, Vanderburgh contends that these measurements do lend support to GR, but this support is weak and defeasible.

Vanderburgh, W. L. (2014a). Quantitative parsimony, explanatory power and dark matter. *Journal for General Philosophy of Science, 45*(2), 317–327. https://doi.org/10.1007/s10838-014-9261-9.

Alan Baker (2003)[2] has argued that quantitative parsimony (the principle that theories which posit fewer entities are superior) is legitimately virtuous since quantitatively parsimonious theories have greater explanatory power. Using dark matter as a case study, Vanderburgh (CSU San Bernardino) challenges Baker's account, and argues more generally that we ought to avoid artificially separating quantitative parsimony from other varieties of parsimony in actual theory choice situations.

Vanderburgh, W. L. (2014b). On the interpretive role of theories of gravity and 'ugly' solutions to the total evidence for dark matter. *Studies in History and Philosophy of Science Part B: Studies in History and Philosophy of Modern Physics, 47*, 62–67. https://doi.org/10.1016/j.shpsb.2014.05.008.

---

[2] Baker, A. (2003). Quantitative parsimony and explanatory power. *British Journal for the Philosophy of Science, 54*, 245–259. https://doi.org/10.1093/bjps/54.2.245

Peter Kosso (2013) has argued that evidence from observations of the Bullet Cluster provides evidential warrant for the equivalence principle in such a way which avoids Vanderburgh's (2001; 2003) "dark matter double bind". Vanderburgh (CSU San Bernardino) responds that even if this is the case, we are still unable to perform the kind of precision tests of general relativity that would confirm its applicability to galactic and supra-galactic scales. Vanderburgh also countenances the possibility that we cannot rule out "ugly" solutions to the dark matter problem which incorporate both dark matter and modified theories of gravity.

17.6   SSK and Social Issues

Collins, H. M. (1985). *Changing Order: Replication and Induction in Scientific Practice*. Sage Publications. ISBN-10: 0226113760.

A classic text in the sociology of scientific knowledge which introduces the notion of the "experimenter's regress" using Joseph Weber's attempts to detect gravitational waves as a case study. According to Collins (Cardiff University), it often happens in frontier science that the best or only check on a result is the proper functioning of the apparatus used to generate it, while the best or only check on the proper functioning of the apparatus is the result; this is the experimenter's regress. For Collins, there is usually no rational way out of the regress – instead, scientists resort to heuristics, rhetoric, compulsion, etc. For those interested specifically in the Weber case study and regress, see Chapter 4 (pp. 79–111).

Collins, H. M. (2004). *Gravity's Shadow: The Search for Gravitational Waves*. University of Chicago Press. ISBN: 9780226113791.

The authoritative social history of gravitational waves from the 1960s-2004, Collins' (Cardiff University) book touches on key issues of scientific knowledge, expertise, and consensus. It serves as an important resource for philosophers interested in gravitational waves who seek to understand the scientific process at a more concrete level. For Collins' other books on the science and sociology of gravitational waves, see Collins, H. (2010). *Gravity's Ghost: Scientific Discovery in the Twenty-first Century*. University of Chicago Press; and Collins, H. (2017). *Gravity's Kiss: The Detection of Gravitational Waves*. MIT Press.

Curiel, E. (2019). The many definitions of a black hole. *Nature Astronomy, 3*(1), 27–34. https://doi.org/10.1038/s41550-018-0602-1.

In this article, Curiel (Munich Center for Mathematical Philosophy; Black Hole Initiative) discusses the phenomenon whereby different communities of physicists define black holes in distinct and conflicting ways. The author presents a helpful sample of such definitions in three boxes, the first of which focuses on those offered by astrophysicists of various specializations. Given that physicists across different fields seek to collaborate on questions of mutual interest, Curiel recommends that each investigative team fix an explicit list of properties and phenomena which they take to be characteristic of black holes in order to ensure shared understanding and to avoid miscommunication.

English, J. (2017). Canvas and cosmos: Visual art techniques applied to astronomy data. International *Journal of Modern Physics D, 26*(04), 1,730,010. https://doi.org/10.1142/S0218271817300105.

An extensive overview of the production as well as cultural, aesthetic, and scientific value of astronomical outreach images (such as those created by the Hubble Heritage Team). English (University of Manitoba) contends that such images, which are created by applying techniques from visual art to representations of data, contribute meaningfully to both the "culture of science" and "culture of art", retaining scientific significance despite being crafted to satisfy aesthetic ends (as evidenced by their widespread incorporation into both research papers and popular media).

Greenberg, J. (2004). Creating the 'Pillars': Multiple meanings of a Hubble image. *Public Understandings of Science, 13*, 83–95. https://doi.org/10.1177/0963662504042693.

Using the public reception and interpretation of the HST's original (1995) image of the Eagle Nebula as a case study, Greenberg (A. P. Sloan Foundation) argues that when scientific images are black-boxed (presented as pure, unquestionable scientific objects, usually by the media and/or by scientists' press releases), it becomes easier for lay-people to augment them with additional, "non-scientific" meanings which build on their supposed status as unadulterated representations of reality. This is a helpful article for those interested in the public perception of astronomy and the sociology of astronomical knowledge.

Heidler, R. (2017). Epistemic cultures in conflict: The case of astronomy and high energy physics. *Minerva, 55*(3), 249–277. https://doi.org/10.1007/s11024-017-9315-3.

The discovery of dark energy suddenly increased the mutual dependency between astronomy and high energy physics, such that physicists had to rely on astronomical instruments and data to answer their own questions (functional dependency), while both disciplines integrated and coordinated their scientific goals (strategic dependency). Heidler (German Research Foundation) argues that these dependencies fostered an epistemic conflict between the disciplines, leading to transgression of social and cognitive boundaries, turbulence in epistemic practices, and self-reflection on the scientists' identities and the moral economy of which they are a part.

Jebeile, J. (2018). Collaborative practice, epistemic dependence and opacity: The case of space telescope data processing. *Philosophia Scientiæ. Travaux d'histoire et de Philosophie Des Sciences, 22*(2), 59–78. https://doi.org/10.4000/philosophiascientiae.1483.

Employing Susann Wagenknecht's (2014) distinction between opaque and translucent epistemic dependence,[3] Jebeile (University of Bern) analyzes the social epistemological relationships between collaborators working with the Herschel Space Observatory. Jebeile identifies cases of opaque epistemic dependence therein, and argues that sources of opacity include not only lack of expertise, but also the non-disclosure of data, failure to understand relevant instrumental processes, and epistemic inaccessibility of numerical calculations.

Kennefick, D. (2000). Star crushing: Theoretical practice and the theoretician's regress. *Social Studies of Science, 30*(1), 5–40. https://doi.org/10.1177/030631200030001001.

An important study in the sociology of astrophysical knowledge and simulations which uses the 1990s controversy over "star-crushing" as a case study. Kennefick (University of Arkansas) introduces the notion of the "theoretician's regress" – a play on Harry Collins' "experimenter's regress" (Collins 1985) – to explain why the controversy over the star-crushing effect could not be resolved by strictly "scientific" debate.

McCray, W. P. (2000). Large telescopes and the moral economy of recent astronomy. *Social Studies of Science, 30*(5), 685–711. https://doi.org/10.1177/030631200030005002.

McCray (UC Santa Barbara) contends that the Gemini 8-Meter Telescopes Project illustrates a tension in American optical astronomy between the "haves" (those with access to telescopes through their institutions) and the "have-nots" (those who must compete for time at federally funded national observatories), while also showing how non-scientific political concerns play into the debate around investment in big science.

---

[3] Wegenknecht, S. (2014). Opaque and translucent epistemic dependence in collaborative scientific practice. *Episteme, 11*(4), 475–492. https://doi.org/10.1017/epi.2014.25. According to Wegenknecht: "A scientist is opaquely dependent upon a colleague's labor, if she does not process the expertise necessary to independently carry out, and to profoundly assess, the piece of scientific labor her colleague is contributing. I suggest, however, that if the scientist does possess the necessary expertise, then her dependence would not be opaque, but translucent" (p. 483).

Messeri, L. R. (2010). The problem with Pluto: Conflicting cosmologies and the classification of planets. *Social Studies of Science, 40*(2), 187–214. https://doi.org/10.1177/0306312709347809.

This paper offers a historical and sociological account of the "demotion" of Pluto by the IAU in 2006. Messeri (Yale University) focuses on the relationships between scientific and cultural cosmologies, and the ways that the public influenced the debate surrounding the definition of "planet". Messeri argues that the IAUs definition of "planet" privileged one cosmology over others, thereby fracturing discourse about planets and Pluto in particular.

Metzger, P. T., Grundy, W. M., Sykes, M. V., Stern, A., Bell, J. F., Detelich, C. E., Runyon, K., & Summers, M. (2022). Moons are planets: Scientific usefulness versus cultural teleology in the taxonomy of planetary science. *Icarus, 374*, 114,768. https://doi.org/10.1016/j.icarus.2021.114768.

In this lengthy and detailed article, Metzger et al. contend that the IAUs 2006 definition of "planet" paid too much credence to folk taxonomy while ignoring the importance of scientific taxonomy. They argue that a purely geophysical definition of "planet", according to which a planet is an object with a certain amount of geological complexity, has stronger historical and pragmatic grounding when compared to the current dynamical definition (and that contemporary planetary scientists already use the geophysical definition anyways). As such, the authors claim that we ought to revise our educational materials for the sake of an improved scientific and cultural planetary taxonomy.

Sovacool, B. (2005). Falsification and demarcation in astronomy and cosmology. *Bulletin of Science, Technology & Society, 25*(1), 53–62. https://doi.org/10.1177/0270467604270151.

In this sociological article, Sovacool (University of Sussex; Aarhus University) analyzes how and to what extent contemporary astronomers and cosmologists rely on the ideas of Karl Popper to resolve crises related to methodology, legitimacy, and testability. Sovacool concludes that Popper's ideas play an important implicit and explicit role in such crises, and he argues more broadly for the relevance of philosophy to scientific practice.

Sundberg, M. (2010). Cultures of simulations vs. cultures of calculations? The development of simulation practices in meteorology and astrophysics. *Studies in History and Philosophy of Science Part B: Studies in History and Philosophy of Modern Physics, 41*, 273–281. https://doi.org/10.1016/j.shpsb.2010.07.004.

Sundberg (Stockholm University) uses "modern" and "postmodern" "computer cultures" as a lens for analyzing contemporary numerical simulation practices in meteorology and astrophysics. Sundberg argues that, by and large, there seems to be a shift occurring towards a more "postmodern" culture of simulations which emphasizes the value of surface-level research using black-box simulations, entertaining visualizations of simulation results, and the playful exploration of simulation codes as a learning tool.

Sundberg, M. (2012). Creating convincing simulations in astrophysics. *Science, Technology, & Human Values, 37*(1), 64–87. https://doi.org/10.1177/0162243910385417.

Sundberg (Stockholm University) examines the methods that astrophysicists use to convince themselves and others of the reliability and credibility of their simulations, especially when those simulations deliver outputs which are uncertain or difficult to interpret. In the process, Sundberg analyzes the distinction between "numerical" and "real" effects, arguing that they cannot be distinguished on the basis of what they derive from.

For another article relevant to this category, see Hudson 2007.

17.7 Typicality and Extra-Terrestrials

Ćirković, M. M. (2006). Too early? On the apparent conflict of astrobiology and cosmology. *Biology and Philosophy, 21*(3), 369–379. https://doi.org/10.1007/s10539-005-8305-2.

Olum's problem, a generalized form of Fermi's paradox, relies on a dehistoricized understanding of the universe. In opposition to this dehistoricized stance, Ćirković (Astronomical Observatory of Belgrade; Future of Humanity Institute) argues that the universe becomes more hospitable to life as time passes, meaning that it is *too early* in cosmological history for the absence of detected extraterrestrial life to constitute a genuine paradox.

Lacki, B. (2021). The noonday argument: Fine-graining, indexicals, and the nature of Copernican reasoning [unpublished manuscript]. arXiv:2106.07738v1 [physics.hist-ph].

Lacki (UC Berkeley) offers a new theory of typicality called Fine Graining with Auxiliary Indexicals (FGAI). He argues that it avoids the paradoxes (such as the Doomsday Argument) faced by other theories of typicality by fine-graining our macrotheories with microhypotheses and by separating indexical from physical facts.

Lewis, G. F. & Barnes, L. A. (2021). The trouble with "puddle thinking:" A user's guide to the Anthropic Principle. *Journal & Proceedings of the Royal Society of New South Wales, 154*(1), 6–11. ISSN 0035–9173/21/010006–06.

A short, popular introduction to the anthropic principle and to Douglas Adams' notion of "puddle thinking", Lewis (University of Sydney) and Barnes (Western Sydney University) argue that the problem of fine-tuning does not concern the question of how we find ourselves in a universe with conditions amenable to human life, but the question of *why* a universe with such conditions exists at all. This question is, they concede, a necessarily philosophical one.

Satta, M. (2021). Evil twins and the multiverse: Distinguishing the world of difference between epistemic and physical possibility. *Synthese, 198*(2), 1153–1160. https://doi.org/10.1007/s11229-019-02092-1.

Brian Greene and Max Tegmark have claimed that if the universe is infinite and matter is roughly evenly distributed within it, then every possible material arrangement of particles must exist in an infinite number of instantiations. Satta (Wayne State University) argues that Green and Tegmark's claims rely on a conflation of physical with epistemic possibility, and that they ignore potential macro-level constraints on possibility from psychology, biology, and sociology.

For another article relevant to this category, see Matarese 2022.

17.8 Dark Matter and MOND

Abelson, S.S. (2022). The fate of tensor-vector-scalar modified gravity. *Foundations of Physics, 52*(31). https://doi.org/10.1007/s10701-022-00545-1.

Abelson (Indiana University, Bloomington) reviews the case of the LIGO neutron star merger detection against TeVeS, long considered the most plausible relativistic extension of MOND. Abelson argues that the physicists' use of language of falsification in a strict Popperian sense was unwarranted. However, Abelson offers an alternative interpretation of the result as a corroboration of the null-hypothesis, along the lines of Mayo's error-statistical account.[4]

De Baerdemaeker, S. & Dawid, R. (2022). MOND and meta-empirical theory assessment. *Synthese 200*(344). https://doi.org/10.1007/s11229-022-03830-8.

De Baerdemaeker and Dawid (both Stockholm University) critically examine some of the philosophical arguments that have been offered by defenders of MOND in terms of different views on theory assessment. They argue, first, that on a standard reading of Popper and Lakatos, the arguments fail. Second, they argue that the strongest philosophical defense of MOND takes the form of meta-empirical theory assessment, but that, according to that account as well, the arguments fail to be convincing.

---

[4] Mayo, D. (2018). *Statistical Inference as Severe Testing: How to Get Beyond the Statistics Wars.* Cambridge University Press. https://doi.org/10.1017/9781107286184

Jacquart, M. (2021). ΛCDM and MOND: A debate about models or theory? *Studies in History and Philosophy of Science Part A, 89,* 226–234. https://doi.org/10.1016/j.shpsa.2021.07.001.

Jacquart (University of Cincinnati) extends Massimi's (2018b) analysis of the debate between proponents of ΛCDM and MOND as one about the challenges of multiscale modeling. Instead of interpreting the debate in terms of theoretical disagreement (as is commonly done), Jacquart advocates for a model-based understanding. According to that interpretation, both rivals are successful within their intended domain of application, but there is a need to be critical of attempts to extend them to new domains. Such extension requires justification that the model accurately represent explanatory dependencies in the target system.

Martens N. C. M., Carretero Sahuquillo, M. A., Scholz, E., Lehmkuhl, D., & Krämer, M. (2022). Integrating dark matter, modified gravity, and the humanities, *Studies in History and Philosophy of Science, 91,* A1-A5. https://doi.org/10.1016/j.shpsa.2021.08.015.

This editorial introduces the aims of a Special Issue on dark matter and MOND. The authors all work on an interdisciplinary research project, The Epistemology of the LHC, of which one sub-project is on LHC, dark matter, and gravity. They provide two motivations for interdisciplinary work on the interface between dark matter and modified gravity. The first is to improve communication and reduce the polemics between physicists on either side of the divide. The second is to start extending the philosophical literature on dark matter—despite dark matter being one of the central problems of contemporary fundamental physics. The editorial includes an extensive reference list of philosophical discussions of dark matter and MOND.

Massimi, M. (2018b). Three problems about multi-scale modelling in cosmology. *Studies in History and Philosophy of Science Part B: Studies in History and Philosophy of Modern Physics, 64,* 26–38. https://doi.org/10.1016/j.shpsb.2018.04.002.

Massimi (University of Edinburgh) argues that the debate between ΛCDM and MOND in contemporary cosmology can best be understood as a debate about challenges to multi-scale modeling. Massimi argues for five claims (i) ΛCDM and MOND work best at different scales, i.e., the macro-scale and the meso-scale, respectively; (ii) Both face challenges when modeling across more than one scale; (iii) The downscaling problem for ΛCDM is one of explanatory power, while the upscaling problem is one of consistency with general relativity; (iv) Hybrid models, which try to unify the best of ΛCDM and MOND, face a problem of predictive novelty; (v) Ultimately, a successful cosmology, in order to be successful cannot avoid having to solve these problems.

McGaugh, S. (2015). A tale of two paradigms: The mutual incommensurability of ΛCDM and MOND. *Canadian Journal of Physics, 93*(2), 250–259. https://doi.org/10.1139/cjp-2014-0203.

McGaugh (Case Western University) reviews the ongoing disagreement between ΛCDM and MOND in terms of a Kuhnian picture of incommensurable paradigms. McGaugh submits that both have significant empirical support, but nonetheless offer inconsistent worldviews. However, McGaugh is concerned about the detectability of dark matter: without a positive dark matter particle detection, there are concerning parallels between the dark matter hypothesis and aether theory.

Merritt, D. (2017). Cosmology and convention. *Studies in History and Philosophy of Science Part B: Studies in History and Philosophy of Modern Physics, 57*, 41–52. https://doi.org/10.1016/j.shpsb.2016.12.002.

Merritt (Rochester Institute of Technology) assesses the current concordance model of cosmology according to Popper's definition of conventionalist stratagems. According to Merritt, dark matter and dark energy were ad hoc auxiliary hypotheses introduced to save the concordance model from falsifying evidence. Moreover, the usual convergence arguments offered in support of the concordance model are argued to fail. As such, cosmology has, plausibly, entered the phase of what Lakatos called a degenerative problem shift.

Merritt, D. (2020). *A Philosophical Approach to MOND: Assessing the Milgromian Research Program in Cosmology*. Cambridge University Press. https://doi.org/10.1017/9781108610926

Merritt (Rochester Institute of Technology) offers a book-length review of the history of MOND, from when it was first proposed in the early 1980s, until today. The philosophical framing of the book is in terms of Lakatos' theory of progressive and degenerative research programs. Merritt argues that especially the earlier theories of the research program were highly successful at predicting novel facts (assessed according to different proposed philosophical accounts of novelty). While the very latest theories are potentially less successful qua novel prediction, the overall research program is deemed to be progressive.

Merritt, D. (2021). Feyerabend's rule and dark matter. *Synthese 199*, 8921–8942. https://doi.org/10.1007/s11229-021-03188-3.

Building on Feyerabend's work, Merritt's (Rochester Institute of Technology) starting point is the claim that, under specific circumstances, the lack of an experimental result can refute a theory while confirming another. Merritt applies this to the current concordance model of cosmology, and argues that there are several examples of such refuting negative results, including the failure of dark matter particle detection, the failure to detect primordial dwarf galaxies, and dynamical friction.

Milgrom, M. (2020). MOND vs. dark matter in light of historical parallels. *Studies in History and Philosophy of Science Part B: Studies in History and Philosophy of Modern Physics, 71*, 170–195. https://doi.org/10.1016/j.shpsb.2020.02.004.

Milgrom (Weizmann Institute) reviews the case for MOND, from its initial proposal, up until its scientific status today. The paper draws multiple comparisons between the history of MOND and well-known episodes in the history of science, including the Copernican revolution and the development of quantum theory. It also draws parallels between dark matter and aether-theory.

For further articles relevant to this category, see De Baerdemaeker 2021, De Baerdemaeker and Boyd 2020, Horvath 2009, Hudson 2007, 2009, 2013, Kosso 2013, Martens 2022, Martens and Lehmkuhl 2020a,b, Sus 2014, and Vanderburgh 2001, 2003, 2005, 2014a,b, Wilson 2021.

Printed in the United States
by Baker & Taylor Publisher Services

Printed in the United States
by Baker & Taylor Publisher Services